Edited by
Sebastian Schlücker

Surface Enhanced Raman Spectroscopy

Related Titles

Andrews, D. L. (ed.)

Encyclopedia of Applied Spectroscopy

2010
ISBN: 978-3-527-40773-6

Salzer, R., Siesler, H. W. (eds.)

Infrared and Raman Spectroscopic Imaging

2009
ISBN: 978-3-527-31993-0

Amer, M. S. (ed.)

Raman Spectroscopy for Soft Matter Applications

2009
ISBN: 978-0-470-45383-4

Wartewig, S.

Materials Characterization
Introduction to Microscopic and Spectroscopic Methods

2008
ISBN: 978-0-470-82298-2

Sasic, S. (ed.)

Pharmaceutical Applications of Raman Spectroscopy

2008
ISBN: 978-0-8138-1013-3

Aroca, R.

Surface-Enhanced Vibrational Spectroscopy

2006
ISBN: 978-0-471-60731-1

Edited by Sebastian Schlücker

Surface Enhanced Raman Spectroscopy

Analytical, Biophysical and Life Science Applications

WILEY-VCH Verlag GmbH & Co. KGaA

The Editor

Prof. Dr. Sebastian Schlücker
University of Osnabrück
Department of Physics
Barbarastraße 7
49076 Osnabrück
Germany

All books published by Wiley-VCH are carefully produced. Nevertheless, authors, editors, and publisher do not warrant the information contained in these books, including this book, to be free of errors. Readers are advised to keep in mind that statements, data, illustrations, procedural details or other items may inadvertently be inaccurate.

Library of Congress Card No.: applied for

British Library Cataloguing-in-Publication Data
A catalogue record for this book is available from the British Library.

Bibliographic information published by the Deutsche Nationalbibliothek
The Deutsche Nationalbibliothek lists this publication in the Deutsche Nationalbibliografie; detailed bibliographic data are available on the Internet at <http://dnb.d-nb.de>.

© 2011 WILEY-VCH Verlag & Co. KGaA, Boschstr. 12, 69469 Weinheim, Germany

All rights reserved (including those of translation into other languages). No part of this book may be reproduced in any form – by photoprinting, microfilm, or any other means – nor transmitted or translated into a machine language without written permission from the publishers. Registered names, trademarks, etc. used in this book, even when not specifically marked as such, are not to be considered unprotected by law.

Cover   Adam Design, Weinheim
Typesetting   Laserwords Private Ltd., Chennai, India
Printing and Binding   Fabulous Printers Pte Ltd, Singapore

Printed in Singapore
Printed on acid-free paper

ISBN: 978-3-527-32567-2
ePDF ISBN: 978-3-527-63276-3
oBook ISBN: 978-3-527-63275-6
ePub ISBN: 978-3-527-63306-7

Foreword

More than 80 years since the discovery of the Raman effect have passed and Raman spectroscopy has become one of the most important methods within the various methods of analysis and structural determinations. Certainly, the discovery of the laser in 1960 has opened up new horizons for Raman spectroscopy and brought several new useful techniques. One of the most interesting and significant findings in this field is undoubtedly surface-enhanced Raman scattering (SERS) which was discovered in 1977. Within this phenomenon, molecules adsorbed onto metal surfaces under certain conditions exhibit an anomalously large interaction cross section for the Raman effect. It might be thought that a subject originated more than three decades ago would be virtually exhausted by now, but nothing could be farther from truth. The recent developments in SERS have led to large increases in the sensitivity of SERS measurements and have enabled new phenomena to be observed and applied. SERS measurements are expected to become increasingly important in chemistry, biochemistry, and biophysics.

In the 14 chapters of this book, an authoritative, up-to-date account of the principles and fundamentals of SERS is given including many examples for its applications. The book includes the basic theory for SERS; summarizes the various SERS substrates; discusses quantitative SERS methods with emphasis on reproducibility, stability and sensitivity up to single molecule detection; and describes SERS microscopy, electrochemical SERS, surface enhanced resonance Raman scattering (SERRS), and surface-enhanced hyper Raman scattering (SEHRS), as well as surface- and tip-enhanced coherent anti-Stokes Raman scattering (SE-CARS, TE-CARS). Applications of SERS include the detection of organic pollutants and pharmaceuticals; studies of electron transfer of proteins at membrane models; investigations of microfluidics, quantitative DNA analysis, biomedical applications by means of SERS microscopy, SERS as an intracellular probe; and coupling of SERS with various separation methods (e.g. liquid or gas chromatography).

The abundant references provide ready access to the original research literature. As the field of SERS has sufficiently matured during the past decades, the danger of rapid obsolescence for this book is less. The subject matter, however, still offers plenty of opportunity for further exploration and exploitation. In my opinion this book, which clearly expresses the current excitement in this extremely active

Surface Enhanced Raman Spectroscopy: Analytical, Biophysical and Life Science Applications. Edited by Sebastian Schlücker
Copyright © 2011 WILEY-VCH Verlag GmbH & Co. KGaA, Weinheim
ISBN: 978-3-527-32567-2

research area, will make a substantial contribution to the further growth of an increasingly important subfield of vibrational spectroscopy.

Professor Schlücker, editor of this volume, is one of the leading researchers working currently in the SERS field. As chemist at the University of Würzburg, Germany, and the National Institutes of Health, Bethesda, USA, and now as physicist at the University of Osnabrück, Germany, he has played a major role in introducing a few important new experimental techniques of SERS (e.g. the direct and label-free SERS detection of solid-phase bound compounds; immuno-SERS microscopy with nanoparticle probes). He is well qualified to present this book to the scientific community.

Wolfgang Kiefer
University of Würzburg and Eisingen
Laboratory for Applied Raman Spectroscopy (ELARS)

Contents

Preface *XV*
List of Contributors *XVII*

1 Basic Electromagnetic Theory of SERS *1*
Pablo G. Etchegoin and Eric C. Le Ru
1.1 Introduction *1*
1.2 Plasmon Resonances and Field Enhancements *2*
1.2.1 Optical Properties of Simple Metals *2*
1.2.2 Planar Surfaces *4*
1.2.3 The Metallic Cylinder (2D) and Sphere (3D) *7*
1.2.3.1 The Electrostatic Approximation *7*
1.2.3.2 Localized Surface Plasmon Resonances of the Cylinder *9*
1.2.3.3 Localized Surface Plasmon Resonances of the Sphere *10*
1.2.3.4 Local Field Enhancements *11*
1.2.4 Size Effects *12*
1.2.5 Shape Effects *12*
1.2.6 Interacting Objects and Gaps *15*
1.2.6.1 Coupled Plasmon Resonances *15*
1.2.6.2 Tip-Enhanced Raman Scattering (TERS) *17*
1.2.7 Choice of Metal *18*
1.2.7.1 Gold versus Silver *18*
1.2.7.2 Other Coinage and Transition Metals *19*
1.3 Field Enhancement Distribution and Localization *20*
1.3.1 Electromagnetic Hot Spots *20*
1.3.2 Long-Tail Distribution of Enhancements *21*
1.4 Electromagnetic Model for the SERS and Fluorescence Enhancement Factors *23*
1.4.1 Enhanced Absorption *23*
1.4.2 Comparison of Raman and Fluorescence Processes *24*
1.4.3 The $|E|^4$ Approximation to SERS Enhancement Factors *27*
1.4.4 Fluorescence Quenching and Enhancement *28*
1.4.5 Comparison of SERS and Fluorescence Enhancements *29*
1.4.6 Other Forms of Enhancements *31*

Surface Enhanced Raman Spectroscopy: Analytical, Biophysical and Life Science Applications. Edited by Sebastian Schlücker
Copyright © 2011 WILEY-VCH Verlag GmbH & Co. KGaA, Weinheim
ISBN: 978-3-527-32567-2

1.5	The Magnitude of the SERS Enhancement Factor in Typical Cases 32
1.6	Conclusions 33
	References 34

2	**Nanoparticle SERS Substrates** 39
	Yuling Wang and Erkang Wang
2.1	Introduction 39
2.2	Preparation and Stability of Metal Nanoparticle Colloidal SERS Substrates 40
2.2.1	Colloidal Spherical Metal Nanoparticles 41
2.2.1.1	Chemical Reaction for Metal Nanoparticle Preparation 41
2.2.1.2	Laser Ablation and Photoreduction for Metal Nanoparticle Preparation 44
2.2.1.3	Size Effect of SERS Signal on Metal Nanoparticles 45
2.2.1.4	Near-Infrared (NIR) Excitation for SERS on Metal Nanoparticles 46
2.2.1.5	Stability of the Metal Nanoparticle Colloids 47
2.2.2	Aggregation of Metal Nanoparticles 47
2.2.3	Bimetallic Nanoparticle SERS Substrate 50
2.2.4	Nanoparticles with Various Shapes 52
2.3	Characterization of Nanoparticle-Based SERS Substrates 57
2.4	Nanoparticles on the Unfunctionalized Solid Surface as SERS Substrates 58
2.5	Conclusion and Outlook 60
	References 60

3	**Quantitative SERS Methods** 71
	Steven E.J. Bell and Alan Stewart
3.1	Introduction 71
3.2	SERS Media 71
3.3	Stability and Shelf Life 73
3.4	Reproducibility and Internal Standards 74
3.5	Selectivity 78
3.6	Conclusion 82
	References 83

4	**Single-Molecule- and Trace Detection by SERS** 87
	Nicholas P.W. Pieczonka, Golam Moula, Adam R. Skarbek, and Ricardo F. Aroca
4.1	Introduction 87
4.1.1	SERS 87
4.1.2	The Two Regimes: Ensemble and Trace/SM 88
4.1.3	Requirements for SM-SERS 89
4.2	Experiments and Results 90
4.2.1	The Langmuir–Blodgett Method for SM-SERRS 90

4.2.2	LB SM-SERRS to Biologically Relevant Systems	91
4.2.3	Experimental Details	93
4.2.4	Single-Molecule Examples	94
4.2.4.1	Tagged Phospholipid	94
4.2.4.2	R18, Octadecyl Rhodamine B	97
4.3	Conclusions	99
	References	99

5 Detection of Persistent Organic Pollutants by Using SERS Sensors Based on Organically Functionalized Ag Nanoparticles 103

Luca Guerrini, Patricio Leyton, Marcelo Campos-Vallette, Concepción Domingo, José V. Garcia-Ramos, and Santiago Sanchez-Cortes

5.1	Introduction	103
5.2	Inclusion Hosts	106
5.2.1	Calixarenes	106
5.2.2	α,ω-Aliphatic Diamines	112
5.3	Contact Hosts	115
5.3.1	Viologens	115
5.3.2	Carbon Nanotubes	118
5.4	Occlusion Hosts	120
5.4.1	Humic Substances	120
5.5	Conclusions	122
	Acknowledgements	124
	References	124

6 SERS and Pharmaceuticals 129

Simona Cîntă Pînzaru and Ioana E. Pavel

6.1	Introduction	129
6.2	SERS of Antipyretics and Analgesics	130
6.3	SERS of Antimalarials	139
6.4	SERS of Anticarcinogenics and Antimutagenics	142
6.4.1	5-Fluorouracil	142
6.4.2	β-Carotene	150
6.5	Concluding Remarks	152
	References	152

7 SERS and Separation Science 155

Alison J. Hobro and Bernhard Lendl

7.1	Introduction	155
7.2	SERS and Capillary Electrophoresis (CE)	157
7.3	SERS and Liquid Chromatography (LC)	161
7.4	SERS and Gas Chromatography (GC)	164
7.5	SERS and Thin Layer Chromatography (TLC)	165
7.6	Other Separation Methods	166

7.7	Conclusions 169
	References 169
8	**SERS and Microfluidics** 173
	Thomas Henkel, Anne März, and Jürgen Popp
8.1	Introduction 173
8.2	Lab-on-a-chip Technology 174
8.3	Microfluidic Platforms and Application for SERS 176
8.3.1	Capillary-Driven Test Stripes 176
8.3.2	Microfluidic Large-Scale Integration and PDMS Microchannels 178
8.3.3	Centrifugal Microfluidics 180
8.3.4	Electrokinetic Platform 181
8.3.5	Droplet-Based Microfluidics 183
8.3.5.1	Straight Plug-Flow Concept 184
8.3.5.2	Surfactant-Stabilized Sample Droplets 184
8.3.5.3	Processing as Foams in Microchannel Systems 185
8.3.5.4	Conclusion 185
8.4	Summary 187
	References 188
9	**Electrochemical SERS and its Application in Analytical, Biophysical and Life Science** 191
	Bin Ren, Yan Cui, De-Yin Wu, and Zhong-Qun Tian
9.1	Electrochemical Surface-Enhanced Raman Spectroscopy 191
9.2	Features of Electrochemical Surface-Enhanced Raman Spectroscopy 192
9.2.1	Electrochemical Double Layer of EC-SERS Systems 193
9.2.2	Potential-Dependent SERS Spectral Characters 194
9.2.3	Electrode Materials and Excitation Energy Dependence 194
9.2.4	Electrolyte Solutions and Solvent Dependence 195
9.2.5	The Electrochemically Influenced SERS Enhancement 195
9.3	Experimental Techniques of EC-SERS 197
9.3.1	Experimental Setup 197
9.3.2	EC-SERS Cell Design 198
9.3.3	Improving the Detection Sensitivity 199
9.3.4	Preparation of SERS-Active Electrode Surfaces 199
9.3.4.1	Electrochemical Oxidation and Reduction Cycles (ORCs) 199
9.3.4.2	Preparation of SERS Substrates Using Metal Nanoparticles 200
9.3.5	SERS Substrate Cleaning 201
9.3.6	An Approach to Reliable SERS Measurement on Bio-related Systems by the Defocussing Method 203
9.4	Applications of EC-SERS 204
9.4.1	Model System – Benzene Adsorption and Reaction on Transition Metal Surfaces 204
9.4.2	SERS for Studying Biological Molecules 206

9.4.2.1	SERS Study of the Adsorption Behaviour of NADH 207
9.4.2.2	SERS Study of Single-Stranded and Double-Stranded DNA on Gold Surfaces 208
9.4.2.3	EC-SERS Study of Cytochrome c on a DNA-Modified Gold Surface 208
9.4.3	EC-SERS as a Method to Improve the Detection Sensitivity of Dopamine 211
9.4.4	Discrimination of Mutations in DNA Sequences by Electrochemical Melting Using SERS as Probing Signal 212
9.5	Perspectives 213
9.5.1	Ordered Nanostructured Electrode Surfaces 214
9.5.2	EC-SERS Study of Cell under Culturing Condition 215
9.5.3	Integration of EC-SERS with Microfluidic Devices 216
9.5.4	Applications of EC-SERS in Biosciences and Biosensors 216
	Acknowledgements 216
	References 217

10   Electron Transfer of Proteins at Membrane Models 219
Peter Hildebrandt, Jiu-Ju Feng, Anja Kranich, Khoa H. Ly, Diego F. Martín, Marcelo Martí, Daniel H. Murgida, Damián A. Paggi, Nattawadee Wisitruangsakul, Murat Sezer, Inez M. Weidinger, and Ingo Zebger

10.1	Introduction 219
10.2	Model Membranes and Membrane Models 221
10.3	Methods for Probing Electron Transfer Processes of Cytochrome c at Coated Electrodes 225
10.4	The Unusual Distance Dependence of the Interfacial Electron Transfer Process 228
10.5	Electron Transfer and Protein Orientational Dynamics 231
10.6	Electric Field Effects on the Electron Transfer Dynamics 232
10.7	Electron Transfer and Protein Structural Changes 234
10.8	Overall Description of the Mechanism and Dynamics of the Interfacial Processes 235
10.9	Interfacial Electric Fields and the Biological Functions of Cytochrome c 237
	References 239

11   Quantitative DNA Analysis Using Surface-Enhanced Resonance Raman Scattering 241
Ross Stevenson, Karen Faulds, and Duncan Graham

11.1	Introduction 241
11.2	SERRS Surfaces 242
11.3	Raman Reporters 245
11.4	SERRS DNA Probes 248
11.5	Sensitivity 252

11.6	Multiplexing	253
11.7	Assays	256
11.8	Conclusion	259
	References	259

12 SERS Microscopy: Nanoparticle Probes and Biomedical Applications 263
Sebastian Schlücker

12.1	Introduction	263
12.2	SERS Nanoparticle Probes	264
12.2.1	Components of a SERS Label	264
12.2.2	Choice of Metal Colloid	265
12.2.3	Choice of Raman Reporter	267
12.2.4	Protection and Stabilization	269
12.3	Biomedical Applications of SERS Microscopy	272
12.3.1	Immunohistochemistry	273
12.3.2	Methodologies in Raman Microspectroscopy	274
12.3.3	Immuno-SERS Microscopy for *In vitro* Tissue Diagnostics	275
12.3.4	Applications *In vivo*	278
12.4	Summary and Outlook	279
	Acknowledgement	280
	References	281

13 1-P and 2-P Excited SERS as Intracellular Probe 285
Janina Kneipp

13.1	From Tags to Probes: Challenges in Intracellular Probing	285
13.1.1	Localization and Targeting	286
13.1.2	Influence of Surroundings on Nanoprobe Aggregation and Stability	287
13.1.3	Probe Identification	290
13.2	Probing of Intracellular Parameters	292
13.3	Surface-Enhanced Hyper Raman Scattering and Its Potential in Studies of Cells	297
	Acknowledgements	301
	References	301

14 Surface- and Tip-Enhanced CARS 305
Taro Ichimura and Satoshi Kawata

14.1	Introduction	305
14.2	CARS : Coherent Anti-Stokes Raman Scattering	305
14.3	Local Enhancement of CARS by Metallic Nanostructures	307
14.4	Surface-Enhanced CARS	309
14.4.1	Experimental System for SECARS Measurements	309
14.4.2	SECARS of Adenine Nanocrystals	310
14.4.3	SECARS of Single-Walled Carbon Nanotubes	314

14.5	Tip-Enhanced CARS	*315*
14.5.1	Experimental System for TECARS Microscopy	*315*
14.5.2	TECARS Imaging of DNA Molecules	*316*
14.5.3	TECARS Imaging of CNTs	*318*
	References	*319*

Index *323*

Preface

The field of surface-enhanced Raman scattering (SERS) is currently undergoing a very dynamic development and many novel directions are rapidly emerging. The aim of this book is to provide an overview of current exciting topics in SERS, focusing on analytical, biophysical and life science applications. International leaders in their respective research areas have contributed to this volume. Their original scientific background and training is quite diverse, ranging from bioorganic chemistry to physical chemistry and solid state physics – in my opinion, this directly reflects the highly multidisciplinary nature of SERS applications, a prerequisite for original and pioneering research between the boundaries of traditionally distinct disciplines. The selection of the scientific topics covered in the 14 chapters is naturally subjective and I must certainly apologize to those who have not received the opportunity to contribute to this edition.

This monograph is intended to be useful for both the newcomer with no or little background in Raman/SERS spectroscopy as well as for the experts in the field who are interested in achieving a quick overview as well as in-depth information on specific subjects.

The first part of this book (Chapters 1–3) lays the foundation for the entire book by providing important theoretical and practical background. Topics are the basic electromagnetic theory of SERS, various aspects of metal colloids as plasmonic nanostructures and practical considerations for quantitative SERS. The second part (Chapters 4–14) covers various analytical, biophysical and life science applications of SERS. Chapters 4 through 8 describe analytical applications of SERS, including single-molecule and trace detection, sensors for detecting organic pollutants based on host–guest systems as well as the detection of pharmaceuticals. Two chapters describe the promising combination of SERS with other analytical techniques such as separation methods and microfluidic platforms for lab-on-a chip detection. Chapters 9 and 10 cover spectroelectrochemistry as a classical and important topic in SERS. After an introduction into the theory and experimental setups for combining SERS with electrochemistry, applications to several biological molecules are summarized. Biophysical applications of spectroelectrochemistry with SERS are focused on the electron transfer in membrane models, in particular cytochrome c on coated electrodes. Chapters 11 through 14 report on recent life science applications of SERS. Quantitative DNA analysis with immense multiplexing and ultrasensitive detection capabilities is demonstrated by surface-enhanced resonance Raman scattering (SERRS). Selective protein localization in cells and tissue specimens via SERS microscopy requires the design and fabrication of functionalized metal colloids for labeling target-specific ligands such as antibodies. Information on intracellular biochemical composition and physiological conditions is accessible via one- or two-photon excited SERS in a label-free approach in conjunction with microscopy. Surface- and tip-enhanced coherent anti-Stokes Raman scattering (CARS) as advanced microspectroscopic techniques with sub-diffraction limited spatial resolution together with first applications to DNA are discussed in the last chapter.

I would like to thank all authors for their hard work and commitment to contribute their chapters. This international and multidisciplinary book project would not have been possible without their dedication. The support from Lesley Belfit (Wiley-VCH) and Manfred Köhl (now Thieme) is greatly appreciated. Thanks to Wolfgang Kiefer for his foreword – many of his former students including the editor have contributed to this book. Finally, I would like to thank my wife Uta-Maria, our sons Jan and Henrik, my parents Marianne and Eberhard as well as my group members for their continuous support.

Osnabrück, August 2010   *Sebastian Schlücker*

List of Contributors

Ricardo F. Aroca
University of Windsor
Department of Chemistry
and Biochemistry
Material and Surface
Science Group
373-3401 Sunset Avenue
Windsor, ON
Canada

Steven E. J. Bell
Queen's University
School of Chemistry and
Chemical Engineering
Innovative Molecular
Materials Group
Belfast BT9 5AG
UK

Marcelo Campos-Vallette
University of Chile
Faculty of Sciences
PO Box 653
Santiago
Chile

Yan Cui
Xiamen University
Department of Chemistry
State Key Laboratory of
Physical Chemistry of
Solid Surfaces
College of Chemistry and
Chemical Engineering
Xiamen 361005
China

Concepción Domingo
Instituto de Estructura de la
Materia
CSIC, Serrano 121
28006 Madrid
Spain

Pablo G. Etchegoin
Victoria University of Wellington
School of Chemical and Physical
Sciences
The MacDiarmid Institute for
Advanced Materials and
Nanotechnology
Kelburn Parade
Gate 7, PO Box 600
Wellington
New Zealand

Surface Enhanced Raman Spectroscopy: Analytical, Biophysical and Life Science Applications. Edited by Sebastian Schlücker
Copyright © 2011 WILEY-VCH Verlag GmbH & Co. KGaA, Weinheim
ISBN: 978-3-527-32567-2

Karen Faulds
University of Strathclyde
Department of Pure and
Applied Chemistry
Centre for Molecular
Nanometrology
WestCHEM
295 Cathedral Street
Glasgow, G1 1XL
UK

Jiu-Ju Feng
Technische Universität Berlin
Institut für Chemie
Sekretariat PC14
Straße des 17. Juni 135
10623 Berlin
Germany

José V. Garcia-Ramos
Instituto de Estructura de la
Materia
CSIC, Serrano 121
28006 Madrid
Spain

Duncan Graham
University of Strathclyde
Department of Pure and
Applied Chemistry
Centre for Molecular
Nanometrology
WestCHEM
295 Cathedral Street
Glasgow, G1 1XL
UK

Luca Guerrini
Instituto de Estructura de la
Materia
CSIC, Serrano 121
28006 Madrid
Spain

Thomas Henkel
Institute of Photonic
Technology e.V.
Nano Biophotonics Department
Albert-Einstein-Str. 9
07745 Jena
Germany

Peter Hildebrandt
Technische Universität Berlin
Institut für Chemie
Sekretariat PC14
Straße des 17. Juni 135
10623 Berlin
Germany

Alison J. Hobro
Vienna University of Technology
Institute for Chemical
Technologies and Analytics
Getreidemarkt 9/164AC
1060 Vienna
Austria

Taro Ichimura
Osaka University
Photonics Advanced
Research Center
2-1 Yamadaoka
Suita, Osaka 565-0871
Japan

Satoshi Kawata
Osaka University
Photonics Advanced
Research Center
2-1 Yamadaoka
Suita, Osaka 565-0871
Japan

Janina Kneipp
Humboldt-Universität zu Berlin
BAM Federal Institute for
Materials Research and Testing
and Institute of Chemistry
Richard-Willstätter-Straße 11
Berlin
Germany

Anja Kranich
Technische Universität Berlin
Institut für Chemie
Sekretariat PC14
Straße des 17. Juni 135
10623 Berlin
Germany

Bernhard Lendl
Vienna University of Technology
Institute for Chemical
Technologies and Analytics
Getreidemarkt 9/164AC
1060 Vienna
Austria

Eric C. Le Ru
Victoria University of Wellington
School of Chemical and Physical
Sciences
The MacDiarmid Institute for
Advanced Materials and
Nanotechnology
Kelburn Parade
Gate 7, PO Box 600
Wellington
New Zealand

Patricio Leyton
Universidad Católica de
Valparaíso
Facultad de Ciencias Básicas y
Matemáticas
Valparaíso
Chile

Khoa H. Ly
Technische Universität Berlin
Institut für Chemie
Sekretariat PC14
Straße des 17. Juni 135
10623 Berlin
Germany

Marcelo Martí
Universidad de Buenos Aires
Ciudad Universitaria
Departamento de Química
Inorgánica, Analítica y Química
Física/INQUIMAE-CONICET
Facultad de Ciencias Exactas y
Naturales
Pab. 2, piso 1
C1428EHA Buenos Aires
Argentina

Diego F. Martín
Universidad de Buenos Aires
Ciudad Universitaria
Departamento de Química
Inorgánica, Analítica y Química
Física/INQUIMAE-CONICET
Facultad de Ciencias Exactas y
Naturales
Pab. 2, piso 1
C1428EHA Buenos Aires
Argentina

Anne März
Friedrich Schiller
University of Jena
Institute of Physical Chemistry
Helmholzweg 4
07743 Jena
Germany

Golam Moula
University of Windsor
Department of Chemistry
and Biochemistry
Material and Surface
Science Group
373-3401 Sunset Avenue
Windsor, ON
Canada

Daniel H. Murgida
Universidad de Buenos Aires
Ciudad Universitaria
Departamento de Química
Inorgánica, Analítica y Química
Física/INQUIMAE-CONICET
Facultad de Ciencias Exactas y
Naturales
Pab. 2, piso 1
C1428EHA Buenos Aires
Argentina

Damián A. Paggi
Universidad de Buenos Aires
Ciudad Universitaria
Departamento de Química
Inorgánica, Analítica y Química
Física/INQUIMAE-CONICET
Facultad de Ciencias Exactas y
Naturales
Pab. 2, piso 1
C1428EHA Buenos Aires
Argentina

Ioana E. Pavel
Wright State University
Department of Chemistry
3640 Colonel Glenn Hwy.
Dayton, OH 45435-0001
USA

Nicholas P.W. Pieczonka
University of Windsor
Department of Chemistry
and Biochemistry
Material and Surface
Science Group
373-3401 Sunset Avenue
Windsor, ON
Canada

Simona Cîntă Pînzaru
Babeş-Bolyai University
Molecular Spectroscopy
Department
Kogălniceanu 1
400084 Cluj-Napoca
Romania

Jürgen Popp
Institute of Photonic
Technology e.V.
Nano Biophotonics Department
Albert-Einstein-Str. 9
07745 Jena
Germany

and

Friedrich Schiller
University of Jena
Institute of Physical Chemistry
Helmholzweg 4
07743 Jena
Germany

Bin Ren
Xiamen University
Department of Chemistry
State Key Laboratory of
Physical Chemistry of
Solid Surfaces
College of Chemistry and
Chemical Engineering
Xiamen 361005
China

Santiago Sanchez-Cortes
Instituto de Estructura de la
Materia
CSIC, Serrano 121
28006 Madrid
Spain

Sebastian Schlücker
University of Osnabrück
Department of Physics
Barbarastraße 7
49076 Osnabrück
Germany

Murat Sezer
Technische Universität Berlin
Institut für Chemie
Sekretariat PC14
Straße des 17. Juni 135
10623 Berlin
Germany

Adam R. Skarbek
University of Windsor
Department of Chemistry
and Biochemistry
Material and Surface
Science Group
373-3401 Sunset Avenue
Windsor, ON
Canada

Ross Stevenson
University of Strathclyde
Department of Pure and
Applied Chemistry
Centre for Molecular
Nanometrology
WestCHEM
295 Cathedral Street
Glasgow, G1 1XL
UK

Alan Stewart
Queen's University
School of Chemistry and
Chemical Engineering
Innovative Molecular
Materials Group
Belfast BT9 5AG
UK

Zhong-Qun Tian
Xiamen University
Department of Chemistry
State Key Laboratory of
Physical Chemistry of
Solid Surfaces
College of Chemistry and
Chemical Engineering
Xiamen 361005
China

Erkang Wang
Changchun Institute of
Applied Chemistry
Chinese Academy of Science
State Key Laboratory of
Electroanalytical Chemistry
Changchun 130022
Jilin
China

Yuling Wang
Changchun Institute of
Applied Chemistry
Chinese Academy of Science
State Key Laboratory of
Electroanalytical Chemistry
Changchun 130022
Jilin
China

Inez M. Weidinger
Technische Universität Berlin
Institut für Chemie
Sekretariat PC14
Straße des 17. Juni 135
10623 Berlin
Germany

Nattawadee Wisitruangsakul
Technische Universität Berlin
Institut für Chemie
Sekretariat PC14
Straße des 17. Juni 135
10623 Berlin
Germany

De-Yin Wu
Xiamen University
Department of Chemistry
State Key Laboratory of
Physical Chemistry of
Solid Surfaces
College of Chemistry and
Chemical Engineering
Xiamen 361005
China

Ingo Zebger
Technische Universität Berlin
Institut für Chemie
Sekretariat PC14
Straße des 17. Juni 135
10623 Berlin
Germany

1
Basic Electromagnetic Theory of SERS
Pablo G. Etchegoin and Eric C. Le Ru

1.1
Introduction

This chapter is aimed at introducing the newcomer to the field of surface-enhanced Raman spectroscopy (SERS), and is not intended to supplant the already available exhaustive literature in the field either in the form of review articles [1, 2] or books [3, 4]. As a technique, SERS is relatively exposed to the dangers of specialization due to its (intrinsic) multidisciplinary nature. The technique is becoming widespread and is finding new and exciting horizons in analytical chemistry [5–7], biology and biotechnology [8–12], forensic science [13, 14] and in the study of artistic objects [15–17]. While this is in many ways an advantage, it is also a handicap in the sense that scientists approaching the technique from a more 'biological' or 'applied' aspect might not have the appropriate background (or predisposition) to venture into the depths of electromagnetic theory and to understand the basic concepts of the theory of plasmon resonances in metallic nanostructures. This could be particularly true for students in the biotechnology field, who might find it desirable to have access to the elementary concepts (with a bare minimum of mathematics) but with enough insight to understand what they are actually doing in the lab. We believe that the success and use of the technique – in an environment which is by nature multidisciplinary – will be more effective if accessible presentations of the basic principles aimed at broader audiences are available at all times (and reviewed over prudent periods of time). This chapter (hopefully) fulfils part of that requirement.

This chapter is organized as follows: in Section 1.2, we introduce the basic principles of plasmon resonances and their associated field enhancements. Section 1.3, on the other hand, looks at the field enhancement distribution and localization produced by these plasmon resonances, while Sections 1.4 and 1.5 study the origin of the enhancement factor (EF) and its characteristic magnitude. Finally, Section 1.6 presents some conclusions and summarizes several main concepts.

Surface Enhanced Raman Spectroscopy: Analytical, Biophysical and Life Science Applications. Edited by Sebastian Schlücker
Copyright © 2011 WILEY-VCH Verlag GmbH & Co. KGaA, Weinheim
ISBN: 978-3-527-32567-2

1.2
Plasmon Resonances and Field Enhancements

1.2.1
Optical Properties of Simple Metals

None of the modern optical techniques such as surface-enhanced fluorescence (SEF) [18–20], surface plasmon resonance spectroscopy [21–23] or SERS itself [1, 4] would exist without the particular optical properties of *coinage metals* (with silver (Ag) and gold (Au) standing out as the most useful ones). The first obvious question is then what is it that makes the optical properties of metals so interesting? Hence, it is worth spending a few paragraphs on the topic of the optical properties of *bulk metals* such as Ag and Au to understand why they are so interesting, and why we use them in the aforementioned techniques.

The optical properties of bulk materials are characterized by their *dielectric function* $\epsilon(\omega)$. Most students from scientific disciplines would have come across the related *index of refraction* $n(\omega)$, which is linked to the former by $n(\omega) = \sqrt{\epsilon(\omega)}$. Both $n(\omega)$ and $\epsilon(\omega)$ depend on the frequency (ω) of the light (with $\omega = 2\pi c/\lambda$, where c is the speed of light and λ the wavelength), due to the fact that most materials respond differently to electromagnetic waves at different frequencies (wavelengths). The dielectric function can therefore be considered indistinctly as either a function of ω ($\epsilon(\omega)$) or λ ($\epsilon(\lambda)$). We shall use one or the other according to convenience. In the most elementary treatments of the optics of material objects (lenses, prisms, etc.) [24], both the dielectric function and the index of refraction are *positive real numbers* (more precisely ϵ, $n \geq 1$). More often than not, however, the dielectric function of materials at a given wavelength will be a *complex* (rather than real) number, and the material will not be transparent. In fact, this is more the rule than the exception, since the list of transparent materials constitutes a really small fraction of the materials we see around us. Metals are amongst the list of materials in which $\epsilon(\omega)$ is complex. The ultimate reason for the optical properties of materials is their *electronic structure*, and this is a canonical topic in solid-state theory [25, 26]. We shall not dwell too much on the details of the connection between the dielectric function of metals and their electronic structure (see Appendix D of Ref. [4] for a slightly more in-depth discussion), but rather take the properties of $\epsilon(\omega)$ of metals as given.

Figure 1.1 shows the dielectric functions of Ag and Au with their real and imaginary parts spanning from the near-UV (~300 nm) to the near-IR (NIR) range (~900 nm). These are analytical representations that interpolate rather well a collection of experimental results for $\epsilon(\lambda)$ obtained with different techniques. The accuracy and limitations of these fits are discussed in more detail in Refs. [4, 27]; here, we shall take these results as the starting point of our discussion on why the optical properties of metals are interesting. The main characteristics of the real and imaginary parts of the bulk $\epsilon(\lambda)$ for both metals can be summarized as follows:

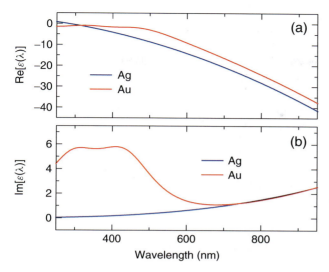

Figure 1.1 The real (a) and imaginary (b) parts of $\epsilon(\lambda)$ for the two most useful metals in SERS; that is, Ag and Au. Note the different vertical scales in (a) and (b); the imaginary parts of $\epsilon(\lambda)$ span over a smaller range and they are always *positive*. The real parts are *negative* across the visible range ($\sim$400–750 nm) and show the overall (expected) characteristic of the simplest description of the dielectric function of metals (the lossless Drude model [25]), which predicts a $\sim -\lambda^2$ dependence for real part of $\epsilon(\lambda)$ at long wavelengths. See the text and Ref. [4] for further details.

- The real part of the dielectric function of both metals, for most of the visible range, is both *large* (in magnitude) and *negative*. Later, this will turn out to be one of the most important properties of these metals as far as their optical properties are concerned, and one of the main reasons for their usefulness as plasmonic materials. Furthermore, ignoring the imaginary parts of $\epsilon(\lambda)$ momentarily, we can claim that the real parts follow at long wavelengths one of the simplest models for the dielectric function of a (lossless) metal, which is the *lossless Drude model*. The latter predicts a dielectric function of the form [4, 25, 26]:

$$\epsilon = \epsilon_\infty \left(1 - \frac{\omega_p^2}{\omega^2}\right) = \epsilon_\infty \left(1 - \frac{\lambda^2}{\lambda_p^2}\right) \tag{1.1}$$

where $\omega_p = 2\pi c/\lambda_p$ is the so-called *plasma frequency*[1]) of the metal (proportional to the square root of the density of free electrons in it). The first expression on the right-hand side in Equation (1.1) holds if we want to express the dielectric function ϵ as a function of ω, while the last expression holds if ϵ is expressed as a function of $\lambda (= 2\pi c/\omega)$. Figure 1.1a reveals that both Ag and Au have actually very similar electronic densities, since the real parts of their dielectric functions are not too far away from each other. This is the approximate quadratic

---

1) For both Au and Ag, $\lambda_p = 2\pi c/\omega_p$ is around $\sim$280 nm; that is, in the UV range.

downturn of the real part of $\epsilon(\lambda)$ seen in Figure 1.1a for longer wavelengths. We can see that, to a good approximation, the simplest lossless Drude model describes already a good fraction of the experimental results for the real parts seen in Figure 1.1a.

- Real bulk metals are *not* lossless, and this is where the imaginary part of $\epsilon(\lambda)$ comes into play. Even though when the Im[$\epsilon(\lambda)$] for both metals are smaller than their real counterparts for most of the visible range, their effects are important and – in some cases – crucial. The imaginary part is always related to the *absorption* of the material (a material with Im[$\epsilon(\lambda)$] = 0 does not absorb light, and has a real index of refraction $n(\lambda) = \sqrt{\text{Re}[\epsilon(\lambda)]}$). It turns out that the imaginary part of $\epsilon(\lambda)$ for Ag can be obtained by a relatively easy generalization of the lossless Drude model (Equation 1.1). For Au, the situation is slightly more complicated; $\epsilon(\lambda)$ has additional contributions (in addition to that from the free electrons) from other electronic transitions in its electronic band structure [27]. This is the reason for the relatively higher absorption of Au (with respect to Ag) for $\lambda \leq 600$ nm, with a 'double hump' structure in the imaginary part (~400 nm), which comes from the so-called interband electronic transitions. Note, however, that for $\lambda \geq 600$ nm, the imaginary parts of $\epsilon(\lambda)$ for both Ag and Au become completely comparable (Figure 1.1b) and – with their real parts being comparable too in this range – both materials are similar (from the viewpoint of their electromagnetic response). Their surface chemistries are of course different, and one material might be preferred over the other for specific chemical reasons. But, as far as the electromagnetic response is concerned, Au is comparable to Ag in the near- and far-IR range.

1.2.2
Planar Surfaces

Once the complex dielectric function $\epsilon(\lambda)$ is known, all the electromagnetic properties of the material can be calculated in different geometries. The normal reflectance R (in the direction perpendicular to the surface) arises as a natural consequence of matching the boundary conditions of the fields at the interface.[2] The reflectance is plotted for Ag and Au in Figure 1.2b using the complex dielectric functions shown in Figure 1.1a and b. Silver has a very high reflectivity ~100% across the entire visible range. Gold, on the contrary, has ~50% for $\lambda \leq 600$ nm (from the yellow-green region towards shorter wavelengths in the UV). This is the reason for the 'yellowish/reddish' colour of flat gold when compared to silver. The overall high reflectivity of Ag does not come as a surprise; this is the reason why Ag

---

2) The standard boundary conditions for all electromagnetic problems require that the components of the electric field *parallel* to the surface (on both sides of the interface) are equal, as well as the *perpendicular* components of the displacement vector $\mathbf{D} = \epsilon(\lambda)\mathbf{E}$. In standard notation [4, 28, 29] for an interface between medium 1 and 2: $E_1^{\|} = E_2^{\|}$, and $\epsilon_1(\lambda)E_1^{\perp} = \epsilon_2(\lambda)E_2^{\perp}$. The normal reflectance at a planar surface between the two media is given by $R = |(n_2 - n_1)/(n_2 + n_1)|^2$, with $n_1 = \sqrt{\epsilon_1}$ and $n_2 = \sqrt{\epsilon_2}$.

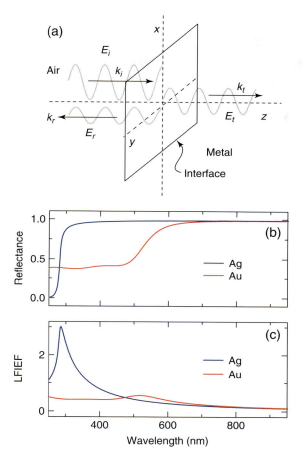

Figure 1.2 (a) An incident electromagnetic wave (with electric field E_i and wavevector k_i) impinges from the left (along z) onto a sharp interface with a (bulk) metal lying in the x–y plane, and transmitted and reflected waves result. The amplitude of the reflected (E_r) and transmitted (E_t) waves result from the matching of boundary conditions for the field at the interface, and depend only on the (complex) index of refraction of the metal ($n = \sqrt{\epsilon(\lambda)}$) [4, 28, 29]. (b) Reflectance at normal incidence for Au and Ag, using the dielectric functions shown in Figure 1.1. Note that Ag has a reflectance close to ~100% across the entire visible range, while for Au the reflectivity decreases from ~500 nm towards the UV range. (c) Local field intensity enhancement factor (LFIEF) at the surface of the metal [(x–y) plane] for Au and Ag (at normal incidence). Note that, in general, the LFIEF is <1 across the visible, meaning that the intensity is typically 'quenched' at the (flat) surface of the metal.

is used as a mirror in the visible. Gold mirrors, on the other hand, are preferred for NIR applications where it reflects as much as Ag, but it is more stable with respect to effects caused by long-term exposure to ambient conditions.

Another aspect of interest, while we dwell on the simplest of examples, is the Local Field Intensity Enhancement Factor (LFIEF) at the surface (i.e. by how much the intensity of the electromagnetic field is changed with respect to the intensity we would have had at the place without the metal). The local field intensity at a specific point is proportional to the square of the electric field amplitude at that point: $|E(r)|^2$.[3] The LFIEF at a specific point is then the normalized value of $|E(r)|^2$ with respect to the intensity of the incoming field at that point: $|E_0(r)|^2$. Explicitly,

$$\text{LFIEF}(r) = |E(r)|^2/|E_0(r)|^2 \tag{1.2}$$

The LFIEF is, therefore, an adimensional magnitude expressing the (normalized) change in local intensity at a specific point produced by the presence of objects (which perturb the electric field of the light). Any optical technique that depends on the intensity of the light at a specific point will hence be linked to the LFIEF and, in general, depending on whether the LFIEF is >1 or <1 the optical process involved will be enhanced or quenched. The LFIEF will also depend on ω (or, equivalently, on λ), simply because the local field at a specific point depends on ω. We can formally write

$$\text{LFIEF}(r, \omega) = |E(r, \omega)|^2/|E_0(r, \omega)|^2 \tag{1.3}$$

In general, however, we will simplify the notation and emphasize only the most important dependence for the explanation of a specific aspect. We will refer, for example, to the LFIEF at a well-specified point in a geometry and at given frequency ω simply as LFIEF(ω).

The LFIEF at a flat surface for normal incidence (which results from the interference between E_i and E_r on the surface, see Figure 1.2) is another aspect of the classical problems in basic electromagnetic theory (and optics) [4, 28, 29], and (like R) is solely determined by ϵ_1 and ϵ_2.

The LFIEF on the surface – for both an interface of Ag and Au with air – are plotted in Figure 1.2c.[4] As can be appreciated, the LFIEF is in general for normal incidence <1 at the surface of a planar interface separating a bulk metal (like Au or Ag) from air; that is, the intensity is 'quenched' at the surface compared to what we would have had in its absence. An ideal (100%-reflective) lossless metal will create a field on the surface, which cancels exactly the incoming one $((E_i + E_r) = 0)$, thus cancelling exactly the transmitted field too ($E_t = 0$) and sending the impinging electromagnetic wave back in the opposite direction from where it came. Therefore, a low LFIEF at the surface (achieved by the condition $E_i \sim -E_r$) is a natural consequence of having a very high reflectivity. In reality, the cancellation is not complete, but it is efficient enough to guarantee a low LFIEF on the surface and a concomitant high reflectance. The LFIEF is only >1 for Ag when $\lambda \leq 400$ nm, but this is the region where it actually stops being a good reflector. Larger (but, nevertheless moderate) LFIEFs may also be obtained at other angles of incidence (different from normal incidence shown here) and it is then also dependent on the incident polarization [4]. The reflection process for an arbitrary angle of incidence

---

3) We shall avoid vector notations throughout for simplicity.
4) The LFIEF immediately above the flat surface for normal incidence is given by $\text{LFIEF} = \left|\frac{4n_1}{n_1+n_2}\right|^2$ [4, 28, 29], where (as before) $n_1 = \sqrt{\epsilon_1}$ and $n_2 = \sqrt{\epsilon_2}$.

and arbitrary polarization results in Fresnel formulas, which is (again) another aspect of the classic topics in the basic electromagnetic theory of optics [4, 28, 29].

It might appear up to this point that metals do not present any major advantage with respect to other types of materials as far as SERS is concerned. Except for their highly reflective properties (most familiar to everybody), it appears that molecules spread over a flat interface on the metal will not have much to win in terms of surface enhancement of the electromagnetic field. The key point to the usefulness of metals as photonic materials starts once we start considering the effects of *shapes*. This is the subject of the next section.

1.2.3
The Metallic Cylinder (2D) and Sphere (3D)

Let us consider now a different problem; the case of a (long) metallic rod (Au or Ag) embedded in a non-absorbing dielectric medium (with $\epsilon_M \geq 1$) being impinged by an electromagnetic wave of wavelength λ with polarization E_i perpendicular to the main axis of the rod. For all practical purposes, we can consider the problem to be two dimensional (2D) as shown in Figure 1.3a, for as long as the aspect ratio of the cylinder (basically its length divided by its diameter) is $\gg 1$.

1.2.3.1 The Electrostatic Approximation

If we want to know what happens to the electromagnetic field around the cylinder now, we have no other option but to actually solve Maxwell's equations subject to appropriate boundary conditions. This can be, in general, a rather difficult undertaking [4, 28, 29]. Full analytical solutions of Maxwell's equations exist in a handful of simple geometries, and these are useful to underpin basic concepts and ideas. More often than not, however, one has to resort to numerical solutions within some approximation scheme [4, 28, 29].

One useful approximation scheme – widely used in the literature – is the *electrostatic approximation*, explained in full detail in Refs. [4, 30]. This is schematically represented in Figure 1.3b for the aforementioned cylinder. In this approximation, the problem is solved as in electrostatics, but at different ω's (λ's) using the complex dielectric function of the material at that frequency (wavelength). Accordingly, we solve in this case *Poisson's equation* for the electrostatic field. This is much easier than solving Maxwell's equations in full, for it involves only an equation for the scalar electric potential $\phi(\mathbf{r})$. Nevertheless, at the time of satisfying boundary conditions we do use the complex dielectric function $\epsilon(\lambda)$.

The electrostatic approximation corresponds then to ignoring the presence of the wavevector k (or, equivalently, the wavelength $\lambda = 2\pi/k$) in Figure 1.3a. The applied electric field does not have a 'wavelength', therefore, but rather it is a uniform field oscillating up and down with frequency ω. Obviously, this is an approximation and it is bound to fail in many cases. It is not too difficult to imagine that the electrostatic approximation works well when the size of the object is much smaller than the wavelength. In this case, the electric field of the light will be in any case

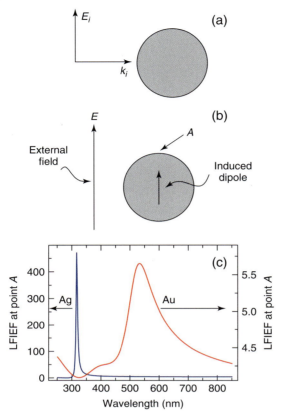

Figure 1.3 (a) A cylindrical metallic object is impinged by an electromagnetic wave coming from the side with wavevector k_i and polarization E_i (on the plane of the page). (b) When the object is small compared to the wavelength ($\leq$ 10–20 nm), it is possible to gain some insight into the situation by solving the problem *electrostatically* [4, 30] (i.e. in a constant electric field in the direction of the polarization and with no wavevector present). The boundary conditions at different wavelengths ($\lambda = 2\pi c/\omega$) are still specified by $\epsilon(\lambda)$, and can be fulfilled exactly at the surface of the cylinder by considering the superposition of an *induced dipole* and the external field (Equation 1.4). (c) The local field intensity enhancement factor (LFIEF) at point A (immediately above the surface) for Ag or Au cylinder and for different wavelengths. Note the different scales for Au and Ag in (c). Silver achieves a much higher LFIEF (at a much shorter wavelength). This is due to the much less absorption of Ag compared to Au in the place where the resonance condition for a cylinder is achieved: $\mathrm{Re}[\epsilon(\lambda)] = -1$ in air.

approximately constant over distances comparable to the size of the object, and it will look like a uniform field oscillating up and down at a frequency ω. In practice – with typical wavelengths in the visible being in the range $\sim$500–600 nm – it means that the electrostatic approximation will be mostly valid for objects of typical sizes in the range of $\sim$10 nm or smaller. However, it is sometimes used for even larger objects and, in some cases, in limits where it would be clearly invalid [30].

With these limitations in mind, it is always possible to use the electrostatic approximation for *nano-optics* and, indeed, it gives truly valuable insights into the electromagnetic properties of many systems of interest and (more than that) yardstick values for LFIEFs.

1.2.3.2 Localized Surface Plasmon Resonances of the Cylinder

The exact electrostatic solution of a 2D cylinder with dielectric function $\epsilon(\lambda)$ turns out to be analytically tractable. It turns out that boundary conditions on the surface of the cylinder can be *exactly* satisfied by considering the superposition of the external field E with an *induced dipole* centred at the origin, as depicted in Figure 1.3b. As far as points outside the cylinder is concerned then, the electric field looks like the superposition of this dipole (p) at the origin and the external field. We shall not go into the details of the solution of the electrostatic problem to keep the mathematical aspects to a bare minimum, but rather only mention that the magnitude of the induced dipole that satisfies the boundary conditions and solves the electrostatic problem is proportional to

$$p \propto \left(\frac{\epsilon(\lambda) - \epsilon_M}{\epsilon(\lambda) + \epsilon_M} \right) \qquad (1.4)$$

Dipoles in 2D are more complicated than standard dipoles in 3D. Note that it should be described more formally as a 'dipolar line' (in the direction perpendicular to the page as shown in Figure 1.3) rather than a dipole. The 2D solution has certain peculiarities that we are not going to analyse in detail here, but rather concentrate on a few salient aspects. The most important detail of the proportionality in Equation 1.4 is the presence of the denominator $(\epsilon(\lambda) + \epsilon_M)$. With $\epsilon(\lambda)$ being a complex number, it is obviously not possible to satisfy in full the condition $\epsilon(\lambda) = -\epsilon_M$ exactly (which would imply $p \to \infty$). But it is evident, at the same time, that an interesting situation will happen when the real part, at least, satisfies the condition $\text{Re}[\epsilon(\lambda)] = -\epsilon_M$; in particular, if the imaginary part of $\text{Im}[\epsilon(\lambda)]$ at that λ is small. This is, indeed, the case for metals (with Ag being a better example of this than Au). At the wavelength where $\text{Re}[\epsilon(\lambda)] = -\epsilon_M$, the magnitude of p will only be limited by how small the imaginary part of $\epsilon(\lambda)$ is, and this will show as a *resonance* (i.e. a large response of the system), called the dipolar localized surface plasmon (LSP) resonance of the cylinder. A very important point to note here is that this resonance is *purely induced by geometrical aspects* (i.e. the shape of the object, a cylinder in this case), and the fact that we need to satisfy boundary conditions. The denominator $(\epsilon(\lambda) + \epsilon_M)$ is purely a consequence of satisfying the specific boundary conditions for this particular geometry (a cylinder). Note, however, that the condition $\text{Re}[\epsilon(\lambda)] = -\epsilon_M$ introduces a small dependence of the resonance wavelength on the embedding medium (characterized by ϵ_M). As a result, the LSP resonance is red shifted in media with a larger ϵ_M (for example, in water compared to air).

Objects with different shapes will have different resonances (sometimes more than one), and this is one of the most important properties why metals are so interesting for nano-optics. The fact that metals have negative $\text{Re}[\epsilon(\lambda)]$ spanning a wide range in magnitude from $\text{Re}[\epsilon(\lambda)] \sim 0$ when $\lambda \sim \lambda_p$ (Equation 1.1) to very large (and negative) values when $\lambda \to \infty$ ($\omega \to 0$, see Figure 1.1) makes them ideal to satisfy a wide rage of *resonance conditions* that appear in the solutions of the electromagnetic responses of many objects. The latter is aided by the fact that these conditions for the real part happen at λ's where $\text{Im}[\epsilon(\lambda)]$ is either small or at least not too large (Figure 1.1). Geometry-induced resonances are at the heart of plasmonics and the usefulness of coinage metals.

1.2.3.3 Localized Surface Plasmon Resonances of the Sphere

We give an additional example to show how the resonance changes with geometry with the standard case of the *metallic sphere* in Figure 1.4. The sphere is one of the problems where we can still resort to exact solutions of Maxwell's equations (Mie theory [4]) or approximate (electrostatic approximation) solutions of the electromagnetic problem. It has been treated in full detail in the literature [4] and we, therefore, only mention here some essential aspects in the simplest of approximations. As before (for the cylinder) the electrostatic problem of a sphere in a uniform field can be solved analytically [28, 29]. It turns out that (as in the case of the cylinder) the electrostatic boundary conditions on the sphere can be satisfied *exactly* by considering the superposition of an induced dipole at the origin p with the external applied field E. However, the magnitude of the induced dipole is now (in the 3D problem) proportional to

$$p \propto \left(\frac{\epsilon(\lambda) - \epsilon_M}{\epsilon(\lambda) + 2\epsilon_M} \right) \tag{1.5}$$

The most important change with respect to the previous case is the fact that the resonance condition in the denominator has changed to $\epsilon(\lambda) = -2\epsilon_M$. As before, however, this condition cannot be satisfied exactly because of the presence

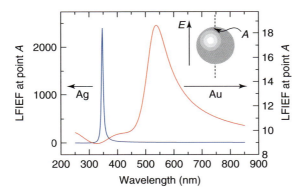

Figure 1.4 Local field intensity enhancement factor (LFIEF) at point *A* (inset) on a sphere of either Au or Ag in the electrostatic approximation [4, 30]. The points of largest LFIEFs on the surface are along the main symmetry axis in the direction defined by the polarization of the electric field *E*. Note that maximum LFIEFs are higher than in the case of cylinders from the same materials (Figure 1.3c) and also that the maxima of the LFIEFs happen at slightly different frequencies; accounting for a different resonance condition (with respect to cylinders) given by $\text{Re}[\epsilon(\lambda)] = -2$ (Equation 1.5) in air.

of the imaginary part of $\epsilon(\lambda)$, but a shape-induced resonance condition will arise when $\text{Re}[\epsilon(\lambda)] = -2\epsilon_M$ and will be limited only by how small $\text{Im}[\epsilon(\lambda)]$ is at that particular λ.

1.2.3.4 Local Field Enhancements

On the surface of the sphere, the places with the largest local fields are the two points along the axis that goes through the centre of the sphere, and is oriented in the direction of the external field (one of them labeled as 'point A' in Figure 1.4). This can be easily understood if we look at the superposition of the induced dipole and the external field, for these are the two points where the two add up (on the surface) in the same direction. In 2D (because of the peculiarities of dipoles in two dimensions) it turns out that the intensity is constant over the surface of the cylinder! We still evaluate, though, the intensity at one point on the surface (also labeled as point A in Figure 1.3b). In Figure 1.3c, we calculate LFIEF at point A as a function of wavelength for the dielectric functions of Ag and Au (given in Figure 1.1). Note the different scales (on the left and right) in Figure 1.3c for the Ag or Au cases. The LFIEF then tells us how bigger or smaller the intensity at point A (Figure 1.3b) will be due to the presence of the cylinder. This would be the *intensity enhancement* that a molecule would experience if it were located at that position. As can be appreciated from Figure 1.3c, the individual LFIEFs have a characteristic peak. This peak appears at the wavelength λ where the condition $\text{Re}[\epsilon(\lambda)] = -\epsilon_M$ is satisfied (Figure 1.1). Ag has a stronger (and narrower) LFIEF resonance. This can be again easily understood by the fact that the condition $\text{Re}[\epsilon(\lambda)] = -\epsilon_M$ is satisfied in Ag at a wavelength λ where the imaginary part is much smaller (comparatively speaking) than that of Au at its corresponding resonance frequency. This makes the resonance in Au lossy and, accordingly, broad. An important concept to realize here is that these resonances happen at λ's where there is no intrinsic feature (or peak) in the *bulk* dielectric function of the materials themselves. In other words, these resonances appear as purely *geometrical* aspects of the problem.

On the other hand, Figure 1.4 displays the LFIEF at point A on a sphere as a function of wavelength for Au and Ag. Figure 1.4 shows again the clear presence of resonance peaks where the LFIEF is large for both Ag or Au spheres. Note that the resonances occur at slightly different wavelengths from those in the 2D cylinder, accounting for the new resonance condition $\text{Re}[\epsilon(\lambda)] = -2\epsilon_M$ (instead of $\text{Re}[\epsilon(\lambda)] = -\epsilon_M$ for the cylinder). Note also that the LFIEFs are larger than in the case of the cylinder. In the case of Ag, a molecule sitting at point A on the sphere will experience (at resonance) more than three orders of magnitude intensity compared to what it would have experienced otherwise.

Despite their simplicity, these truly basic examples show already why plasmon resonances in metals are interesting and important. Intensities can be boosted by large factors and resonance can be 'tuned' by geometrical aspects of the problem. These two basic topics are further explored in the next section.

1.2.4
Size Effects

Note that in the electrostatic approximation, the problem becomes *scale invariant*; that is, it does not really matter what the actual size of the sphere is. If we increase or decrease the size of the sphere by a certain factor, the LFIEF will still be the same in the electrostatic approximation. Nevertheless, the conditions under which the electrostatic approximation will represent most faithfully the *real* solution of the problem (which is *not* electrostatic [4]) is when the size of the sphere is a few tens of nanometers (at most) in size. In all the examples, therefore, we always assume objects in the tens-of-nanometers size range, even though the actual solution of the problem is independent of this assumption.

But *size does matter*, and for objects in the range of typical dimension $\sim$30–100 nm there will be, in general, size effects (see also Section 2.2.1.3). The fact that the size of the object is now a substantial fraction of the wavelength cannot be ignored anymore and the electrostatic approximation fails. The effect on the plasmon resonances of different sizes is very often really complicated to analyse in simple terms and relies mostly on the numerical solution of Maxwell's equations. But the size effects can be qualitatively summarized as follows:

- LSP resonances red shift as the size increases.
- LSP resonances are strongly damped as the size increases, mostly as a result of increased radiation losses. This results in the broadening of the resonance, and more importantly in a dramatic decrease in the associated LFIEF. The resonance (and any substantial LFIEF) ultimately disappears for large sizes, typically 100 nm for dipolar LSP in spheres, but possibly at larger sizes for other geometries.
- Another typical consequence of size is the appearance of resonances that do not exist in the small size limit (where the electrostatic approximation was sufficient). These 'new' size-related resonances (for a fixed shape) are typically related to the activation of *multipolar resonances* (with the quadrupolar resonance playing typically the most important role) that do not couple to light very effectively in the limit of small sizes. Size-induced resonances add yet another layer to the diversity (and complexity) of optical phenomena in metallic nanostructures.

1.2.5
Shape Effects

The two examples given earlier highlight the concept of *geometry-induced resonance* which is a defining characteristic of LSP resonances in metallic nanostructures (see also Section 2.2.4). The obvious question that now arises is what happens with other geometries? Unfortunately, a rule of thumb is that the simplest examples of geometries are at the same time the *only* ones that can be typically solved analytically. Except for a handful of exceptions, even some of the simplest geometries beyond cylinders and spheres (like triangular shapes or prisms) are not analytically soluble (not even in the electrostatic approximation). Numerical solutions come here as an aid to these cases and, as a rule of thumb, we have no option but to resort to them in order to obtain the solution of the electromagnetic problem.

Figure 1.5 shows an example of shape-induced resonances for a 2D shape with a triangular cross section. Figure 1.5a and b shows the spatial distribution of the LFIEF at two different wavelengths, while Figure 1.5c shows the spectral dependence of the LFIEF on the surface at two different points (labeled A and B in Figure 1.5a and b). Note that in an equilateral triangle all vertices are, in principle, equivalent, but the presence of an electric field in the vertical direction in this case breaks the symmetry of the problem and makes points like A and B in Figure 1.5a and b inequivalent. Figure 1.5c shows the spectral (wavelength) dependence of the LFIEF at points A and B (on the surface) of the triangular shape. The problem was solved numerically in the electrostatic approximation [4]. A few obvious conclusions arise from the calculation in Figure 1.5c:

- Unlike what happens in the examples of the cylinder and the sphere, there will be, in general, more than one resonance condition associated with a given shape. Some of these resonances have complicated spatial distributions of the enhancement.
- Different points on the surface can have their maximum LFIEFs at different wavelengths (as it is the case for points A and B in Figure 1.5a and b). The LFIEF is strongly position dependent in most cases.
- The maxima of the LFIEFs shown in Figure 1.5a and b are simple examples of what is normally dubbed as *the lightning rod effect* (i.e. electric fields concentrating at sharp ends). One should keep in mind though that the real lightning rod effect makes reference to a truly electrostatic situation ($\omega = 0$, or $\lambda = +\infty$), while the cases shown in Figure 1.5a and b are resonances that occur at frequencies in the visible range ($\omega \sim 10^{15}$ Hz).
- The LFIEF in more complicated shapes than the simplest cases of the cylinder or sphere can be really high in some circumstances (in particular, if the shape has sharp corners as in Figure 1.5). In the specific case of the triangular shape, the LFIEF at point B felt by a molecule can achieve (at its maximum) well over four orders of magnitude the intensity it would have felt if it had been in free space under the influence of the same field.
- In general, the resonances (in their wavelength position and intensity) will depend not only on the shape but also on the orientation with respect to the field. If the direction of the electric field is changed in Figure 1.5 while keeping the triangular shape in the same position, the LFIEFs at points A and B will change accordingly (in both intensity and frequency position). In the *real* solution of the electromagnetic problem (not in the electrostatic approximation), there is also the direction imposed by the wavevector of the light k. While the latter has in general less importance than the direction of E (in particular, for really small objects of the order of ~ 10 nm in size), there are many subtle details of the electromagnetic field distribution that does depend on it.

The dependence of the LFIEFs on shape and polarization/wavevector directions is what gives plasmon resonances in metallic nano-objects their vast richness and

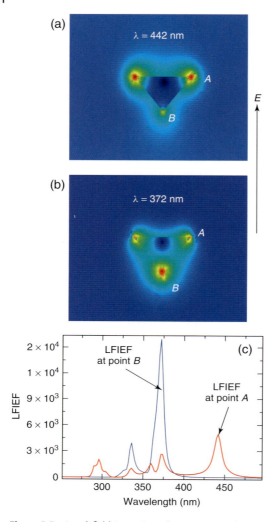

Figure 1.5 Local field intensity enhancement factor (LFIEF) at two different wavelengths for a triangular Ag shape in the electrostatic approximation. The direction of the polarization is vertical. The LFIEF is shown at two different wavelengths: (a) 442 nm and (b) 372 nm, where the highest LFIEFs occur at points A and B, respectively. The plots are in a logarithmic (false) colour scale, where red is the highest and blue the lowest. The wavelength dependence of the LFIEF at A and B is shown in c. These are examples of shape-induced resonances, affecting the enhancement factor at different points on the surface of the shape. Note that point A achieves the maximum LFIEF at a different wavelength from point B. Even when the triangular cross section is equilateral and all vertices are equivalent, the electric field breaks the symmetry of the problem and points A and B are no longer equivalent. The resonances at A and B are simple examples of the 'lightning rod effect' in the optical range (see the field distribution in (a) and (b)).

complex variety of phenomena. In addition, once interactions among objects are introduced and the concept of coupled plasmon resonances appears, these are *not* the only factors that can produce them.

1.2.6
Interacting Objects and Gaps

1.2.6.1 Coupled Plasmon Resonances
The induced dipoles appearing in the cylinder and sphere cases (Figures 1.3 and 1.4) provide the necessary mind frame to introduce another extremely important effect in the metallic nanostructures; to wit, the existence of *coupled plasmon resonances* for two or more closely spaced objects (see also Sections 2.2.2 and 2.4). We give a somewhat oversimplified description here emphasizing the qualitative aspects.

Imagine the presence of two (instead of one) cylinders, as depicted in Figure 1.6a. If the cylinders are far apart from each other (several diameters) they can be considered as two independent problems of single cylinders. However, as they approach each other, the field produced by their respective induced dipoles start to interact. This interaction can reinforce or weaken the field in certain regions of space. In a manner reminiscent of atomic orbital bonding in atoms (in the H_2 molecule, for example), the interaction of the induced dipoles changes the spatial configuration of the fields (wavefunctions in the case of orbital bonding) and shifts the intrinsic energy of the resonances. In fact, something somewhat reminiscent of the creation of a *bonding* and anti-bonding resonance happens, with the 'bonding' resonance being concentrated in the middle of the two cylinders and being red shifted with respect to the individual (isolated) resonances at far distances. The analogy with orbital bonding is mostly semantic, for they are indeed very different problems. But it helps to understand the qualitative picture of interaction and red shifting of the resonance.

A much more complicated picture arises in the electromagnetic case though [31, 32], with resonances coming from higher-order multipoles being activated by the interaction. Be it as it may, the fact remains that there will be a red shifted plasmon resonance with its intensity mainly concentrated in the middle of the two cylinders, and that this *coupled resonance* comes (primarily) from the interaction of the induced dipoles in each individual cylinder. This can be illustrated by the examples in Figure 1.6b–d, where the LFIEF at the centre of the axis separating the two cylinders in Figure 1.6a is calculated for different separations (gaps). These results are of course limited by the validity of the electrostatic approximation, but the qualitative features do not change with more sophisticated methods. At a separation of 20 nm in Figure 1.6b, the interaction between the two cylinders is weak and we can still see the main individual resonance of the cylinders, slightly affected by the interaction with a small shoulder at shorter wavelengths. When the cylinders are drawn together, a clear red shift of the strongest peak (the dipolar interaction coupled plasmon resonance) can be seen. The strongest coupled plasmon resonance is indicated with an arrow in Figure 1.6b–d. Note that not only the resonance shifts but the LFIEF also becomes larger. At separations of

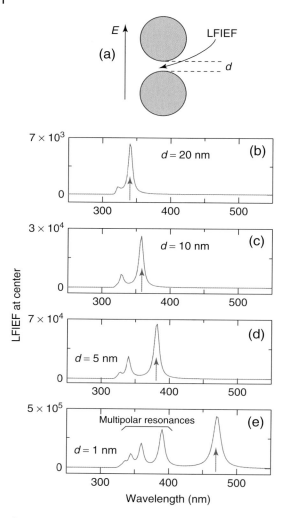

Figure 1.6 Gap effects: (a) the local field intensity enhancement factor (LFIEF) at the centre of a dimer formed by two identical (50 nm radius) Ag metallic cylinders separated by a distance d and with the electric field pointing along the axis joining the two cylinders. The LFIEF as a function of λ (in the electrostatic approximation) is plotted for different separations of (b) 20 nm, (c) 10 nm, (d) 5 nm, and (e) 1 nm. The peak labeled with an arrow is the interacting coupled (dipolar) plasmon resonance between the two cylinders, which red shifts and increases the LFIEF at the centre as the cylinders get closer (note the different vertical scales in (b)–(e)). The additional resonances contributing to the LFIEF at shorter wavelengths (clearly visible in (e), for example) are higher-order multipolar resonances. Note that the maximum LFIEF in (e) is about ∼3 orders of magnitude larger than the maximum value for a single cylinder (Figure 1.3).

the order of ~1 nm, LFIEFs can reach (for Ag) values of the order of ~10^5. These values (when used within the framework of SERS) allow the observation of single molecules, as we shall discuss later. Note that SERS EF is approximately the square of the LFIEF, as explained later.

Coupled plasmon resonances provide some of the highest EFs for optical spectroscopy available. It is difficult to emphasize enough their importance in the fields of SERS and related technique. By the same token, they add an additional level of complexity to the pehenomena described already in terms of shape and size. The complexity of plasmon resonances when the issues of shape, size, and interactions are included is enough to justify an entire field or research by itself: *plasmonics*. One of the main aims of plasmonics (and plasmonic engineering) is to precisely tailor-make and understand nanostructures that can benefit the most from plasmon resonances for applications in optical spectroscopy. Plasmonics (as a field) is, in fact, more general and does not only include the study of nanoparticles but also the properties of propagating plasmons, meta-materials, near-field effects, and so on.

1.2.6.2 Tip-Enhanced Raman Scattering (TERS)

Undoubtedly, a major breakthrough in SERS in the last few years has been the introduction of the related technique tip-enhanced Raman scattering (TERS) [33–35]. Needless to say, coupled plasmon resonances are not limited to gaps between objects with the same geometries, but rather exist (to a larger or lesser degree) for any pair of metallic interacting objects. A particularly important case of coupled plasmon resonances happens between a flat metallic surface and a tip, as displayed schematically in Figure 1.7. Metallic tips of different kinds can be used in this technique, thus opening (simultaneously) the possibility of combining TERS with other types of microscopy (AFM, STM, etc.). Figure 1.7 shows an example in the spirit of the previous cases studied in this chapter; that is, a calculation in the simplest case of the electrostatic approximation. If we set the external field direction along the axis of the tip (as actually shown in Figure 1.7), we are modelling an experimental situation often encountered in which the laser is delivered from the side at almost grazing incidence with the surface and with the polarization along the tip. This choice of electric field is due to the fact that this is the polarization that couples most efficiently to the plasmon resonance resulting from the interaction between the surface and the tip. Figure 1.7a shows a LFIEF map (in a false-colour log-scale) at 620 nm where the clear presence of a hot spot in between the tip and the surface can be seen. In Figure 1.7b, on the other hand, we show the LFIEF at point A (in Figure 1.7a) that is located ~0.5 nm away from the surface and directly below the tip. This is the position that a deposited molecule on the substrate (of typical size ~1 nm) would be occupying under the tip. As can be seen from Figure 1.7b, a clear coupled surface plasmon resonance develops under the tip. The position of the resonance (and its maximum LFIEF) can be tuned to some degree by both the geometrical aspects of the tip and the separation distance from the surface d. The latter can be very efficiently controlled, for TERS systems normally use piezo-controllers (of the same type used in AFM and STM) to position

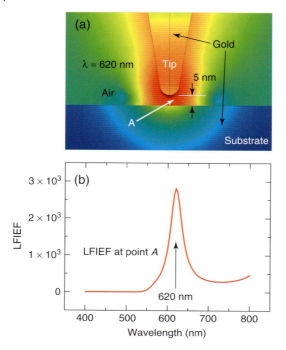

Figure 1.7 (a) A typical TERS geometry: a flat substrate and a tip (both made of gold in this case) are brought together with a gap of a few nanometers in between. The tip in this case has a conical body with a hemisphere termination (10 nm in diameter), and it is separated by 5 nm from the surface. A 3D simulation in the electrostatic approximation at 620 nm excitation (with the field E in the vertical direction) shows the displayed LFIEF map (in a false-colour log-scale as shown in the previous figures). If we monitor the LFIEF for different λ's at point A (which is 0.5 nm above the surface and immediately below the tip), we obtain the result displayed in (b). This would be approximately the position that could be occupied by a molecule lying on the surface. The peak at $\sim$620 nm is the coupled plasmon resonance between the tip and the surface.

the tip and scan over the surface. TERS (and other tip-related techniques; see also Chapter 14) could be considered as a branch of SERS, even though it has a certain life of its own. In the last few years, impressive advances have been made on the technique from both experimental and theoretical aspects, as any brief detour into the current literature can demonstrate (see, for example, Refs. [33–38] to acquire a flavour of current topics in tip-related techniques).

1.2.7
Choice of Metal

1.2.7.1 Gold versus Silver
In the examples given above, it is clear that Ag outperforms Au in most cases, and this can be tracked down (ultimately) to the higher absorption (which is proportional

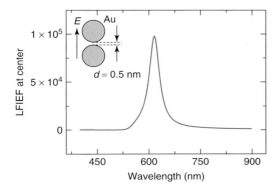

Figure 1.8 Local field intensity enhancement factor (LFIEF) caused by a coupled plasmon resonance at the centre of two 50 nm radius Au cylinders separated by a small gap of 0.5 nm. The coupled plasmon resonance is red shifted to a wavelength where the absorption of Au is comparable to that of Ag (see the comparison imaginary parts of $\epsilon(\lambda)$ in Figure 1.1(b)), thus achieving a maximum LFIEF which is comparable to those obtained in Ag. For coupled resonances in the red and NIR ($\lambda > 600$ nm), Au is as efficient as Ag for SERS applications and preferable in many cases due to its intrinsic stability, well-known surface chemistry, and biocompatibility.

to Im[$\epsilon(\lambda)$]) that Au has at the frequencies where the resonances occur. This is basically the origin of the results seen in Figures 1.3c and 1.4. However, the red shift induced by object interaction and/or shape-and-size effects can push the resonance in Au to the wavelength region $\lambda > 600$ nm, where (as can be seen in Figure 1.1) Im[$\epsilon(\lambda)$] becomes comparable for both metals. In this case, *gold can be as good as silver* as far as LFIEFs are concerned. This is illustrated explicitly in Figure 1.8 for a coupled plasmon resonance in Au. Irrespective of the oversimplified nature of the example in Figure 1.8 (2D cylinders and electrostatic approximation), the fact remains that coupled plasmon resonances in Au nanostructures can be as efficient as those in Ag. In fact, the range $\lambda > 600$ nm is really important for many practical applications of SERS. Many biological [39] (and forensic) applications of the technique are based on NIR lasers (typical examples being diode lasers at ~750 or ~830 nm; see also Section 2.2.1.4). In biological applications, therefore, Au will probably be the most preferred plasmonic substrate (see also Chapters 12 and 13). To the fact that LFIEFs can be comparable to the best values in Ag, we add the advantage of the greater (chemical) stability of Au surfaces in the long run, and the better biocompatibility with many molecules of interest. Examples of EFs in Au comparable to the best values found in Ag have been extensively studied in the literature [4], with more realistic geometrical models for hot spots (a dimer of spheres, for example) and exact solutions (Mie theory) of the electromagnetic problem.

1.2.7.2 Other Coinage and Transition Metals

It is worth mentioning that other coinage metals (besides Ag and Au; see also Sections 2.2.1.1 and 2.2.3) can also be potentially used for enhancing electromagnetic fields. The reason why they are not as useful as Ag and Au is mainly because of their specific dielectric functions [4]. For example, for aluminium (Al) [40, 41], in the region where $\text{Re}[\epsilon(\lambda)]$ is negative, – and can satisfy resonance conditions such as those imposed by Equations 1.4 and 1.5 – the imaginary part is at least an order of magnitude larger than that of Ag or Au, peaking at $\text{Im}[\epsilon(\lambda)] \sim 50$ at ~ 800 nm. Compared to Ag and Au, therefore, resonances experience a much larger damping and are broader and weaker. The use of other coinage metals and transition metals (Pt, Ru, Rh, Pd, Fe, Co and Ni and their alloys [42–44]) is sometimes pursued as an academic interest, but occasionally also as important substrates for specific applications (SERS in the UV range, for example).

1.3
Field Enhancement Distribution and Localization

1.3.1
Electromagnetic Hot Spots

The calculation shown in Figure 1.6 is an example of how the LFIEF can reach really high values ($\sim 10^5$) through coupled resonances at the 'gaps' between metallic nano-objects. These large enhancements are informally dubbed *hot spots* in the SERS literature. Hot spots play a fundamental role in techniques such as SERS, for they provide in many cases enough enhancement to detect single molecules [47, 48] (see also Chapter 4). But, there are other aspects associated with hot spots that are important and worth highlighting besides the magnitude of the LFIEF. The large increase in the LFIEF at hot spots is normally associated also with a strong *spatial localization* of the resonance in the gap. This effect has been studied in full detail in the specialized literature [4] with realistic geometries (a dimer of spheres, for example) and more sophisticated methods to solve Maxwell's equations (generalized Mie theory [4], for example), but – as before – we shall provide here one of the simplest demonstrations with two cylinders in 2D in the electrostatic approximation. Figure 1.9a shows the spatial distribution of the LFIEF at 471 nm under the conditions used previously in Figure 1.6e. At 471 nm we are at the wavelength where the LFIEF is maximum due to the coupled (dipolar) plasmon resonance between the two cylinders in Figure 1.6e. As can be seen in Figure 1.9a, the LFIEF is highly localized in the gap separating the two cylinders (the LFIEF is represented in a false-colour logarithmic scale in Figure 1.9a). This is further reinforced by the data in Figure 1.9b, in which the LFIEF on the surface of the bottom cylinder is shown as a function of Θ. The intensity falls already by a factor of two when $\Theta \sim 2.66°$ from the central axis. How much this actually represents in terms of distance along the surface of the cylinder depends on the radius of the cylinders, but a rule of thumb is that it is typically enough to move by a few nanometers (say ~ 5 nm) from the maximum LFIEF at a hot spot to have a decrease in the LFIEF by an order of magnitude (or more). This is a defining characteristic of the hot spots and one that results in a probability distribution of enhancements

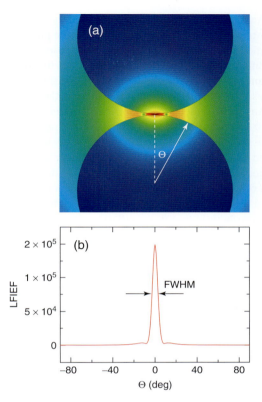

Figure 1.9 (a) Spatial distribution of the LFIEF between two (50 nm radius) Ag cylinders at 471 nm (the position of the peak labeled with an arrow in Figure 1.6e). The LFIEF is plotted in a logarithmic (false-colour) intensity scale with red being the most intense and blue being the weakest. The highly concentrated (in the gap) nature of the dipolar coupled plasmon resonance at 471 nm is clearly seen in the LFIEF map. Localized coupled plasmon resonances like this one are normally responsible for the so-called *hot spots* in SERS substrates. Hot spots are highly localized regions in space. In (b) we show the angular variation of the LFIEF on the surface of the cylinder as a function of Θ. The LFIEF decays to half of the value at the maximum at $\Theta \sim 2.6°$. In real (experimental) hot-spots used for single-molecule SERS, the intensity can decrease by an order of magnitude by moving a few nanometres from the maximum, thereby resulting in some extreme statistical behaviour that has been the topic of study (and of contradictions in the literature) for a very long time [45, 46].

that is *long tail*, for molecules randomly distributed on the surface of the metallic nano-objects.

1.3.2
Long-Tail Distribution of Enhancements

Imagine we have two metallic objects producing a hot spot at a gap (coupled plasmon resonance). For the sake of argument, let us take the example of a dimer formed by two 30 nm in radius Au spheres separated from each other by a gap of 2 nm. As in all the previous examples, we can gain some insight into the problem by solving it within the electrostatic approximation. Let us further suppose that we take a typical experimental condition; for example, we are using a 633 nm laser (HeNe) to illuminate the dimer and we have molecules (randomly) distributed on the surface of the metallic spheres. A pertinent question at this point is then what is the probability of having a certain enhancement if a random point on the surface of a sphere is chosen? Figure 1.10 shows a simulation in which 2×10^5 points are randomly selected over one of the spheres in the dimer, the LFIEF is calculated at each point and a histogram (which can be normalized to the total number of points to represent a probability) is created. The most defining characteristic of this result is that the probability of having a certain LFIEF is described by a *power law* at high enhancements (which looks like a 'line' on a double logarithmic plot). This is one example of the so-called *long-tail distributions*, and it is a reflection of the fact that the chances of a molecule (distributed at random) to find a hot spot are increasingly rare for the highest enhancements. Moreover, this is directly linked to the strong spatial localization of the hot spot and the fact that they represent a really small fraction of the typical total area available to the molecule.

Examples of long-tail distributions of enhancement at SERS hot spots have been given in the specialized literature (with more sophisticated methods to solve the

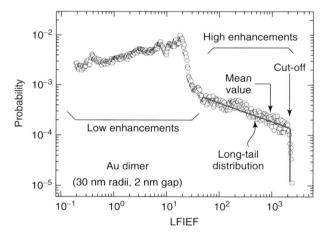

Figure 1.10 Random points on the surface (2×10^5) are chosen for a dimer of Au spheres (30 nm radii) separated by a gap of 2 nm. The LFIEF is obtained in the electrostatic approximation and the histogram displayed here is the resulting distribution (not that it is plotted on a log–log plot). The 'slope' in the *high enhancement region* defines a long-tail distribution (with a cut-off at the hot spot) that completely dominates the statistics of LFIEF. All the details of what happens in the *low enhancement region* can be mostly ignored. Note, for example, that the average value of the enhancement is right towards the end of the tail. Long-tail distributions can also have variances that are larger than the mean! The extreme fluctuations and the dominance of rare events with high enhancements turn out to be a defining property of the SERS enhancement factor (which is linked to the LFIEF).

electromagnetic problem) [45, 49], and here we will only mention that they are fairly universal and always present when hot spots arise. Long-tail distributions have very interesting consequences and produce some extreme statistical characteristics that are a trademark of single-molecule SERS. The topic of the spatial localization of hot spots is therefore central to the understanding of single-molecule SERS statistics, as well as some of its historical contradictions [46, 49].

In the example shown in Figure 1.10, there is a cut-off at the maximum enhancement attainable on the surface, which is right on the main axis joining the two spheres. It is worth mentioning that the tail of the distribution at high enhancements completely dominates the statistics of LFIEFs. For example, Figure 1.10 shows the position of the average LFIEF (mean value) in the distribution. All the details of what happens at low LFIEFs (the region labeled as 'low enhancements' in Figure 1.10) are mostly irrelevant. For all practical purposes, we can replace the probability distribution by the effect of the 'tail' at high LFIEFs (the region labeled as 'high enhancements' in Figure 1.10). More advanced descriptions of these probability distributions have also been studied in the literature [50], to account for the fact that hot spots are imperfect and geometrical parameters vary among certain ranges in real systems. Distributions that do not end abruptly at a cut-off, for example, have been examined and proposed in the literature [50, 51]. Irrespective of their details, the fact remains that the EF distribution amid the presence of hot spots is long-tail and that this a defining property (and a typical feature) of many of the techniques that are associated with the LFIEF, like SERS.

1.4
Electromagnetic Model for the SERS and Fluorescence Enhancement Factors

We are now in a position to address the issue of how much the presence of plasmon resonances affects (enhances or quenches) the Raman or fluorescence signals of molecules. While the main focus of this chapter is on SERS, the mechanism of fluorescence enhancement and quenching is intimately intertwined with it, and it is very useful to understand and highlight the differences. In particular, we can now address the question of what is the connection between the enhancement of Raman and/or fluorescence signals and the LFIEFs discussed in the previous sections.

1.4.1
Enhanced Absorption

The LFIEF represents the increase in intensity a molecule would experience at a specific point with respect to the intensity it would have had in the absence of a metallic nanostructure. Accordingly, all optical magnitudes that depend directly on the intensity are modified (typically enhanced) by the LFIEF. For example, if the molecule absorbs light at a specific wavelength (λ) with a cross section given by σ_{abs} [4], and the field is now affected by an LFIEF, we expect the absorption cross section to increase to LFIEF $\times \sigma_{abs}$; that is, the absorption is LFIEF times the one of the bare molecules. In most of these cases, we talk about LFIEFs that are $\gg 1$ (which are the cases of interest), but we should keep in mind that occasionally the LFIEF can be <1 (and sometimes even $\ll 1$), in which case it is a *quenching* effect rather than an enhancement. But, if the LFIEF is $\gg 1$, this is equivalent to have the intensity of the incoming light increased at the position of the molecule; hence, the increased absorption.

However, Raman and fluorescence are more complicated optical effects involving at least *two* photons (one taken from the incoming laser beam and the other re-emitted by the molecule) and, moreover, these photons are at different wavelengths. A necessary step to understand the application and use of plasmon resonances in the framework of SERS (or SEF) is, therefore, what the modification of the electromagnetic field does to the efficiency of these processes. It turns out that the effect of the local field enhancement by plasmon resonances is very different for SEF (fluorescence) and SERS (Raman), and it pays off to dwell on the topic shortly and reflect on the two very different scenarios that they present.

1.4.2
Comparison of Raman and Fluorescence Processes

A (Stokes) vibrational Raman scattering event in a molecule is an *instantaneous* optical scattering process in which an incoming photon from the laser at ω_L excites a molecular vibration (with frequency ω_v) while emitting a scattered photon at $\omega_S = (\omega_L - \omega_v)$ [52, 53]. The incident photon does not need to be absorbed and induce electronic transitions in the molecule, since Raman processes are usually excited in the *transparency region* of the optical properties of a molecule (otherwise the process is called *resonance Raman scattering*). In the electronically non-resonant case, it can be considered (from a quantum mechanical point of view) as an interaction with a 'virtual state' as depicted in Figure 1.11 [54]. The scattering is *instantaneous* for both photons and they are directly connected through the scattering process. This makes a fundamental difference with fluorescence, for if we 'enhance' the rate (photons per unit time) at which inelastically scattered photons (red in Figure 1.11) are produced and detected in the far field (as a Raman signal), we force that increased rate on the incoming (green in Figure 1.11) photons too, and vice versa (i.e. more photons are drawn from the laser).

Fluorescence, on the other hand, is a stepwise process (not instantaneous) that evolves over time through a series of intermediate steps briefly summarized in Figure 1.11 [52, 53]. The initial step involves the absorption of a photon from the ground singlet state S_0 to a state in the vibrational substructure of the first singlet state S_1. The first 'leg' of fluorescence is, therefore, an absorption process. Unlike Raman, the photon must have enough energy to reach S_1 in order to start a fluorescence event and that can only happen for energies above a certain value. This is schematically shown in Figure 1.11a and b. Once the molecule is left in the excited state, it undergoes a series of (rapid) vibrational relaxation processes, reaching (typically) the vibrational ground state of S_1 after a few picoseconds. By

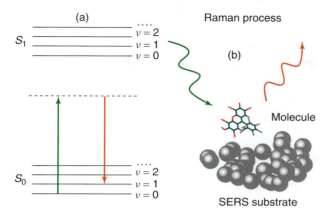

Figure 1.11 A (Stokes) Raman scattering process (under both normal and SERS conditions) is an *instantaneous* process in which the scattered photon is directly linked to the incoming one. The incoming photon does not have to be absorbed at all by the molecule (non-resonant Raman scattering). This is depicted in (a), where a schematic Jablonski diagram for the electronic structure of the molecule is shown. From a quantum mechanical point of view, the Raman process can be considered as an incoming photon (green) interacting with a 'virtual state' (dashed line) and emitting instantaneously a scattered photon (red) that leaves the molecule in an excited vibrational state ($v = 1$ in (a)). The same is depicted schematically in (b): both photons are simultaneous and benefit from the enhancement provided by the SERS substrate.

the time the molecule has reached the vibrational ground state of S_1, it remains there for a few nanoseconds (a typical lifetime before emission for a bare molecule). The main point to realize here is that (unlike Raman) the emission process is now completely independent of the initial absorption; that is, both photons are not linked to each other in a coherent (and instantaneous) way as they are in Raman. For example, if we increase (by some external means) the rate (photons per unit time) at which photons are emitted from the vibrational ground state of S_1, we cannot *force* more photons per unit time to be absorbed as a result. If one photon has been 'taken' from the laser beam to produce a Raman process, then there will be a scattered photon (one cannot exist without the other). In fluorescence, on the contrary, we have situations in which some of the potentially emitted photons (from the ground state of S_1) go 'missing' (in non-radiative recombination, for example). Once the molecule is exited to the ground state of S_1, the best we can do is to recover everything that has been excited in the initial absorption step (in general a fraction will be missing through processes that allow the molecule to relax back to the ground state of S_0 without emitting a photon). But this is independent of the initial absorption process. Therefore, the two processes are effectively 'disconnected' in fluorescence (unlike Raman) and this has important consequences for the different way the EFs work in Raman or fluorescence.

One could argue (hopefully without running the risk of straying too much into semantics) that the fluorescence emission (Figure 1.12e and f) has lost the 'handle' to control the absorption process (Figure 1.12a and b). Crucial to this

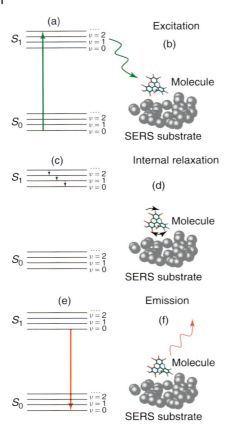

Figure 1.12 A fluorescence process (unlike Raman) can be described as a sequence of events that evolve in time. Fluorescence starts with the absorption of a photon (depicted in (a) and (b)). This process is favoured if the LFIEF is large, resulting in *enhanced absorption*. In (c) and (d), the molecule undergoes vibrational relaxation in the first electronically excited state (S_1). After a certain amount of time (a few nanoseconds typically), the molecule relaxes to the vibrational levels of the ground state thus emitting a photon (as shown in (e) and (f)). When we increased the rate at which the emission is produced, for example ((e) and (f)), this *did not* force more absorption processes ((a) and (b)) to happen, because the two processes have become disconnected from each other in the relaxation step ((c) and (d)).

'disconnection' of the two effects is the vibrational relaxation of the molecule within the electronically excited state S_1. To summarize these very important points then (for the forthcoming discussion) we can say that both Raman and fluorescence are two photon processes; however, they are fundamentally different from each other: The Raman process is the inelastic scattering of light, while fluorescence involves electronic transitions and, prior to the emission, vibrational relaxation in the electronically excited state. These differences in the microscopic details of two processes will actually result in very different EFs, as we shall see later.

1.4.3
The $|E|^4$ Approximation to SERS Enhancement Factors

Imagine that we have an isolated molecule in which Raman processes are occurring, and we are detecting that as a Raman signal in the far field. Furthermore, imagine now that we put the molecule in an environment where the laser field is enhanced by a certain amount through an LFIEF, such as the ones studied in the previous sections. Evidently, we will produce more Raman processes since, after all, an increase in the LFIEF at ω_L is equivalent to increasing the laser power. We will therefore observe more scattered photons at ω_S. Let us imagine, on the other hand, that we put the molecule in a place that enhances the emitted field at ω_S (Figure 1.11) (in a cavity tuned at ω_S, for example). This will increase the Raman intensity too, for the emission is directly coupled to the excitation and we can always draw more photons from the laser to feed the increased emission rate favoured by the presence of the cavity.

Hence, the Raman process benefits from *both* the emission and excitation enhancements, that is, the LFIEF at both ω_L and ω_S, and this leads to an EF for SERS of the form.[5]

$$\text{EF} = \text{LFIEF}(\omega_L) \times \text{LFIEF}(\omega_S) \qquad (1.6)$$

This formula includes a series of implicit approximations, one of which is the fact that it ignores any polarization issues between the incoming and scattered fields. The real Raman process is mediated by a tensor [4, 54], and Equation (1.6) is only an approximation to the real expression of EF [55]. Moreover, further approximations are normally possible. The difference between the LFIEF at ω_L and that at ω_S can sometimes be ignored in many cases. This is due to the fact that $\omega_L - \omega_S$, that is the Raman shift ($\hbar\omega_v = \hbar\omega_L - \hbar\omega_S \sim 0\text{--}200$ meV), is sometimes small compared to the typical frequency ranges where the LFIEF shows substantial changes. This is *not* always the case, and there are important cases in SERS where this approximation is actually not valid [56, 57]. But, when it does hold, it provides the simplest possible version of the SERS EF. Taking into account that the LFIEF at a given point is given by Equation (1.2), the SERS EF at r with all the above approximations included reads

$$\text{EF} \sim \text{LFIEF}^2(\omega_L) = \frac{|E(r)|^4}{|E_0(r)|^4} \qquad (1.7)$$

This is the so-called $|E|^4$ approximation for the SERS enhancement. Despite its many approximations and simplifications, it provides a very useful yardstick estimate for the actual experimental SERS enhancements in a single molecule located at r. It is also used as a typical figure of merit to evaluate and to compare theoretical models with experiments [45].

We conclude that the EF represented by Equation 1.7 (with all its implicit approximations) represents then a good estimate of the SERS enhancement of a

---

5) A more rigorous approach using the *optical reciprocity theorem* shows that, indeed, both the incident and emitted photons are favoured by their respective LFIEFs at the two frequencies.

single molecule. In many experimental situations, however, we measure not one but many molecules that can spread across different places with very different EFs. This is, for example, exemplified in the distribution of the LFIEF in Figure 1.9 close to a hot spot; molecules that differ in position only by a few nanometres can have *very* different EFs. In such cases, a surface-averaged SERS EF is a more appropriate measure of the overall SERS substrate performance.

Moreover, it is necessary occasionally to reintroduce some of the polarization effects that are washed out in the $|E|^4$ approximation. As a result, a large number of different EFs can be defined for different experimental situations: they all arise ultimately from different averages (over polarization, position, etc.) of the single-molecule EFs. There is not a *single* EF in SERS, but rather several different versions of it. A whole list of different EFs that can be defined for different experimental purposes is given in Ref. [58]. Single-molecule EFs are useful only in some situations (when single molecules are studied!), but there is a natural need to have definitions of the EF that include explicitly the averaging over EFs or orientations. This is particularly true for analytical applications of the technique, which typically imply the measurement of a large number of molecules spread over a substrate with a distribution of EFs according to its geometry and characteristics.

1.4.4
Fluorescence Quenching and Enhancement

Since our emphasis here is on SERS, we shall not dwell into all the details of the fluorescence enhancement, but rather highlight the main points to stress the differences with the SERS case. As stated before, fluorescence begins by an absorption process. The LFIEF at ω_L enhances the intensity of the laser and, therefore, it enhances the absorption. Fluorescence benefits hence from this LFIEF in the initial 'leg' of the process (Figure 1.11). For the emission, however, we have a very different situation than that is in SERS. The emission process is 'disconnected' from the absorption (it can neither stimulate it nor quench it) because there has been an irreversible interaction (with a concomitant delay time) in the vibrational relaxation process in the excited state S_1. Therefore, the only possible effect of the environment on the 'emission leg' of the process is to modify the decay rate and the relative contribution of radiative to non-radiative decays. This results in a modified quantum yield or *radiative efficiency*. In simple terms, the best we can do on the emission leg is to recover all the possible radiation that the molecule would have emitted. Dyes with a very high quantum yield (which are the majority) have a very high efficiency and produce one photon for each excited molecule to S_1 with a very high efficiency ($\sim$100%). In general, however, the presence of a metallic surface nearby provides channels for the absorption of the emitted radiation and the radiation efficiency (η) – related to the modified quantum yield – can be smaller (and sometimes much smaller) than one.

This implies that on the 'emission leg' fluorescence can be lost only by the presence of a metal. Combining the effect of enhanced absorption and modified quantum yield, we obtain an EF for fluorescence (for good fluorophores with intrinsic bare quantum yields ~100%) given by

$$EF_{fluo} \sim LFIEF(\omega_L)\eta \tag{1.8}$$

where η is the radiation efficiency ($0 \leq \eta \leq 1$). In general, the LFIEF at ω_L will be much larger than one, while the radiation efficiency η can be $\ll 1$ due to non-radiative processes (i.e. emission that ends up absorbed in the metal rather than being radiated to the far field). The competition between the LFIEF in absorption and η is what produces the wide variety of effects observed in the experiments – from fluorescence enhancements (when the LFIEF at the absorption dominates) to fluorescence quenching (where the effect of η is predominant). A more complete description of the SERS and fluorescence EFs has been given in Ref. [4].

1.4.5
Comparison of SERS and Fluorescence Enhancements

At very short distances from the surface (~1 nm), most metallic nanostructures will look basically like a 'plane' from the point of view of the emitter. The problem of emission close to a plane is well studied in several approximations [4, 59].[6] Equations 1.7 and 1.8 actually contain all the basic phenomenology observed in SERS probes under *resonant* excitation for typical substrates, to wit:

- Far away from the surface, the fluorescence signal is typically much larger (by many orders of magnitude than the Raman signals, which will be typically swamped in the fluorescence background.
- Closer to the surface (meaning typically ~10 nm), there is a mild enhancement of the local field intensity, but the difference between the much more efficient fluorescence process (Figure 1.12) and Raman (Figure 1.11) cannot be compensated; that is, fluorescence may be enhanced (SEF regime) and will again completely dominate the spectrum.
- At much closer distances (a few nanometers), Raman begins benefiting from higher enhancements at both the incoming and outgoing frequencies, while the emission leg of the fluorescence begins experiencing the effect of a reduced η. Even though fluorescence benefits from the increased LFIEF at the incoming field (enhanced absorption), the condition $\eta \ll 1$ starts to dominate and the fluorescence appears as 'partially quenched'. In this regime, the Raman (SERS) signals begin to 'pop out' above the fluorescence background. Both SERS and SEF are observed.
- At really short distances from the surface (~1 nm), the quenching of the fluorescence emission due to η is really effective, and fluorescence may disappear from the spectrum. It is important to realize that this does not mean that the molecule itself is not fluorescing, but rather that its emission is being mainly

---

6) At very short distances (d) from a surface the non-radiative components of the emission scale like ~$1/d^3$. Therefore, at sufficiently close distances from the surface (subject to the geometrical constraints imposed by the molecular size and shape of the chromophore) the fluorescence emission can always be quenched.

channelled into non-radiative processes in the metal. At the same time, the Raman process observed in the far field is still benefiting from the LFIEF at both the excitation and emission wavelengths. Accordingly, this is the regime where Raman (SERS) signals dominate and fluorescence is quenched. It is worth highlighting, however, that a residual fluorescence signal may still be present, albeit strongly spectrally modified, and this can then be the origin of the so-called SERS continuum in resonant conditions [20]. The case when the laser excitation wavelength is not only in resonance with the plasmonic substrate but also with an electronic absorption of the adsorbed molecule is called surface-enhanced resonance Raman scattering (SERRS; see also Chapters 10 and 11).

Most of the phenomenologies observed in SERS and SEF under the presence of plasmon resonances is actually contained in combinations of the above given situations, and this is schematically displayed in Figure 1.13.

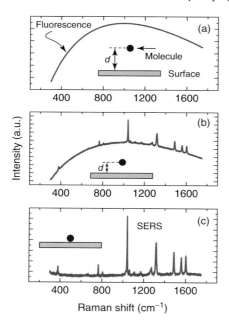

Figure 1.13 Schematic representation of the relative intensities of fluorescence and SERS for molecules at different distances from a metallic surface. We assume the case of a resonant excitation of a dye that produces fluorescence under normal conditions at the chosen laser wavelength. In (a) the molecule is far away from the surface and fluorescence dominates. In (b), the molecule has come closer to the surface and the LFIEF increases. SERS benefits from the higher LFIEF, and fluorescence (even when it is benefiting from the higher LFIEF at the absorption leg of the process) begins experiencing the quenching of the radiative efficiency η. In (c), the molecule has come even closer to the surface and the fluorescence has been quenched, thus revealing the SERS spectrum which is at its maximum efficiency. Note that this does *not* imply that the molecule is not fluorescing, but rather that the emission is quenched (i.e. η dominates).

The 'competition' between the efficiencies of both processes refers here to spectrally integrated magnitudes (the total spectrally integrated amount of fluorescence or the integrated signal of a Raman peak). Additionally, if we look at the spectral domain (i.e. not only the overall intensity but also the shape of the spectrum), there are other phenomena that arise besides the general phenomenology described above such as *spectrally modified fluorescence* [20]. These are, however, more advanced topics that can be introduced in a second stage and are very case dependent.

1.4.6
Other Forms of Enhancements

Besides the electromagnetic enhancement based on plasmon resonances (described up to here in this chapter), there are other known forms of enhancements in SERS, which are normally grouped under the general denomination of 'chemical enhancement' [60–63]. The EFs described so far in this chapter (all linked in one way or another to the LFIEF) would exist even if the molecule were not there at all! They are an intrinsic electromagnetic property of the substrate, not the analyte. But once the molecule is introduced into the problem, its simultaneous interaction with the metal and the electromagnetic field can induce additional contributions to the SERS enhancement. The existence of chemical enhancements is well established and documented since the early days of SERS [2]. It is normally divided into different categories depending on the strength of the interaction between the electronic structure of the molecule and that of the metal [4, 60]. It can contain, in addition, a number of very interesting and complex aspects, such as photo-induced electron transfer between the molecule and the metal [61] or the subtler effects of ionic species (in particular, chlorine [62]). From the standpoint of this introduction, we only mention its existence and refer the reader to the more specialized literature for further details [4]. An important point to highlight, however, is the fact that in all known SERS cases the chemical contribution to the enhancement can only account for an additional factor of ~ 10 (in the best of cases). Standard SERS EFs in the wide range of $\sim 10^3 – 10^{10}$ are, hence, primarily electromagnetic in nature and produced by surface plasmon resonances in metals (of the kind described in this chapter). SERS – as we know it – would not exist if the chemical enhancement were the only source boosting the signals. The chemical enhancement plays a crucial role in the understanding of the finer details of the effect and in its basic science. But, any realistic application of the technique to boost the Raman signals will always start by tailoring the electromagnetic enhancements through surface plasmon resonances in different geometries and configurations [64, 65]. It is only now that the ever-increasing power of computers has also started to unveil some of the details pertaining to the electronic interaction of molecules with metals [63]; the ultimate microscopic source of the chemical enhancement.

1.5
The Magnitude of the SERS Enhancement Factor in Typical Cases

Before concluding, a brief comment is appropriate on the magnitude of the EF observed experimentally under different conditions. This has been, in fact, a contentious issue for a very long time in SERS, and a topic that has been plagued for many years by experimental uncertainties and diverging views [45, 58]. As mentioned earlier, part of the problem is that there is a myriad of different EFs that one can define (spatially averaged, polarization averaged, etc.). But, this is only part of the problem, for some of the issues relate to different experimental practices and lack of consensus on how to actually measure the EFs. We shall not dwell here on all the various details of the problem, which have been thoroughly studied in Ref. [58] and further expanded in Ref. [4]. We shall concentrate, however, on the 'simplest' case: single-molecule EFs in SERS.

When we are measuring a single molecule, we are in a situation in which many of the uncertainties of *spatial* averaging (the fact that molecules at different positions experience different enhancements) are ruled out. There are still a few pending issues that can affect the definition of the EF (or what is actually being measured) such as surface selection rules [66, 67]. We only mention here that of all possible cases related to the quantification of SERS EFs, the single-molecule case posses the simplest one with the smallest number of assumptions. Needless to say, the price is paid here in a different way: through the experimental challenge of making sure that we are actually measuring *one* molecule. The rest of the task is to be able to normalize the signal coming from a single molecule with respect to a (non-SERS) spectrum of a reference compound with a known Raman differential cross section. But the latter is a relatively simpler experimental procedure, and the real difficulty lies in making sure that single molecules are indeed being observed reliably.

It is a relatively recent development though [58] that SERS EFs have been quantified and measured with techniques that allow the identification of single-molecule cases. The conditions needed to ensure that the SERS signals we are measuring come from single-molecules can be achieved through a variety of methods (Langmuir–Blodgett films [68–71], TERS [33–35], and bi-analyte SERS [5, 72–74]), but direct quantifications of single-molecule SERS cross sections and EFs have been done almost exclusively with the bi-analyte SERS method [58, 72]. Several situations of single-molecule EFs have been considered in the literature, including the possible effects of photobleaching [51] (which limits the maximum observed EFs [75]). What follows is a very brief (and, accordingly, necessarily incomplete) description of the main experimental findings to date:

- Single-molecule SERS EFs up to a maximum value of $\sim 10^{10}$ [76] have been observed in many different experimental conditions. The minimum value to observe single molecules is around $\sim 10^7-10^8$ for resonant or pre-resonant molecules, and even though these values are strongly dependent on the exact experimental conditions [45] the fact remains that they are (at least) up to a million times smaller than what was originally thought to be necessary to observe single molecules in SERS. These enhancements are in perfect accord

with what is expected theoretically from electromagnetic calculations (in the $|E|^4$ approximation, for example).
- For cases involving many molecules, the EF is invariably an average over the different situations of the individual molecules over the substrate. Average EFs play an important role in applications, even if their microscopic origin is obscured by the averaging process. But they provide sometimes the only mean to compare the performance of different substrates. For non-optimized conditions, average EFs for SERS can be as low as $\sim 10-10^3$. More typical values for useful applications will be in the range of EF $\sim 10^4-10^6$, and these values should be considered standard for the technique.

Enhancements in the range 10^7-10^{10} are obtained in many situations and can easily provide single-molecule sensitivity (including cases of non-resonant molecules [76]). These latter cases appear more often in 'disordered' substrates, where it is difficult to control the geometrical parameters of the problem and we rely solely on 'accidental' formation of hot spots. An alternative to produce hot spots with EFs in this range (in a controlled manner) is the TERS described earlier [33–35], albeit with the proviso in mind that it is very difficult to position the molecule at the right place. The vast majority of single-molecule SERS experiments is performed in disordered substrates and relies heavily on the statistics of events. There is, nevertheless, an increasing number of single-molecule experiments with TERS, but reliable values for the actual EFs that are achieved have not yet appeared in the literature. The 'normalization' procedure (with respect to a reference compound of known differential cross section) using TERS is a lot more difficult.

A brief comment on our actual experimental ability to control the EF is opportune at this stage. The more our desire to control the EF, the more it becomes necessary to control the geometrical aspects of the problem at the nanoscale. This is particularly true for high enhancements, which normally come from coupled plasmon resonances at gaps (as shown in Figure 1.9) with typical dimensions $\sim 1-2$ nm. This has been informally dubbed 'the SERS uncertainty principle' by Natan [77], in the sense that the higher the enhancement we desire, the more uncontrollable the geometrical variables of the problem become (from an experimental point of view). While this is merely a figure of speech (and not an actual principle), it does highlight the fact that highest possible enhancements are always limited by our ability to control the nanoscale world to a precision comparable to molecular dimensions (~ 1 nm).

1.6
Conclusions

We hope that somebody who has not heard about plasmon resonances and SERS/SEF enhancements before would have obtained at this stage a clear idea of the basic concepts underlying the EF in SERS and fluorescence, at least at a qualitative level. *Shape, size and interactions* are at the heart of the immensely rich and complex optical response of metallic nanostructures and the stunning variety of optical phenomena that arise in plasmonics. Needless to say, a much deeper understanding of the details of the electromagnetic response of metallic nanostructures requires dwelling on the details of the solutions and the multitude of analytical and numerical methods to obtain them. While this is left to more specialized literature [4], we believe that a very basic conceptual understanding can be obtained through the topics highlighted in this chapter. More advanced theoretical concepts on the theory of plasmon resonances can be obtained from Ref. [78], or from a more SERS-oriented point of view in Refs. [3, 4].

References

1. Moskovits, M. (1985) Surface-enhanced spectroscopy. *Rev. Mod. Phys.*, **57**, 783–826.
2. Otto A. (1984) Surface enhanced Raman scattering: classical and chemical origins, in *Light Scattering in Solids IV*, Topics in Applied Physics, vol. 103 (eds M. Cardona. and G. Guntherodt.), Springer, Berlin, pp. 289–418.
3. Aroca, R. (2006) *Surface Enhanced Vibrational Spectroscopy*, Wiley, Chichester.
4. Le Ru, E.C. and Etchegoin, P.G. (2009) *Principles of Surface-enhanced Raman Spectroscopy and Related Plasmonic Effects*, Elsevier, Amsterdam.
5. Goulet, P.J.G. and Aroca, R.F. (2007) Distinguishing individual vibrational fingerprints: single-molecule surface-enhanced resonance Raman scattering from one-to-one binary mixtures in Langmuir-Blodgett monolayers. *Anal. Chem.*, **79**, 2728–2734.
6. McCabe, A.F., Eliasson, C., Prasath, R.A. et al. (2006) SERRS labeled beads for multiplex detection. *Faraday Discuss.*, **132**, 303–308.
7. Smith, W.E. (2008) Practical understanding and use of surface enhanced Raman scattering/surface enhanced resonance Raman scattering in chemical and biological analysis. *Chem. Soc. Rev.*, **37**, 955–964.
8. Lyandres, O., Shah, N.C., Yonzon, C.R. et al. (2005) Real-time glucose sensing by surface-enhanced Raman spectroscopy in bovine plasma facilitated by a mixed decanethiol/mercaptohexanol partition layer. *Anal. Chem.*, **77**, 6134–6139.
9. Faulds, K., Barbagallo, R.P., Keer, J.T. et al. (2004) SERRS as a more sensitive technique for the detection of labelled oligonucleotides compared to fluorescence. *Analyst*, **129**, 567–568.
10. Sabatte, G., Keir, R., Lawlor, M. et al. (2008) Comparison of surface-enhanced resonance Raman scattering and fluorescence for detection of a labeled antibody. *Anal. Chem.*, **80**, 2351–2356.
11. Koo, T.W., Chan, S., Sun, L. et al. (2004) Specific chemical effects on surface-enhanced Raman spectroscopy for ultra-sensitive detection of biological molecules. *Appl. Spectrosc.*, **58**, 1401–1407.
12. Vo-Dinh, T., Allain, L.R. and Stokes, D.L. (2002) Cancer gene detection using surface-enhanced Raman scattering (SERS). *J. Raman Spectrosc.*, **33**, 511–516.
13. Haynes, C.L., Yonzon, C.R., Zhang, X. and Van Duyne, R.P. (2005) Surface enhanced Raman sensors: early history and the development of sensors for quantitative biowarfare agent and glucose detection. *J. Raman Spectrosc.*, **36**, 471–484.
14. Sägmüller, B., Schwarze, B., Brehm, G. and Schneider, S. (2001) Application of SERS spectroscopy to the identification of (3,4-methylenedioxy)amphetamine in forensic samples utilizing matrix stabilized silver halides. *Analyst*, **126**, 2066–2071.
15. Chen, K., Leona, M., Vo-Dinh, K.C. et al. (2006) Application of surface-enhanced Raman scattering (SERS) for the identification of anthraquinone dyes used

in works of art. *J. Raman Spectrosc.*, **37**, 520–527.
16. Centeno, S.A. and Shamir, J. (2008) Surface enhanced Raman scattering (SERS) and FTIR characterization of the sepia melanin pigment used in works of art. *J. Mol. Struct.*, **873**, 149–159.
17. Chen, K., Leona, M. and Vo-Dinh, T. (2007) Surface-enhanced Raman scattering for identification of organic pigments and dyes in works of art and cultural heritage material. *Sens. Rev.*, **27**, 109–120.
18. Dulkeith, E., Morteani, A.C., Niedereichholz, T. *et al.* (2002) Fluorescence quenching of dye molecules near gold nanoparticles: radiative and nonradiative effects. *Phys. Rev. Lett.*, **89**, 203002–203005.
19. Lakowicz, J.R. (2005) Radiative decay engineering 5: metal-enhanced fluorescence and plasmon emission. *Anal. Biochem.*, **337**, 171–194.
20. Le Ru, E.C., Etchegoin, P.G., Grand, J. *et al.* (2007) The mechanisms of spectral profile modification in surface enhanced fluorescence. *J. Phys. Chem. C*, **44**, 16076–16079.
21. Haes, A.J. and Van Duyne, R.P. (2005) A unified view of propagating and localized surface plasmon resonance biosensors. *Anal. Bioanal. Chem.*, **379**, 920–930.
22. Zayatsa, A.V., Smolyaninov, I.I. and Maradudin, A.A. (2005) Nano-optics of surface plasmon polaritons. *Phys. Rep.*, **408**, 131–314. (and references therein).
23. Novotny L. (2000) Forces in optical near-fields, in *Near-field Optics and Surface Plasmon Polaritons, Topics in Applied Physics*, vol. 81 (ed. S. Kawata.), Springer-Verlag, Berlin, pp. 123–141.
24. Born, M. and Wolf, E. (1999) *Principles of Optics*, Cambridge University Press, Cambridge.
25. Madelung, O. (1978) *Introduction to Solid-State Theory*, Springer-Verlag, Berlin.
26. Kittel, C. (1986) *Introduction to Solid State Physics*, John Wiley & Sons, Inc, New York.
27. Etchegoin, P.G., Le Ru, E.C. and Meyer, M. (2006) An analytic model for the optical properties of gold. *J. Chem. Phys.*, **125**, 164705–164707.
28. Griffiths, D.J. (1999) *Introduction to Electrodynamics*, Prentice Hall, New Jersey.
29. Jackson, J.D. (1998) *Classical Electrodynamics*, Wiley, New York.
30. Rojas, R. and Claro, F. (1993) Theory of surface enhanced Raman-scattering in colloids. *J. Chem. Phys.*, **98**, 998–1006.
31. Prodan, E., Radloff, C., Halas, N.J. and Nordlander, P. (2003) A hybridization model for the plasmon response of complex nanostructures. *Science*, **302**, 419–422.
32. Etchegoin, P., Cohen, L.F., Hartigan, H. *et al.* (2003) Electromagnetic contribution to surface enhanced Raman scattering revisited. *J. Chem. Phys.*, **119**, 5281–5289.
33. Pettinger, B., Picardi, G., Schuster, R. and Ertl, G. (2002) Surface-enhanced and STM-tip-enhanced Raman spectroscopy at metal surfaces. *Single Mol.*, **5** (6), 285–294.
34. Pettinger, B., Ren, B., Picardi, G. *et al.* (2005) Tip-enhanced Raman spectroscopy (TERS) of malachite green isothiocyanate at Au(111): bleaching behavior under the influence of high electromagnetic fields. *J. Raman Spectrosc.*, **36**, 541–550.
35. Steidtner, J. and Pettinger, B. (2008) Tip-enhanced Raman spectroscopy and microscopy on single dye molecules with 15 nm resolution. *Phys. Rev. Lett.*, **100**, 236101–236104.
36. Anderson, M.S. and Pike, W.T. (2002) A Raman-atomic force microscope for apertureless-near-field spectroscopy and optical trapping. *Rev. Sci. Instrum.*, **73**, 1198–1203.
37. Downes, A., Salter, D. and Elfick, A. (2006) Finite element simulations of tip-enhanced Raman and fluorescence spectroscopy. *J. Phys. Chem. B*, **110**, 6692–6698.
38. Ward, D.R., Halas, N.J., Ciszek, J.W. *et al.* (2008) Simultaneous measurements of electronic conduction and Raman response in molecular junctions. *Nano Lett.*, **8**, 919–924.
39. Hudson, S.D. and Chumanov, G. (2009) Bioanalytical applications of SERS

(surface-enhanced Raman spectroscopy). *Anal. Bioanal. Chem.*, **394**, 679–686.

40. Vial, A. (2007) Implementation of the critical points model in the recursive convolution method for modeling dispersive media with the finite-difference time domain method. *J. Opt. A: Pure Appl. Opt.*, **9**, 745–748. (and references therein).

41. (1985) *Handbook of Optical Constants of Solids* (ed. E.D. Palik.), Academic Press.

42. Tian, Z.Q., Ren, B. and Wu, D.Y. (2002) Surface-enhanced Raman scattering: from noble to transition metals and from rough surfaces to ordered nanostructures. *J. Phys. Chem. B*, **106**, 9463–9483.

43. Tian, Z.Q., Gao, J.S., Li, X.Q. et al. (1998) Can surface Raman spectroscopy be a general technique for surface science and electrochemistry? *J. Raman Spectrosc.*, **29**, 703–711.

44. Tian, Z.Q. and Ren, B. (2004) Adsorption and reaction at electrochemical interfaces as probed by Surface-Enhanced Raman Spectroscopy. *Annu. Rev. Phys. Chem.*, **55**, 197–229.

45. Etchegoin, P.G. and Le Ru, E.C. (2008) A perspective on single molecule SERS: current status and future challenges. *Phys. Chem. Chem. Phys.*, **10**, 6079–6089.

46. Etchegoin, P.G., Meyer, M. and Le Ru, E.C. (2007) Statistics of single molecule SERS signals: is there a Poisson distribution of intensities? *Phys. Chem. Chem. Phys.*, **9**, 3006–3010.

47. Nie, S. and Emory, S.R. (1997) Probing single molecules and single nanoparticles by surface-enhanced Raman scattering. *Science*, **275**, 1102–1106.

48. Kneipp, K., Wang, Y., Kneipp, H. et al. (1997) Single molecule detection using surface-enhanced Raman scattering (SERS). *Phys. Rev. Lett.*, **78**, 1667–1670.

49. Le Ru, E.C., Etchegoin, P.G. and Meyer, M. (2006) Enhancement factor distribution around a single surface-enhanced Raman scattering hot spot and its relation to single molecule detection. *J. Chem. Phys.*, **125**, 204701–204714.

50. Le Ru, E.C. and Etchegoin, P.G. (2009) Phenomenological local field enhancement factor distributions around electromagnetic hot spots. *J. Chem. Phys.*, **130**, 181101–181104.

51. Fang, Y., Seong, N.-H. and Dlott, D.D. (2008) Measurement of the distribution of site enhancements in surface-enhanced Raman scattering. *Science*, **321**, 388–392.

52. Haken, H., Wolf, H.C. and Brewer, W.D. (2004) *Molecular Physics and Elements of Quantum Chemistry: Introduction to Experiments and Theory*, Springer-Verlag, Berlin.

53. Demtröder, W. (2002) *Laser Spectroscopy*, Springer-Verlag, Berlin.

54. Long, D.A. (2002) *The Raman Effect, A Unified Treatment of the Theory of Raman Scattering by Molecules*, John Wiley & Sons, Ltd, Chichester.

55. Le Ru, E.C. and Etchegoin, P.G. (2006) Rigorous justification of the $|E|^4$ enhancement factor in surface enhanced Raman spectroscopy. *Chem. Phys. Lett.*, **423**, 63–66.

56. Le Ru, E.C., Etchegoin, P.G., Grand, J. et al. (2008) Surface enhanced Raman spectroscopy on nanolithography-prepared substrates. *Curr. Appl. Phys.*, **8**, 467–470.

57. Le Ru, E.C., Grand, J., Félidj, N. et al. (2008) Experimental verification of the SERS electromagnetic model beyond the $|E|^4$ approximation: polarization effects. **112**, 8117–8121.

58. Le Ru, E.C., Blackie, E., Meyer, M. and Etchegoin, P.G. (2007) SERS enhancement factors: a comprehensive study. *J. Phys. Chem. C*, **111**, 13794–13803.

59. Ford, G.W. and Weber, W.H. (1984) Electromagnetic interactions of molecules with metal surfaces. *Phys. Rep.*, **113**, 195–287.

60. Tian, Z.Q. (2006) General discussions section. *Faraday Discuss.*, **132**, 309–319.

61. Xie, Y., Wu, D.Y., Liu, G.K. et al. (2003) Surface-enhanced Raman scattering study of the surface coordination of porphyrins adsorbed on silver. *J. Electroanal. Chem.*, **554**, 417–420.

62. Otto, A., Bruckbauer, A. and Chen, Y.X. (2003) On the chloride activation in

SERS and single molecule SERS. *J. Mol. Struct.*, **661**, 501–514.
63. Wu, D.Y., Duan, S., Ren, B. and Tian, Z.Q. (2005) Density functional theory study of surface-enhanced Raman scattering spectra of pyridine adsorbed on noble and transition metal surfaces. *J. Raman Spectrosc.*, **36**, 533–540.
64. Banholzer, M.J., Millstone, J.E., Quin, L. and Mirkin, C.A. (2008) Rationally designed nanostructures for surface-enhanced Raman spectroscopy. *Chem. Soc. Rev.*, **37**, 885–897.
65. Lal, S., Grady, N.K., Kundu, J. *et al.* (2008) Tailoring plasmonics substrates for surface-enhanced spectroscopies. *Chem. Soc. Rev.*, **37**, 898–911.
66. Moskovits, M. (1982) Surface selection rules. *J. Chem. Phys.*, **77**, 4408–4416.
67. Le Ru, E.C., Meyer, M., Blackie, E. and Etchegoin, P.G. (2008) Advanced aspects of electromagnetic SERS enhancement factors at a hot-spot. *J. Raman Spectrosc.*, **39**, 1127–1134.
68. Pieczonka, N.P.W. and Aroca, R.F. (2008) Single molecule analysis by surfaced-enhanced Raman scattering. *Chem. Soc. Rev.*, **37**, 946–954.
69. Constantino, C.J.L., Lemma, T., Antunes, P.A. and Aroca, R. (2001) Single-molecule detection using surface-enhanced resonance Raman scattering and Langmuir-Blodgett monolayers. *Anal. Chem.*, **73**, 3674–3678.
70. Goulet, P.J.G., Pieczonka, N.P.W. and Aroca, R.F. (2005) Mapping single-molecule SERRS from Langmuir-Blodgett monolayers on nanostructured silver island films. *J. Raman Spectrosc.*, **36**, 574–580.
71. Goulet, P.J.G., Pieczonka, N.P.W. and Aroca, R.F. (2003) Overtones and combinations of single-molecule surface-enhanced resonance Raman scattering spectra. *Anal. Chem.*, **75**, 1918–1923.
72. Le Ru, E.C., Meyer, M. and Etchegoin, P.G. (2006) Proof of single-molecule sensitivity in surface enhanced Raman scattering (SERS) by means of a two-analyte technique. *J. Phys. Chem. B*, **110**, 1944–1948.
73. Dieringer, J.A., Lettan II, R.B., Scheidt, K.A. and Van Duyne, R.P. (2007) A frequency domain existence proof of single-molecule surface-enhanced Raman spectroscopy. *J. Am. Chem. Soc.*, **129**, 16249–16256.
74. Blackie, E., Le Ru, E.C., Meyer, M. *et al.* (2008) Bi-analyte SERS with isotopically edited dyes. *Phys. Chem. Chem. Phys.*, **10**, 4147–4153.
75. Etchegoin, P.G., Lacharmoise, P.D. and Le Ru, E.C. (2009) Influence of photostability on single-molecule surface enhanced Raman scattering enhancement factors. *Anal. Chem.*, **81**, 682–688.
76. Blackie, E.J., Le Ru, E.C. and Etchegoin, P.G. (2009) Single-molecule surface-enhanced Raman spectroscopy of nonresonant molecules. *J. Am. Chem. Soc.*, **131**, 14466–14472.
77. Natan, M.J. (2006) Concluding remarks: surface enhanced Raman scattering. *Faraday Discuss.*, **132**, 321–328.
78. Novotny, L. and Hecht, B. (2006) *Principles of Nano-optics*, Cambridge University Press, Cambridge.

2
Nanoparticle SERS Substrates
Yuling Wang and Erkang Wang

2.1
Introduction

Since the discovery that high-intensity Raman scattering of small molecules could be obtained on electrochemical roughened silver surface by Fleischmann *et al.* [1] in 1974, who attributed the high enhancement to the large number of molecules on the roughened surface, and Jeanmaire *et al.*'s [2] and Creighton *et al.*'s [3] independent discovery in 1977 that the enhancement of the Raman scattering is related to an intrinsic surface enhancement effect, marking the beginning of surface-enhanced Raman scattering (SERS) spectroscopy [4], substantial interest has been focussed on the research of the fabrication of SERS-active substrates and on the applications of SERS to many fields, including surface, analytical and life sciences. Meanwhile, investigations of the enhancement mechanisms responsible for the extraordinarily large enhancement of Raman signal of the adsorbate on the roughened metal surfaces have never stopped, and now it is widely accepted that there are two separate mechanisms that describe the overall SERS effect: the electromagnetic effect (EM) and chemical effect (CM). The EM mechanism is based on the interaction of the transition moment of an adsorbed molecule with the electric field of surface plasmons induced by the incoming light on the metal [5–9], which is an effect independent of the probe molecules. For the EM mechanism, the localized surface plasmon resonance (LSPR) plays a key and dominant role in the overall enhancement; whereas the CM is due to the interaction of the adsorbed molecules on the metal with the metal surface, mostly from the first layer of the charge-transfer resonance between the adsorbate and the metal [10–12].

More than 30 years have passed since the initial discovery of SERS and the research activity in this field has dramatically increased due to the improvement in techniques resulting from advances in nanotechnology and improved instrumental capabilities. Therefore, SERS has now become a very useful tool in various fields including chemistry, physics and biology. Among these, the basic focus as well as the key step for practical SERS applications is still the successful fabrication of the SERS substrate because the application really depends on the activity and reproducibility of the substrate. The first used SERS substrate was a roughened electrode obtained

Surface Enhanced Raman Spectroscopy: Analytical, Biophysical and Life Science Applications. Edited by Sebastian Schlücker
Copyright © 2011 WILEY-VCH Verlag GmbH & Co. KGaA, Weinheim
ISBN: 978-3-527-32567-2

by electrochemical cyclic voltammetry. Subsequently, metal films produced by vacuum deposition were employed. The most used SERS substrates, however, are metallic nanoparticles with a large size distribution and various shapes prepared by wet chemical methods, benefitting from the development of nanoscience and nanotechnology. Compared to that on the electrode and the vacuum deposited film, the molecular structure of the analyte on the nanoparticle SERS substrate is not influenced by the oxidation–reduction cycle during pretreatment and does not require the use of expensive vacuum evaporation chambers. Therefore, nanoparticles as SERS substrates have great advantages such as low cost as well as simple and easy manipulation.

Nanoparticles are artificially created structures with dimensions between 1 and 100 nm. According to the SERS enhancement mechanism (see also Chapter 1 for the EM enhancement mechanism), SERS phenomena will happen on metal surfaces with a certain degree of roughness. Metal nanoparticles with dimensions between 10 and 100 nm have the specific surface plasmon resonance property compared to their bulk materials – which is basic for the production of SERS – and there will be very great enhancement of the Raman signal for molecules on or near their surface. From this point, SERS is indeed a 'nanostructure-enhanced Raman scattering'. Therefore, the preparation of metal nanoparticle substrates has been the key issue that dictates signal intensity and reproducibility. As the initial SERS substrate in solution, metal nanoparticle colloid is the most commonly employed SERS substrate because of its easy preparation and manipulation. Many methods have been reported for the preparation of metal nanoparticles with various sizes, shapes and compositions. Owing to the rapid development of this field, some good reviews and books exist that describe nanoparticles as SERS substrates from different aspects [13–16] (see also Chapter 1). In this chapter, we focus on nanoparticle SERS substrates based on colloidal solutions including the preparation of spherical metal nanoparticles, aggregation of metal nanoparticles, bimetallic nanoparticles and metal nanoparticles with other various shapes, mainly with respect to the influence of the size and shape of the nanoparticles and the excitation wavelength on the SERS signals. Finally, we address the technniques for the characterization of these nanoparticles and the nanoparticles on the unfunctionalized surface as the SERS substrates.

2.2
Preparation and Stability of Metal Nanoparticle Colloidal SERS Substrates

There is a large variety of SERS substrates other than colloids, such as roughened electrodes, metal island films, and so on. However, metal colloids are frequently used in SERS because of the simplicity of their preparation and characterization. Some wet chemistry methods have been developed for the preparation of metal nanoparticle colloids, such as chemical reduction, laser ablation and photoreduction, which provide inexpensive and simple manipulation approaches for SERS substrate fabrication. With the advent of nanoscience and nanotechnology, nanoparticle-based SERS substrates have developed from spherical nanoparticles to metal nanoparticles with various shapes and a wide size distribution. Meanwhile, the composition of the nanoparticles has changed from the single composite to multiple composites or alloy nanoparticles to improve the SERS signal. Aggregation of the metal nanoparticles is recognized to be important due to the presence of 'hot spots' for better signals. Since the SERS intensity strongly depends on the excitation wavelength and the strength of the plasmons propagating on the surface of the metal nanoparticles, it is important to engineer the surface plasmon resonances of the nanoparticles to maximize the signal strength. Accordingly, new types of metal nanoparticles have been prepared, some of which can give enhancement factors as high as 10^9, such as the core–shell metal nanoparticles [17] or dimers of nanoparticles [18, 19] reported in the literature. Therefore, in this section, different types of colloidal metal nanoparticles will be introduced as SERS substrates. Compared with solid SERS substrates, there are several advantages using metal nanoparticles, such as the ease of colloid formation, minimization of the burning of the sample which allows the use of higher laser powers and the use of more energetic laser lines. Also, the use of a colloidal solution permits the acquisition of an average spectrum due to the Brownian monition that governs colloidal dispersions, which can be further improved by recirculating the sample [13]. Therefore, in this section, we focus on the preparation of spherical metal nanoparticles, aggregates of metal nanoparticles, bimetallic nanoparticles and metal nanoparticles with various shapes, as well as the stability and SERS properties of these nanoparticles.

2.2.1
Colloidal Spherical Metal Nanoparticles

2.2.1.1 Chemical Reaction for Metal Nanoparticle Preparation

SERS on spherical metal nanoparticles in a colloidal solution was first reported by Creighton in 1979, who used ice-cold $NaBH_4$ solution to reduce $AgNO_3$ and $K[AuCl_4]$ to get the silver and gold nanoparticles and then chose pyridine as the probe molecule to investigate the SERS activity of the metal colloid. In this preliminary work, it was found that the SERS signal has a strong dependence on the excitation wavelength [20]. So far, the most frequently used system to produce SERS-active media in solutions is metal nanoparticles. A typical method for the preparation of silver nanoparticles to be used in SERS is the reduction of $AgNO_3$ by sodium citrate under boiling [21, 22]; the size of the as-prepared silver nanoparticles is about 60–80 nm (Figure 2.1a) and the colloids show a turbid grey-green colour with a surface plasmon absorption maximum at 410 nm (Figure 2.1b). The simplest and most common method of synthesizing spherical metal nanoparticles is by chemical reaction in solution, usually by reducing the metal salts with a variety of reducing and capping agents. Typical chemical reducing agents include sodium citrate [23], sodium borohydride [24], hydrazine [25] and hydroxylamine hydrochloride [26, 27]. Through control of the reaction temperature, pH of the solution and, most importantly, the kind of metal salt, reductant and surfactant, the size and size distribution and even the aggregation state of metal

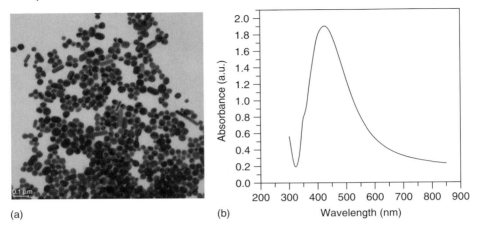

Figure 2.1 (a) TEM image and (b) UV–vis absorption spectrum of Ag colloid obtained by sodium citrate reduction under boiling [21, 22].

nanoparticles can be controlled. Recently, many more different kinds of reducing and capping agents have been developed for the preparation of silver nanoparticles by simple chemical reactions. For example, Nickel *et al.* prepared a very stable silver colloid by the reduction of aqueous silver nitrate with hydrazine dihydrochloride in a weakly alkaline solution. They obtained spherical particles with the size ranging between 40 and 70 nm in diameter, which gives a high SERS signal with Nile Bule A [28]. Li *et al.* also prepared mercaptoacetic acid-capped spherical silver nanoparticles with a diameter of about 17 nm by a simple chemical reaction and found that the SERS signal depends on the sizes and aggregation state of the silver nanoparticles [29]. Kvitek *et al.* used a one-step chemical reduction route to prepare silver colloidal particles with controllable sizes ranging from 45 to 380 nm, which were obtained by the reduction of $[Ag(NH_3)_2]^+$ complex with various reducing sugars (Figure 2.2). They achieved an efficient enhancement for the Raman signal of 1-methyl adenine on the as-prepared silver nanoparticle colloid [30].

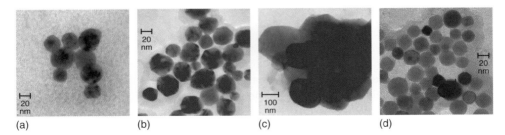

Figure 2.2 TEM images of colloidal silver particles prepared by reduction of $[Ag(NH_3)_2]^+$ using (a) glucose, (b) xylose, (c) fructose and (d) maltose for 0.005 mol dm^{-3} ammonia concentration in the reaction mixture [30].

For the preparation of gold nanoparticles, the commonly employed method is analogous to that with silver by using sodium citrate to reduce chloroauric acid to produce particles with sizes in the range of 20–100 nm, which was developed by Frens [31]. The obtained gold nanoparticles show a red to deep red colour with a surface plasmon absorption maximum between 515 and 540 nm depending on the size of the gold nanoparticles [31]. Recently, Peter *et al.* reported the preparation of gold nanoparticles with different sizes based on the seed-mediated growth method and studied the optical and SERS activity of these gold nanoparticles. In this case, they synthesized gold nanoparticles smaller than 25 nm in diameter by reducing $HAuCl_4$ with sodium acrylate and refluxing for 30 min (Figure 2.3a–c). The reaction produced a deep red solution characteristic of the formation of colloidal gold. Gold nanoparticles with diameters larger than 25 nm were prepared using a seed-mediated growth method: the basic element of the synthesis protocols involves seed formation and seed growth using a combination of reducing and capping agents, including sodium citrate, sodium acrylate and acrylic acid. The seed underwent a growth reaction in the presence of $HAuCl_4$ under controlled concentrations of the reducing and capping agents to form larger Au nanoparticles

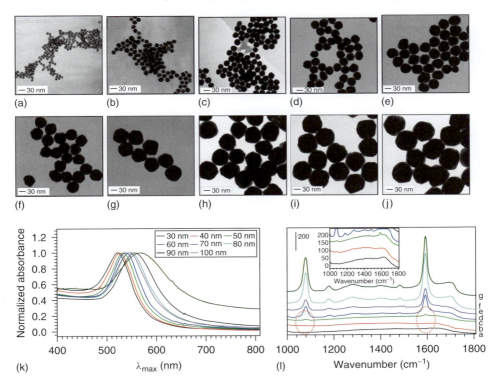

Figure 2.3 (a–j) TEM images, (k) UV–vis absorption spectra and (l) SERS spectra of different sizes of gold nanoparticles generated by seed-mediated growth methods [32].

(up to 100 nm, Figure 2.3d–j). The maximum of the surface plasmon band is shifted from 519 to 569 nm, showing a clear dependence on the particle size (Figure 2.3k). The SERS signal from different sizes of gold nanoparticles reveals that its intensity increases with the particle size (Figure 2.3l) [32]. Another method of preparing gold nanoparticles for efficient SERS enhancement was reported by Schwartzberg *et al.*, who utilized the template method for fabricating hollow spherical gold nanoparticles and got strong SERS signal of the probe molecules (4-mercaptobenzoic acid) on these nanoparticles, which was attributed to the homogeneous structure on the particles [33].

Apart from the synthesis in aqueous solutions, the preparation from organic solvents has also been reported. Shen *et al.* synthesized monodisperse noble-metal nanoparticles through the direct reaction of the related metal salt with oleylamine in toluene and obtained SERS signal from the prepared metal nanoparticles [34].

It is noteworthy that adsorption and self-assembly of organic molecules in metal colloidal sols have been extensively investigated by SERS due to the importance of understanding the processes of self-assembly for manipulating the physico-chemical properties of interfaces for a variety of heterogeneous phenomena. The most widely studied and well-characterized systems include alkanethiols [35–38], dialkyl sulfides [39], aromatic thiols [40] and dicyanobiphenyls [41–44] on gold and silver nanoparticle surfaces. According to the surface selection rule [45], vibrational modes whose polarizability tensor elements are perpendicular to the metal surface should be strongly enhanced in an SERS spectrum. Based on the EM theory, Creighton reported that for aromatic molecules with C_{2v} symmetry, the relative enhancement factors for different modes should be $a_1 : a_2 : b_1 : b_2 = 1-16 : 4 : 4 : 1$ for face-on adsorption of the benzene ring and $a_1 : a_2 : b_1 : b_2 = 1-16 : 1 : 4 : 4$ for perpendicular adsorption [46]. Meanwhile, Aroca's group has investigated the adsorption of small molecules on metal colloids with different laser lines, such as the adsorption of *p*-nitrothiophenol on silver colloids at 514.5, 633 and 780 nm laser excitation. They concluded that the relative intensities of the Raman peaks of the molecules vary with the orientation of the polarizability derivatives and the magnitude of the local field at the surface [47, 48]. Recently, they synthesized new silver nanoparticles with different surface charges by amino acid reduction, and found that the surface charge can be correlated with the degree of adsorption and with the average SERS signal. Some others group also reported the adsorption of small molecules on colloids [49].

2.2.1.2 Laser Ablation and Photoreduction for Metal Nanoparticle Preparation

Laser ablation and photoreduction are the other most used methods for the preparation of metal nanoparticles. There is a big advantage for the nanoparticle SERS substrates prepared by these methods in that they can overcome the problem of the influence of residual ions in the chemical reaction on the SERS signal because the nanoparticles obtained by laser ablation or photoreduction are 'chemically pure'. So far, Ag, Au, Cu and Pt colloids in aqueous solution and organic solvents have been prepared by these methods and successfully applied in SERS and

SERRS spectral studies of several adsorbates [50–53]. For laser ablation, the typical procedure consists of using a pulse laser (1064 nm from a Nd:YAG laser) to ablate metal plates or foils in distilled water or a NaCl solution [51, 52]. The properties of the formed metal colloids are strongly dependent on the laser pulse energy, the time of ablation, the laser beam focussing and the presence of additional ions [51]. Under optimized size distribution of the laser-ablated metal colloids, reproducible SERS spectra can be obtained [52]. However, for the photoreduction method, the common protocol is based on the light (gamma or UV laser irradiation) to reduce the metal salt [54–58]. For example, Tan *et al.* prepared silver nanoparticles via photoreduction of $AgNO_3$ in branched polyethyleneimine (PEI) and 4-(2-hydroxyethyl)-1-piperazineethanesulfonic acid (HEPES) solution for good SERS enhancement for the probe molecules [57]. Ahern *et al.* also have obtained the SERS spectra of pyridine, biotin and citrate from silver colloids generated *in situ* by laser irradiation. They observed that the SERS signal followed a wavelength dependence similar to that from conventional Ag colloids [59].

2.2.1.3 Size Effect of SERS Signal on Metal Nanoparticles

Chapter 1 has addressed the size effect in SERS activity based on theory, and many experiments have also demonstrated that there is a strong relationship between the size of the metal nanoparticles and their SERS activity [60–64], which is due to the variation of the localized surface plasmon (LSP) with nanoparticle size. Jang *et al.* have reported the SERS spectra of 4-biphenylmethanethiolate on gold nanoparticles with different sizes and found that the enhancement is very weak on gold nanoparticles when their size is smaller than 11 nm [65]. However, for the larger nanoparticles (43 and 97 nm), the enhancement is better. Meanwhile, Nie and co-workers have systematically investigated the SERS activity on different sizes of silver nanoparticles to examine the relationship between the optical excitation wavelength and particle size. They chose 488, 568 and 647 nm laser lines to excite the SERS on silver nanoparticles and found that the SERS-active nanoparticles identified at different wavelengths have dramatically different sizes, as shown in Figure 2.4 [66] which gives the correlated SERS spectra and the tapping-mode atomic force microscopy (AFM) images obtained from spatially isolated single silver nanoparticles. It can be seen clearly that the enhancement from the nanoparticles is strongly related to the size of the particles and the excitation wavelength. They also studied the correlation between the size and the optical and spectroscopic properties of gold nanoparticles with diameters ranging from 10 to 100 nm in aqueous solutions. The size correlation with the SERS intensity obtained from the adsorption of 4-mercaptobenzoic acid on the nanoparticles in aqueous solutions revealed that there exists a critical size of the nanoparticles in the solution beyond which particle–particle interaction is operative and responsible for the SERS effect [67].

As also discussed in Chapter 1, the size effect is a more important aspect for nanoparticle-based SERS substrates and, therefore, to get the best SERS signal, it is necessary to choose the preferred size and the corresponding excitation wavelength to excite surface plasmon resonance on the nanoparticles.

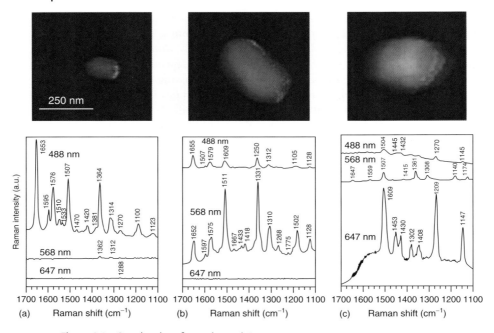

Figure 2.4 Correlated surface-enhanced Raman spectra and tapping-mode AFM images obtained from spatially isolated, single Ag nanoparticles. These particles were selected by wide-field screening for maximum enhancement at (a) 488 nm, (b) 567 nm and (c) 647 nm, respectively [66].

2.2.1.4 Near-Infrared (NIR) Excitation for SERS on Metal Nanoparticles

Several groups have also investigated SERS from metal colloids with near-infrared (NIR) excitation because NIR SERS provides excellent discrimination against fluorescent interference and is out of the electronic resonance with most molecules, allowing the use of higher laser power without photobleaching the analyte. Angel et al. found that the surface enhancement factors in the NIR region on the surface of copper and gold colloidal particles were at least as large as those for SERS excited with radiation in the visible spectral region [68], and Chase et al. reported that the enhancement factor was even higher than that for visible excitation [69]. An enhancement factor of about 5×10^5 was reported for pyridine in grey silver colloids with Fourier transform (FT) techniques by Zhang et al. [70]. Liang et al. used NIR radiation to excite SERS in silver colloids and they found that the enhancement factors for the totally symmetric breathing mode at 1036 cm^{-1} were about 16.5 and 21 times larger, respectively, than those found for excitation at 514.5 nm. More importantly, they come to the conclusion that the chloride ion has some influence on the SERS effect in both NIR and visible regions [71]. But the drawback of NIR excitation is the low sensitivity of the CCD detector used in the instrument. Therefore, one has to choose the exact excitation line and the nanoparticles depending on the application.

2.2.1.5 Stability of the Metal Nanoparticle Colloids

At the end of this part, we address the stability of colloidal metal nanoparticles because it is very important to keep the nanoparticles from aggregating and stabilized in the colloidal solution for different applications. Metal nanoparticles prepared by the conventional citrate reducing methods are reported to be highly stable for months, but because the chemical properties of the nanoparticle surface will change considerably after several days, the SERS signals from old colloids will not be consistent and will be weaker than those obtained from freshly prepared metal nanoparticles. To improve the stability of the metal colloids, stabilizers such as poly(vinyl alcohol), poly(vinylpyrrolidone) and sodium dodecyl sulfate can be added to prevent aggregation [49, 72, 73]. However, the addition of the stabilizer has a disadvantage that it will give interfering signals from the polymer and, more importantly, will increase the electrostatic barrier and lead to a decrease in the adsorption of analytes. Therefore, to improve the stability of the metal nanoparticles and successfully use the nanoparticles in the biological field, silica or bovine serum albumin (BSA) coated SERS-active metal nanoparticles have been developed [74, 75], which has been dealt with in other chapters of this book (see also Chapter 12).

2.2.2
Aggregation of Metal Nanoparticles

Metal nanoparticles, in general, show strong SERS enhancement and are easily prepared and manipulated. To generate much stronger SERS signals, aggregation of the metal colloids is essential because both experimental and theoretical results have indicated that a stronger enhancement effect will be produced when single nanoparticles form aggregates of two or multiple nanoparticles due to the coupling of the electromagnetic field. A common approach for the aggregation is hydrosol activation leading to some 'pre-aggregation', which is induced by adding inorganic salts (e.g. KCl, NaCl, NaBr, $NaNO_3$, $MgSO_4$), surfactants, ethanol, organic amines or mineral acids to the colloids [76–78]. Our group has compared the SERS spectra of Brilliant green in silver colloid before (Figure 2.5A) and after activation by KCl (Figure 2.5B) and found that KCl can enhance the Raman scattering greatly, which can be explained by the increase of electromagnetic field induced by the anion and the reorientation of the molecules on silver nanoparticles [22]. Nickel et al. prepared silver colloids by reduction with hydrazine and aggregation of the nanoparticles by Cl^- for large SERS activity [25]. Apart from the inorganic salts, the analyte itself, most of time, can induce aggregation of the nanoparticles and give different enhancements. For example, Heard et al. found that addition of cetylpyridinium and cetylquinolinium salts result in a colour change of the sol, which is due to the aggregation of the metal nanoparticles into clusters and this aggregation can produce strong enhancement for the Raman signal of the analytes. Therefore, the aggregation of the colloidal substrate is an important factor for observing enhanced Raman scattering from an adsorbed molecule [72]. Smith's group has compared the SERS from unaggregated and aggregated nanoparticles, which was induced by the

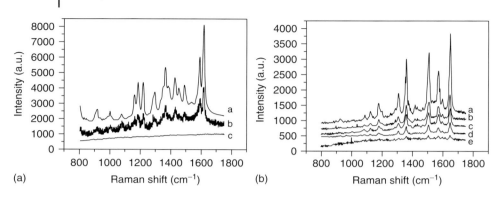

Figure 2.5 SERS spectra of Brilliant Green on AgNPs (A) at different concentrations (a: 10 μM; b: 1 μM; c: 0.1 μM); and that on AgNPs pre-activated by chloride ions (B) at different concentrations (a: 20 nM; b: 5 nM; c: 0.5 nM; d: 0.1 nM; e: 5 fM) [22].

analyte itself. Their research indicated that, if all the clusters could be engineered to give a plasmon resonance frequency that coincides with the excitation frequency, a very large enhancement would be obtained per particle compared to that from isolated single particles [21].

Natan et al. investigated the effect of colloidal gold particle aggregation on SERS by SERS filtration experiments and found that (i) under conditions of minimal aggregation, appreciable SERS intensity derives from aggregates with effective diameters less than 200 nm; (ii) the amount of aggregant clearly controls the average aggregate size; and (iii) similarly aggregated solutions based on colloidal Au particles of different diameters give different distributions of aggregates. These studies provide insight into the dynamics of colloidal gold aggregation, suggest a procedure for signal optimization in colloid SERS experiments and set the stage for controlled surface confinement of SERS-active particle clusters [79].

Many researchers have reported that anions can enhance the Raman scattering in metal nanoparticle colloids and the reasons have been explained from three aspects. One is from the increase of the electromagnetic field by the aggregation of the metal nanoparticles [80–84]. Nie and co-workers have reported that the addition of chloride ions to a silver colloid can detect single rhodamine 6G (R6G) molecules (see also Chapter 4), which is due to the dramatic electromagnetic field produced from the aggregation of silver nanoparticles induced by the chloride ions [80, 81]. Meanwhile, other authors also have supported this view [82–84]. The second view derives from the charge transfer between the metal and the molecules [85–87]. The third view is based on the report by Grochala et al. [88], who considered that the observed increase in intensities through the addition of the chloride ions cannot be exclusively explained in terms of the charge-transfer effect and the 'active site' effect caused by thermodynamically unstable atomic-scale roughness but should be ascribed to anion-induced reorientation of the dye molecule or to the potential-independent molecular resonance caused by the strong influence of co-adsorbed chloride ions on the electronic levels of dyes. Berlin and co-workers also found the phenomena of the anion-induced changes of the Raman spectra and believed that the spectral differences are related not only to silver nanoparticle aggregation but also to coalescence or fusion and new Ag depositions [89]. Smith and co-workers have reported on four different dyes adsorbed on silver nanoparticles and concluded that the complex results of the Raman spectra of dyes can be explained at least in part by changes in the chemistry of surface adhesion [90]. Basu *et al.* investigated the analyte-induced aggregation of gold nanoparticles for SERS and found that the SERS intensity was strongly related to the interparticle spacing [91]. Because of the high enhancement effect observed for aggregated nanoparticles, single-molecule detection has been reported for analytes such as thionine from aggregated gold nanoparticles [92].

Dimers of metal nanoparticles exhibit large SERS enhancements, which are due to the gap effect in the junctions as discussed in Chapter 1. The formation of the dimer is mostly dependent on the nanoparticle assembly technique, such as the use of DNAs to assemble gold nanoparticles into dimers [93] or the rigid multivalent thiol-linkers [94]. Another method for the dimer formation was reported by Shumaker-Parry *et al.*, who produced dimers of gold nanoparticles using organic molecules through an asymmetric functionalization pathway [95]. Chen *et al.* prepared polymer-encapsulated dimers of gold nanoparticles by functionalizing the surface of gold nanoparticles with a thiol-terminated hydrophobic ligand [96]. Recently, Li *et al.* reported a simple one-pot method to generate dimers of silver nanospheres without any additional assembly steps (Figure 2.6a) [97]. Because the dimers consist of 30 nm silver nanospheres, which are separated by a gap

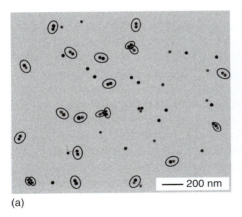

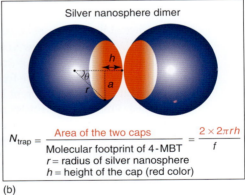

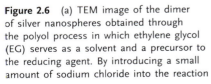

(a) (b)

Figure 2.6 (a) TEM image of the dimer of silver nanospheres obtained through the polyol process in which ethylene glycol (EG) serves as a solvent and a precursor to the reducing agent. By introducing a small amount of sodium chloride into the reaction solution, dimerization is induced due to a change to colloidal stability. (b) Scheme showing the approach to estimate the number of probe molecules trapped in the hot spot (N_{trap}) of a dimer [97].

of 1.8 nm, they obtained an SERS enhancement factor for the dimer as large as 2×10^7 (Figure 2.6b) [97]. Halas also has reported the dimer structure of the metal nanoparticles as being responsible for the huge SERS enhancement [98]. With the development of the dimer structure, some new theoretical calculation have been proposed to explain the enhancement in the so-called hot spot, the gap between a pair of strongly coupled silver or gold nanoparticles. For example, McMahon *et al.* used the finite-element method to calculate the extinction spectra and electromagnetic contributions to SERS on dimers, which revealed that the EM properties depend significantly on the junction region, specifically the distance between the nanoparticles, for a spacing of less than 1 nm [99].

2.2.3
Bimetallic Nanoparticle SERS Substrate

Besides silver or gold nanoparticles with the single composition, bimetallic or core–shell bimetal nanoparticles [100] allow a shift of the resonance frequency as a function of the shell thickness or the ratio between the two metals, which will combine the SERS activities of both metals and provide new or great enhancement for SERS. Simply mixing the as-prepared silver and gold nanoparticle colloids and using them as an SERS substrate was reported by Fang *et al.*, who investigated SERS of dye-coated mixed silver–gold colloids and found that a suitable ratio of mixed silver and gold colloids could form a favourable state of aggregation and significantly increase the SERS activity compared to single silver or gold colloids [101]. But the most typical approach for the fabrication of bimetallic nanoparticles is through the chemical reaction as reported in the literature [102–105]. For example, Jana prepared the Ag-coated Au particles by the seed-mediated method; that is, 12 nm gold particles were first synthesized according to the citrate reduction method and used as the seed. Then ascorbic acid reduction of $AgNO_3$ was carried out in the presence of the 12 nm gold seed. Uniform silver coated gold nanoparticles of 25–50 nm size appears as a new SERS substrate with a very high sensitivity [104]. Natan's group prepared the Ag-coated Au colloidal particles by the reduction of Ag^+ in the presence of pre-formed gold colloids. Through control of the concentration of $AgNO_3$, different kinds of Ag-coated Au colloidal particles can be obtained. From the transmission electron microscopy (TEM) image, there is no evidence for the formation of separate Ag colloids (Figure 2.7a,b); but the different optical property between the Ag-coated Au colloid and a mixture of Ag and Au colloid can be seen clearly from the UV–vis absorption spectra (Figure 2.7c). SERS results (Figure 2.7d) indicated that enhancement behaviour of the Ag-coated Au colloids with 647.1 nm excitation is highly dependent upon the Ag:Au ratio. Very small amounts of Ag ($x \leq 5$) lead to an increase in SERS intensity, but further increases lead to complete loss of signal [102]. Similar research was done by Rivas *et al.*, who prepared Ag-coated Au and Au-coated Ag colloidal particles through chemical reaction and found that the composition of Ag/Au colloidal particles depends on the relative amount of the depositing metal; the SERS activity of the bimetallic nanoparticles was found to be placed between that of Ag

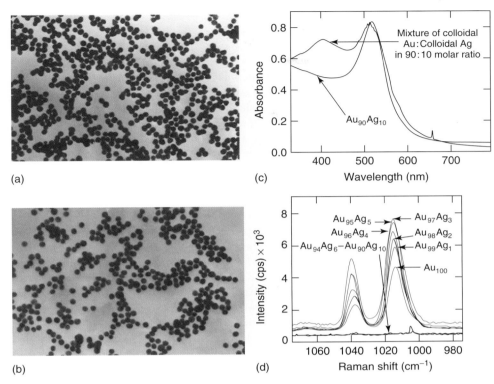

Figure 2.7 TEM image of (a) 12 nm diameter colloidal Au and (b) 12 nm diameter Au90Ag10. (c) Normalized UV–vis spectra of 12 nm Au90Ag10 and of a 90:10 M mixture of 12 nm diameter Au : 12–15 nm diameter colloidal Ag and (d) SERS spectra of at Au100–Au90Ag10 colloids made 20 mM in pyridine. Laser excitation: 647.5 nm [102].

and Au particles [103]. Silver/gold bimetallic nanoparticles with hollow interiors were made by our group and these showed great enhancement for the probe molecules [106].

Composite Au/Ag or Ag/Au core–shell nanoparticles were also fabricated for SERS substrate [107, 108] and used for single-molecule detection [100]. Lu *et al.* fabricated composite Au/Ag nanoshells with different layers of silver nanoparticles and aggregates of composite Au/Ag nanoshells with high SERS enhancement to R6G [109]. Recently, Tian *et al.* proposed the 'borrowing concept' in SERS (see also Chapter 9). They first chemically synthesized gold nanoparticles and then coated them with ultra-thin shells of various transition metals. Because of the long-range effect of the enhanced EM field generated by the highly SERS-active Au core, the original low enhancement effect on the transition metal can be improved greatly, which is very useful for the study of the catalytic reaction on the transition metal surface [110].

2.2.4
Nanoparticles with Various Shapes

Recently, nanoparticles with various shapes, including cubes, prisms, rods and octahedra, depending on the reaction conditions and surface-active agents have been fabricated for use as SERS substrates because theoretical results have shown that metal nanoparticles with a sharp edge and angle will give an extra enhancement owing to the 'lightning rod effect'. Experimental results also indicate that anisotropic nanoparticles have a stronger enhancement effect over spherical nanoparticles. One of the typical methods for the preparation of nanoparticles with various shapes is by seed-mediated growth. So far, silver and gold nanorods (NRs) [111–113] and branched metal nanoflowers [114] have been synthesized through the seed-mediated method and used as active SERS substrates. The most obvious example can be found in the preparation of gold NRs, which involves seed formation and growth (Figure 2.8a) [112]. Gold NRs with different aspect ratios can be obtained by controlling the concentration of $AgNO_3$, which plays an important role in the synthesis. Because gold NRs show transverse and longitudinal surface plasmon resonance modes due to the geometric asymmetry and the longitudinal modes have a stronger absorption band and will red shift and increase in intensity with increasing aspect ratio (Figure 2.8b) [113], SERS activity can be tuned properly on NRs of different aspect ratios according to their surface plasmon bands (Figure 2.8c) [115]. Dong's group has reported the synthesis of branched gold nanoparticles with the assistance of citrate by the seed-mediated method [116]. Different sizes and shapes of gold nanoparticles could be obtained (Figure 2.9a–e) and the surface plasmon bands red-shifted with increasing size (Figure 2.9f). A notable red shift can be clearly seen for the branched gold nanoparticles. Strong SERS signals were generated on the branched gold particles due to this unique structure providing the high electromagnetic field for SERS enhancement (Figure 2.9g) [116]. Similar work was reported by Jeong, who prepared multi-branched gold nanoparticles as efficient SERS substrates, and the SERS activity of the particles was shown to depend on the aspect ratio of their branches, which is most likely related to a great increase in the localized electromagnetic field enhancement from their unique sharp surface features arising from the branches [117]. In addition, silver nanoplates of disc and flower shapes have been prepared as SERS-active substrates using the silver seed-mediated method and it was found that flower-like silver nanoplates exhibit excellent SERS enhancement ability compared to spherical silver nanoparticles and the disc-like silver nanoplates due to the high shape anisotropy of the former [118]. Apart from the seed-mediated method, some work has been reported on the preparation of nanoparticles with other shapes such as urchin-like shape of gold nanoparticles [119, 120], polyhedral gold nanocrystals [121], spinous, multipod microspheres [122] and poly(N-vinyl-2-pyrrolidone) (PVP) capped dendritic gold nanoparticles [123] – which were produced by controlling the experimental conditions including the capping agent, reductant and temperature – which all show SERS enhancement of the analyte due to the LSPR residing in the nanoparticles. It should be noted that light-induced synthesis of nanoparticles with various shapes

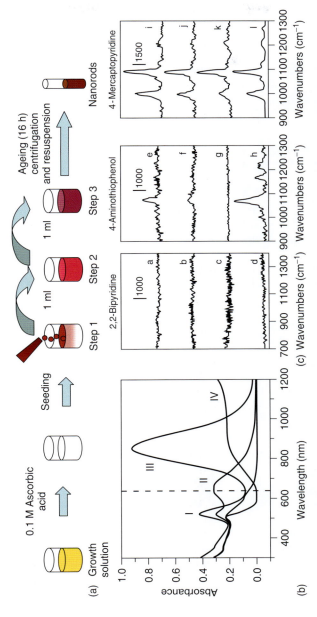

Figure 2.8 (a) General methodology for the generation of gold nanorods; (b) UV–vis absorption spectra of different aspect ratio of nanorods and (c) the SERS spectra of different probe molecules on different aspect ratio of gold nanorods [112, 113, 115].

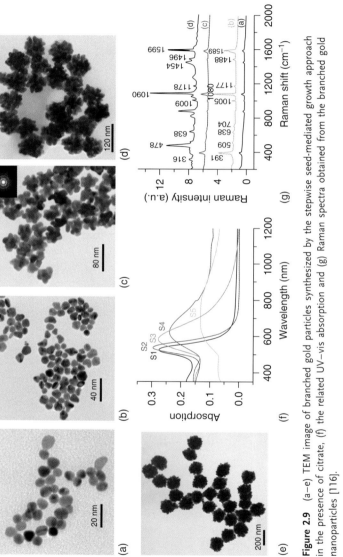

Figure 2.9 (a–e) TEM image of branched gold particles synthesized by the stepwise seed-mediated growth approach in the presence of citrate, (f) the related UV–vis absorption and (g) Raman spectra obtained from the branched gold nanoparticles [116].

is also very effective for the fabrication of SERS-active substrates. Utilizing the light-induced method, tetrahedral silver nanoparticles [124], gold nanolenses [125] and triangular silver nanoparticles [126] have been fabricated to generate strong SERS enhancement for the analyte.

Another method to be mentioned is the polyol synthesis of silver nanoparticles with various shapes as reported by Xia's group [127–129]. This method uses ethylene glycol as both a solvent and a source of reducing agent; through control of the concentration of glycoaldehyde and the temperature of reaction, the shape of the resultant silver nanoparticles can be properly tuned. Most of these nanoparticles can be used as SERS substrates owing to their anisotropic optical properties. For instance, they compared the SERS from sharp and truncated silver nanocubes and found that larger particles (90 and 100 nm) have higher SERS efficiencies (size effect) and particles with shaper corners gave more intense SERS signals than their truncated counterparts [130] owing to the lightning rod effect. They also compared three different molecules on single-crystal nanocubes and nanospheres of silver and concluded that sharp features on the silver nanocubes could greatly increase the contribution of the chemical enhancement to the SERS intensity [131].

Several groups have reported on gold nanostars for efficient SERS, which can be due to the electromagnetic field residing in the nanostars [132, 133]. For example, Hrelescu *et al.* prepared gold nanostars using a one-pot synthesis in aqueous solution with cetyltrimethylammoniumbromide as capping agent and studied the SERS effect on single gold nanostars [132]. Vo-Dinh reported the preparation of gold nanostars with different sizes according to the seed-mediated method. It was found that variations in star size were accompanied by shifts of the long plasmon band in the NIR region, which means the tuning capabilities can be exploited in specific SERS applications [133].

In addition, bimetallic nanoparticles with various shapes, such as NRs [134, 135], nanowires [136, 137] and spindle-shaped core–shell nanoparticles [138], were also prepared and utilized to increase the SERS activity, similar to that of spherical nanoparticles [102, 103].

As is known, inorganic salts can induce the aggregation of the spherical nanoparticles to show high SERS enhancement. Recently, it was reported that inorganic salts can also induce the aggregation of anistropic nanoparticles and give high enhancement. Dong's group prepared four different sizes of the citrate-protected silver nanoplates (Figure 2.10a–d) and induced the aggregation of the nanoparticles by NaCl, and used them as active SERS substrate (Figure 2.10e). Their results demonstrated that highly anisotropic silver nanoplates could also be coupled with each other upon aggregation in aqueous solution to show an enhancement factor of about 4.5×10^5 for the probe molecules under the NIR excitation, which is believed to mainly come from EM enhancement through the interacting silver nanoplates. Since the new plasmon resonance of aggregated silver nanoplates in aqueous solution is more red-shifted than that of the commonly used Lee–Meisel silver colloids (Figure 2.10f), they are a very desirable substrate for NIR Raman measurements [139]. Similar work was also reported by Jana, who tested a wide variety of colloidal solutions of anisotropic metal nanoparticles (NRs, nanoplates) and

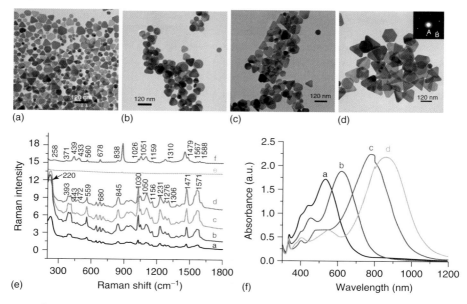

Figure 2.10 TEM images of different size of silver nanoplates (a–d) and the related UV–vis (e) and Raman spectra of 2-ATP on the aggregated silver nanoplates (f) [139].

optimized the length, particle anisotropy, particle composition and experimental conditions for obtaining an enhanced Raman signal [140]. It was found that aggregation of the nanoparticles induced by salts can broaden the plasmon band (as shown in Figure 2.11a and the related TEM images shown in Figure 2.11b) so that a wide range of laser excitation wavelengths can be used for electromagnetic enhancement, and aggregation processes produce coupled plasmon bands that are localized in the junctions between nanoparticle dimers/oligomers/fractals, which can act as electromagnetic hot spots. Meanwhile, during the aggregation processes, the analyte can adsorb on those hot spots to produce a strong SERS signal (Figure 2.11c). All these studies have demonstrated that the aggregation route to colloidal substrates has been the most practical approach for obtaining an intense SERS signal.

For nanoparticles with various shapes, the size effect also plays a great role for the overall SERS enhancement. For instance, Sant'Ana *et al.* reported the size-dependent SERS enhancement of colloidal silver nanoplates in suspension without aggregation [141], and Sabur reported the SERS intensity optimization by controlling the size and shape of faceted gold nanoparticles [142]. However, the shape effect will dominate more for these kinds of nanoparticles [143, 144] due to their shape-dependent and tunable LSPR residing in the nanoparticles (see also Chapter 1). Another important shape-dependent mechanism is the CM related to the crystal structure of the nanoparticles with various shapes as reported by Zheng

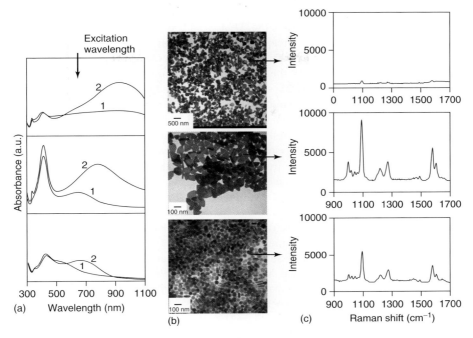

Figure 2.11 Ag platelets as an SERS substrate showing the particle-anisotropy dependence of the plasmon band and Raman signal: (a) Absorption spectra of silver platelets of different aspect ratio in the presence (1) and absence (2) of salt. (b) Representative transmission electron microscopy images (each image corresponds to the spectra to their left and right) of different sized platelets all having a width of 10–20 nm. (c) Surface-enhanced Raman spectra of the platelets in the presence of 1 nM MPy in the presence of 0.1 M NaCl [140].

et al., who attributed the great differences of the SERS spectra of rhodamine B on different shapes of silver colloids to the shapes and crystal planes of silver nanoparticles [145].

2.3
Characterization of Nanoparticle-Based SERS Substrates

As discussed above, the most important properties for nanoparticles as SERS substrates are their morphology, size, monodispersity, shape, crystal structure, dipole plasmon absorption and composition. Electron microscopy, such as TEM and scanning electron microscopy (SEM), can give much more information about the morphology, size and shape of the nanoparticles. Compared with TEM, SEM offers an excellent field of view where large areas of the samples can be imaged simultaneously and with lateral resolution. In addition to TEM and SEM, AFM is another important tool for the characterization of nanoparticles because it can provide much more superior vertical resolution and can be used on conducting or non-conducting surfaces. Therefore, TEM, SEM and AFM are always used together since they can provide complementary information on the nanoparticles. The monodispersity of the nanoparticles has always been characterized by the dynamic light scattering (DLS). For the composition of the nanoparticles, X-ray energy dispersive spectrometry (EDS) is a common technique associated with electron microscopy to give chemical information of the nanoparticles. X-ray photoelectron spectroscopy (XPS) is another frequently used technique to give the elemental composition, chemical state and electronic state of the elements that constitute the nanoparticles. Because the crystal structure is also a very important property of the nanoparticle SERS substrate (recent research has shown the crystal structure of the nanoparticles is related to the chemical enhancement in SERS [145, 146]), X-ray diffraction (XRD) technique can be useful to determine the crystal structure of the nanoparticles. Optical properties of nanoparticles are easily determined in the UV–vis–NIR absorption spectrum, which can provide important information on the excitation wavelength of the nanoparticle SERS substrate. Another important property of the nanoparticle SERS substrate is the surface charge, which can be evaluated by the zeta potential. Smith's group and Aroca's group have investigated the surface charge of the nanoparticles in relation to the SERS intensity, and found that there is some correlation between the surface charge and the absorption of the analyte on the surface of the particles [100, 147]. A more detailed account of the characterization of nanoparticles can be found in the recent reviews by Aroca [13], Wang [148] and Olson [149].

2.4
Nanoparticles on the Unfunctionalized Solid Surface as SERS Substrates

As is known, the relatively weak SERS enhancement and instability are the major issues for SERS detection with colloidal metal nanoparticles. To avoid these problems, some researchers have used unfunctionalized solid substrates, such as glass, silica wafer or metal surface, to disperse the as-prepared metal nanoparticle colloids on them to form a close packed nanoparticle film, and these have been widely adopted at present [133, 150–153]. As shown in the scheme of Figure 2.12a, the manipulation is very simple: just depositing one drop of the colloid on the substrate, then drying naturally in a desiccator for several hours, followed by vigorous rinsing in distilled water and finally drying with a dust blower [154]. Then the nanoparticle SERS film can be formed on the solid surface. The morphology of the film can be examined by AFM (Figure 2.12b,c shows the AFM images of silver and gold nanoparticles on a glass surface, respectively). During this process, the analytes (Raman active molecules) can be mixed with the colloidbreak first [155, 156], or the analyte can be dropped onto the surface after the nanoparticles form the film [154]. In this case, it should be pointed out that thorough cleaning of the substrate and concentrating the colloid are very important to form a uniform surface for a reproducible and stable SERS signal.

Figure 2.12 (a) Scheme of the nanoparticles on an unfunctionalized solid substrate for SERS research [155] and AFM images of (b) Ag nanoparticles ([155]) and (c) Au nanoparticles [154].

For example, Wu et al. have reported depositing a drop of the nanoparticle colloid on the metal surface to get a compact layer of nanoparticles for obtaining a good SERS signal. They found that a more uniform SERS substrate could be obtained by concentrating the nanoparticle sol and controlling the dropping times and the drying rate because a single dispersion is non-uniform and the SERS signals from different areas are quite inhomogeneous [157]. Similar findings were reported by Culha et al., who observed that the aggregates on a glass substrate formed from higher concentrations of the colloidal suspension show a significant improvement in SERS activity [158]. Another approach to get a high SERS activity substrate with great reproducibility and stability is to induce the aggregation of nanoparticles first by different additional reagents such as $NaClO_4$ and NaOH as discussed above, and then, during this process, to place a glass slide at the bottom of the solution. Nanoparticles will precipitate on the glass surface, which is then dried in air to obtain an SERS-active substrate [159]. Ozaki and co-workers investigated the SPR and SERS characteristics of gold nanoaggregates with different morphologies after dropping gold nanoparticle colloids on a glass substrate. They found an SERS enhancement of at least a million times higher in Raman intensity using dye molecules adsorbed on gold nanoaggregates [160]. Similar work was done by Sztainbuch et al., who investigated the effect of gold aggregate morphology on SERS enhancement and concluded that dimensions of the nanoparticle itself or a 'protrusion' emanating from an irregular shaped single nanoparticle must be such that it sustains surface plasmon resonance and enhances SERS generation [161].

Nikoobakht et al. studied the SERS activity of gold NRs with different surface coverages on a silica surface. In their research, they used two approaches to produce silica surfaces covered with NRs. In the first approach, monodispersed NRs are gradually deposited from a solution onto a silica surface and the number of NRs was increased by increasing the deposition time. The other approach is based on the aggregation of NRs in solution, followed by their deposition on the silica surface, similar to that reported for the production of sphere gold nanoparticle aggregation [154, 159]. It was found that stronger SERS signal could be obtained on the surface covered with aggregated NRs, which was attributed to the enhancement of the electric field between the particles in the aggregates [146].

2.5
Conclusion and Outlook

In this review chapter, different nanoparticles and the related preparation methods have been addressed for the SERS substrate, including spherical metal nanoparticles, aggregated nanoparticles, bimetallic nanoparticles and nanoparticles with various shapes. The techniques for the characterization of the nanoparticles have also been introduced. Finally, nanoparticles on unfunctionalized solid surfaces as SERS substrates have been reviewed. The properties of the nanoparticle SERS substrate have been analysed with a view to selecting the proper substrate for the practical application.

The rapid development of nanoscience and nanotechnology has attracted great research efforts for the fabrication of SERS substrates with more activity and reproducibility and brought great opportunities for nanoparticle SERS substrates for practical applications, such as biocompatible SERS nanoprobes composed by the Raman enhancer nanoparticles, Raman reporter molecules and the coating by BSA or a silica layer [74, 162, 163].

References

1. Fleischmann, M., Hendra, P.J. and McQuillan, A.J. (1974) Raman spectra of pyridine adsorbed at a silver electrode. *Chem. Phys. Lett.*, **26**, 163.
2. Jeanmaire, D.L. and Van Duyne, R.P. (1977) Surface Raman spectroelectrochemistry Part I. Heterocyclic, aromatic, and aliphatic amines adsorbed on the anodized silver electrode. *J. Electroanal. Chem.*, **1**, 84.
3. Albrecht, M.G. and Creighton, J.A. (1977) Plasma resonance enhancement of Raman scattering by pyridine adsorbed on silver or gold sol particles of size comparable to the excitation wavelength. *J. Am. Chem. Soc.*, **99**, 5215–5217.
4. Schatz, G.C. (1984) Theoretical studies of surface enhanced Raman scattering. *Acc. Chem. Res.*, **17**, 370–376.
5. Kahl, M. and Voges, E. (2000) Analysis of plasmon resonance and surface-enhanced Raman scattering on periodic silver structures. *Phys. Rev. B*, **61**, 14078–14088.
6. Shalaev, V.M. and Sarychev, A.K. (1998) Nonlinear optics of random metal-dielectric films. *Phys. Rev. B*, **57**, 13265–13288.
7. Moskovits, M., Dilella, D.P. and Maynard, K.L. (1988) Surface Raman spectroscopy of a number of cyclic aromatic molecules adsorbed on silver: selection rules and molecular reorientation. *Langmuir*, **4**, 67–76.
8. Franzen, S., Folmer, J.C.W., Glonmm, W.R. and O'Neal, R. (2002) Optical properties of dye molecules adsorbed on single gold and silver nanoparticles. *J. Phys. Chem. A*, **106**, 6533–6540.
9. Cao, L.Y., Diao, P., Tong, L.M., Zhu, T. and Liu, Z.F. (2005) Surface-enhanced Raman scattering of p-Aminothiophenol on a

Au(core)/Cu(shell) nanoparticle assembly. *ChemPhysChem*, **6**, 913–918.
10. Ueba, H. (1983) Theory of charge transfer excitation in surface enhanced Raman scattering. *Surf. Sci.*, **131**, 347–366.
11. Otto, A., Billmann, J., Eickmans, J., Erturk, U. and Pettenkofer, C. (1984) The "adatom model" of SERS (Surface Enhanced Raman Scattering): the present status. *Surf. Sci.*, **138**, 319–338.
12. Arenas, J.F., Woolley, M.S., Otero, J.C. and Marcos, J.I. (1996) Charge-transfer processes in surface-enhanced Raman scattering. Franck-condon active vibrations of pyrazine. *J. Phys. Chem.*, **100**, 3199–3206.
13. Aroca, R.F., Alvarez-Puebla, R.A., Pieczonka, N., Sanchez-Cortez, S. and Garcia-Ramos, J.V. (2005) Surface-enhanced Raman scattering on colloidal nanostructures. *Adv. Colloid Interface Sci.*, **116**, 45–61.
14. Banholzer, M.J., Millstone, J.E., Qin, L. and Mirkin, C.A. (2008) Rationally designed nanostructures for surface-enhanced Raman spectroscopy. *Chem. Soc. Rev.*, **5**, 885–897.
15. Baker1, G.A. and Moore, D.S. (2005) Progress in plasmonic engineering of surface-enhanced Raman-scattering substrates toward ultra-trace analysis. *Anal. Bioanal. Chem.*, **382**, 1751–1770.
16. Vo-Dinh, T. (1998) Surface-enhanced Raman spectroscopy using metallic nanostructures. *Trends Anal. Chem.*, **17**, 557–582.
17. Oldenburg, S.J., Westcott, S.L., Averitt, R.D. and Halasa, N.J. (1999) Surface enhanced Raman scattering in the near infrared using metal nanoshell substrates. *J. Chem. Phys.*, **111**, 4729.
18. Dadosh, T., Sperling, J., Bryant, G.W., Breslow, R., Shegai, T., Dyshel, M., Haran, G. and Bar-Joseph, I. (2009) Plasmonic control of the shape of the Raman spectrum of a single molecule in a silver nanoparticle dimer. *ACS Nano*, **3**, 1988–1994.
19. Khoury, C.G., Norton, S.J. and Vo-Dinh, T. (2009) Plasmonics of 3-D nanoshell dimers using multipole expansion and finite element method. *ACS Nano*, **3**, 2776–2788.
20. Creighton, J.A., Blatchford, C.G. and Albrecht, M.G. (1979) Plasma resonance enhancement of Raman scattering by pyridine adsorbed on silver or gold sol particles of size comparable to the excitation wavelength. *J. Chem. Soc. Faraday Trans. II*, **75**, 790.
21. Faulds, K., Littleford, R.E., Graham, D., Dent, G. and Smith, W.E. (2004) Comparison of surface-enhanced resonance Raman scattering from unaggregated and aggregated nanoparticles. *Anal. Chem.*, **76**, 592–598.
22. Wang, Y., Li, D., Li, P., Dong, S.J. and Wang, E.K. (2007) Surface enhanced Raman scattering of brilliant green on Ag nanoparticles and applications in living cells as optical probes. *J. Phys. Chem. C*, **111**, 16833.
23. Lee, P.C. and Meisel, D. (1982) Adsorption and surface-enhanced Raman of dyes on silver and gold sols. *J. Phys. Chem.*, **86**, 3391–3395.
24. Weaver, G.C., Norrod, K., (1998) Surface enhanced Raman spectroscopy: a novel physical chemistry experiment for the undergraduate Laboratory. *J. Chem. Educ.*, **75**, 621–624.
25. Nickel, U., Mansyreff, K. and Schneider, S. (2004) Production of monodisperse silver colloids by reduction with hydrazine: the effect of chloride and aggregation on SER(R)S signal intensity. *J. Raman Spectrosc.*, **35**, 101–110.
26. Cañamares, M.V., Garcia-Ramos, J.V., Gómez-Varga, J.D., Domingo, C. and Sanchez-Cortes, S. (2005) Comparative study of the morphology, aggregation, adherence to glass, and surface-enhanced Raman scattering activity of silver nanoparticles prepared by chemical reduction of Ag^+ using citrate and hydroxylamine. *Langmuir*, **21**, 8546–8553.
27. Leopold, N. and Lendl, B. (2003) A new method for fast preparation of highly surface-enhanced Raman scattering (SERS) active silver colloids at room

temperature by reduction of silver nitrate with hydroxylamine hydrochloride. *J. Phys. Chem. B*, **107**, 5723–5727.
28. Nickel, U., Castell, A.Z., Pöppl, K. and Schneider, S. (2000) A silver colloid produced by reduction with hydrazine as support for highly sensitive surface-enhanced Raman spectroscopy. *Langmuir*, **16** (23), 9087–9091.
29. Li, X.-L., Zhang, J.-H., Xu, W.-Q., Jia, H.-Y., Wang, X., Yang, B., Zhao, B., Li, B.-F. and Ozaki, Y. (2003) Mercaptoacetic acid-capped silver nanoparticles colloid: formation, morphology and SERS activity. *Langmuir*, **19**, 4285–4290.
30. Kvitek, L., Prucek, R., Panacek, A., Novotny, R., Hrbac, J. and Zboril, R. (2005) The influence of complexing agent concentration on particle size in the process of SERS active silver colloid synthesis. *J. Mater. Chem.*, **15**, 1099–1105.
31. Frens, G. (1973) Controlled nucleation for regulation of particle-size in monodisperse gold suspensions. *Nat. Phys. Sci.*, **241**, 20–22.
32. Njoki, P.N., Lim, I.S., Mott, D., Park, H.-Y., Khan, B., Mishra, S., Sujakumar, R., Luo, J. and Zhong, C.-J. (2007) Size Correlation of optical and spectroscopic properties for gold nanoparticles. *J. Phys. Chem. C*, **111**, 14664–14669.
33. Schwartzberg, A.M., Oshiro, T.Y., Zhang, J.-Z., Huser, T. and Talley, C.E. (2006) Improving nanoprobes using surface-enhanced Raman scattering from 30-nm hollow gold particles. *Anal. Chem.*, **78**, 4732–4736.
34. Shen, C.-M., Hui, C., Yang, T.-Z., Xiao, C.-W., Tian, J.-F., Bao, L.-H., Chen, S.-T., Ding, H. and Gao, H.-J. (2008) Monodisperse noble-metal nanoparticles and their surface enhanced Raman scattering properties. *Chem. Mater.*, **20**, 6939–6944.
35. Joo, S.W., Han, S.W. and Kim, K. (2000) Adsorption characteristics of 1,3-propanedithiol on gold: surface-enhanced Raman scattering and ellipsometry study. *J. Phys. Chem. B*, **104**, 6218–6224.
36. Joo, T.H., Kim, K. and Kim, M.S. (1986) Surface-enhanced Raman-scattering (SERS) of 1-propanethiol in silver sol. *J. Phys. Chem.*, **90**, 5816–5819.
37. Joo, S.W., Han, S.W. and Kim, K. (2000) Multilayer formation of 1,2-ethanedithiol on gold: surface-enhanced Raman scattering and ellipsometry study. *Langmuir*, **16**, 5391–5396.
38. Joo, S.W., Han, S.W. and Kim, K. (1999) Adsorption characteristics of p-xylene-alpha,alpha′-dithiol on gold and silver surfaces: surface-enhanced Raman scattering and ellipsometry study. *J. Phys. Chem. B*, **103**, 10831–10837.
39. Kim, K.L., Lee, S.J. and Kim, K. (2004) Surface-enhanced Raman scattering of benzyl phenyl sulfide in silver sol: excitation-wavelength-dependent surface-induced photoreaction. *J. Phys. Chem. B*, **108**, 9216–9220.
40. Joo, S.W., Han, S.W. and Kim, K. (2001) Adsorption of 1,4-benzenedithiol on gold and silver surfaces: surface-enhanced Raman scattering study. *J. Colloid Interface Sci.*, **240**, 391–399.
41. Lee, C.R., Bae, S.J., Gong, M.S., Kim, K. and Joo, S.W. (2002) Surface-enhanced Raman scattering of 4,4′-dicyanobiphenyl on gold and silver nanoparticle surfaces. *J. Raman Spectrosc.*, **33**, 429–433.
42. Bae, S.J., Lee, C.R., Choi, I.S., Hwang, C.S., Gong, M.S., Kim, K. and Joo, S.W. (2002) Adsorption of 4-biphenylisocyanide on gold and silver nanoparticle surfaces: surface-enhanced Raman scattering study. *J. Phys. Chem. B*, **106**, 7076–7080.
43. Joo, S.W., Chung, T.D., Jang, W.C., Gong, M.S., Geum, N. and Kim, K. (2002) Surface-enhanced Raman scattering of 4-cyanobiphenyl on gold and silver nanoparticle surfaces. *Langmuir*, **18**, 8813–8816.
44. Lee, C.R., Kim, S.I., Yoon, C.J., Gong, M.S., Choi, B.K., Kim, K. and Joo, S.W. (2004) Size-dependent adsorption of 1,4-phenylenediisocyanide onto gold nanoparticle surfaces. *J. Colloid Interface Sci.*, **271**, 41–46.

45. Moskovits, M. (1982) Surface selection rules. *J. Chem. Phys.*, **77**, 4408–4416.
46. Creighton, J.A. (1983) Surface Raman electromagnetic enhancement factors for molecules at the surface of small isolated metal spheres: the determination of adsorbate orientation from SERS relative intensities. *Surf. Sci.*, **124**, 209–219.
47. Skadtchenko, B.O. and Aroca, R. (2001) Surface-enhanced Raman scattering of p-nitrothiophenol molecular vibrations of its silver salt and the surface complex formed on silver islands and colloids. *Spectrochim. Acta Part A: Mol. Biomol. Spectrosc.*, **57A**, 1009–1016.
48. Menendez, J.R., Obuchowska, A. and Aroca, R. (1996) Infrared spectra and surface enhanced Raman scattering of naphthalimide on colloidal silver. *Spectrochim. Acta Part A: Mol. Biomol. Spectrosc.*, **52A**, 329–336.
49. Siiman, O., Bumm, L.A., Callaghan, R., Blatchford, C.G. and Kerker, M. (1983) Surface-enhanced Raman scattering by citrate on colloidal silver. *J. Phys. Chem.*, **87**, 1014–1023.
50. Besner, S., Kabashin, A.V. and Meunier, M. (2007) Two-step femtosecond laser ablation-based method for the synthesis of stable and ultra-pure gold nanoparticles in water. *Appl. Phys. A-Mater.*, **88**, 269–272.
51. Procházka, M., Mojzeš, P., Štěpánek, J., Vlčková, B. and Turpin, P.-Y. (1997) Probing applications of laser-ablated Ag colloids in SERS spectroscopy: improvement of ablation procedure and SERS spectral testing. *Anal. Chem.*, **69**, 5103–5108.
52. Neddersen, J., Chumanov, G. and Cotton, T.M. (1993) Laser ablation of metals: a new method for preparing SERS active colloids. *Appl. Spectrosc.*, **47**, 1959–1964.
53. Sibbald, M.S., Chumanov, G. and Cotton, T.M. (1996) Reduction of cytochrome c by halide-modified, laser-ablated silver colloids. *J. Phys. Chem.*, **100**, 4672–4678.
54. Sato-Berru, R., Redon, R., Vaquez-Olmos, A. and Saniger, J.M. (2009) Silver nanoparticles synthesized by direct photoreduction of metal salts. application in surface-enhanced Raman spectroscopy. *J. Raman Spectrosc.*, **40**, 376–380.
55. Torreggiani, A., Jurasekova, Z., D'Angelantonio, M., Tamba, M., Garcia-Ramos, J.V. and Sanchez-Cortes, S. (2009) Fabrication of Ag nanoparticles by gamma-irradiation: application to surface-enhanced Raman spectroscopy of fungicides. *Colloids Surf., A: Phys. Eng. Asp.*, **339**, 60–67.
56. Ding, L.-P. and Fang, Y. (2007) An investigation of the surface-enhanced Raman scattering (SERS) effect from laser irradiation of Ag nanoparticles prepared by trisodium citrate reduction method. *Appl. Sur. Sci.*, **253**, 4450–4455.
57. Tan, S., Erol, M., Attygalle, A., Du, H. and Sukhishvili, S. (2007) Synthesis of positively charged silver nanoparticles via photoreduction of $AgNO_3$ in branched polyethyleneimine/HEPES solutions. *Langmuir*, **23**, 9836–9843.
58. Pal, A., Pal, T., Stokes, D.L. and Vo-Dinh, T. (2003) Photochemically prepared gold nanoparticles: a substrate for surface enhanced Raman scattering. *Curr. Sci. India*, **84**, 1342–1346.
59. Ahern, A.M. and Garrell, R.L. (1987) In situ photoreduced silver nitrate as a substrate for surface-enhanced Raman spectroscopy. *Anal. Chem.*, **59**, 2813–2816.
60. Seney, C.S., Gutzman, B.M. and Goddard, R.H. (2009) Correlation of size and surface-enhanced Raman scattering activity of optical and spectroscopic properties for silver nanoparticles. *J. Phys. Chem. C*, **113**, 74–80.
61. dos Santos, D.S., Alvarez-Puebla, R.A., Oliveira, O.N. and Aroca, R.F. (2005) Controlling the size and shape of gold nanoparticles in fulvic acid colloidal solutions and their optical characterization using SERS. *J. Mater. Chem.*, **15**, 3045–3049.
62. Wang, J.-A., Zhu, T., Zhang, X. and Liu, Z.-F. (1999) The SERS intensity vs the size of Au nanoparticles. *Acta Phys. Chim. Sin.*, **15**, 476–480.

63. Glaspell, G.P., Zuo, C. and Jagodzinski, P.W. (2005) Surface enhanced Raman spectroscopy using silver nanoparticles: the effects of particle size and halide ions on aggregation. *J. Clust. Sci.*, **16**, 39–51.
64. Praharaja, S., Janaa, S., Kundua, S., Pandea, S. and Pal, T. (2009) Effect of concentration of methanol for the control of particle size and size-dependent SERS studies. *J. Colloid Interface Sci.*, **333**, 699–706.
65. Jang, S., Park, J.S., Shin, S., Yoon, C., Choi, B.K., Gong, M.S. and Joo, S.W. (2004) Adsorption of 4-biphenylmethanethiolate on different-sized gold nanoparticle surfaces. *Langmuir*, **20**, 1922.
66. Emory, S.R., Haskins, W.E. and Nie, S.-M. (1998) Direct observation of size-dependent optical enhancement in single metal nanoparticles. *J. Am. Chem. Soc.*, **120**, 8009–8010.
67. Krug, J.T., Wang, G.D., Emory, S.R. and Nie, S.M. (1999) Efficient Raman enhancement and intermittent light emission observed in single gold nanocrystals. *J. Am. Chem. Soc.*, **121**, 9208–9214.
68. Angel, S.M., Katz, L.F., Archibald, D.D. and Honigs, D.E. (1989) Near-infrared surface-enhanced Raman spectroscopy. part ii: copper and gold colloids. *Appl. Spectrosc.*, **43**, 367–372.
69. Chase, D.B. and Parkinson, B.A. (1988) Surface-enhanced Raman spectroscopy in the near-infrared. *Appl. Spectrosc.*, **42**, 1186–1187.
70. Zhang, P.-X., Wang, X. and Guo, Z.-G. (1990) Fourier transform infrared surface enhanced Raman scattering from pyridine Ag colloids. *Chin. Phys. Lett.*, **7**, 465.
71. Liang, E.J., Engert, C. and Kiefert, W. (1993) Surface-enhanced Raman scattering of pyridine in silver colloids excited in the near-infrared region. *J. Raman Spectrosc.*, **24**, 775–779.
72. Heard, S.M., Griese, F. and Barraclough, C.G. (1983) Surface-enhanced Raman scattering from amphiphilic and polymer molecules on silver and gold sols. *Chem. Phys. Lett.*, **95**, 154–158.
73. Lee, P.C. and Meisel, D. (1983) Surface-enhanced Raman scattering of colloid-stabilizer systems. *Chem. Phys. Lett.*, **99**, 262–265.
74. Su, X., Zhang, J., Sun, L., Koo, T.W., Chan, S., Sundararajan, N., Yamakawa, M. and Berlin, A.A. (2005) Composite organic-inorganic nanoparticles (COINs) with chemically encoded optical signatures. *Nano Lett.*, **5**, 49–54.
75. Mulvaney, S.P., Musick, M.D., Keating, C.D. and Natan, M.J. (2003) Glass-Coated, analyte-tagged nanoparticles: a new tagging system based on detection with Surface-enhanced Raman scattering. *Langmuir*, **19**, 4784.
76. Sun, L.-L., Song, Y.-H., Wang, L., Guo, C.-L., Sun, Y.-J., Liu, Z.-L. and Li, Z. (2008) Ethanol-induced formation of silver nanoparticle aggregates for highly active SERS substrates and application in DNA detection. *J. Phys. Chem. C*, **112**, 1415–1422.
77. Kneipp, K. and Kneipp, H. (2006) Surface-enhanced Raman scattering on silver nanoparticles in different aggregation stages. *Isr. J. Chem.*, **46**, 299–305.
78. Bell, S.E. and Sirimuthu, N.M. (2006) Surface-enhanced Raman spectroscopy (SERS) for sub-micromolar detection of DNA/RNA mononucleotides. *J. Am. Chem. Soc.*, **128**, 15580–15581.
79. Freeman, R.G., Bright, R.M., Hommer, M.B. and Natan, M.J. (1999) Size selection of colloidal gold aggregates by filtration: effect on surface-enhanced Raman scattering intensities. *J. Raman Spectrosc.*, **30**, 733–738.
80. Nie, S.M. and Emery, S.R. (1997) Probing single molecules and single nanoparticles by surface-enhanced Raman scattering. *Science*, **275**, 1102–1106.
81. Doering, W.E. and Nie, S.M. (2002) Single-molecule and single-nanoparticle SERS: examining the roles of surface active sites and chemical enhancement. *J. Phys. Chem. B*, **106**, 311.
82. Futamata, M. and Maruyama, Y. (2007) Electromagnetic and chemical interaction between Ag nanoparticles and

adsorbed rhodamine molecules in surface-enhanced Raman scattering. *Anal. Bioanal. Chem.*, **388**, 89–102.

83. Sanchez-Cortes, S. and Garcia-Ramos, J.V. (2001) Influence of coverage in the surface-enhanced Raman scattering of cytosine and its methyl derivatives on metal colloids: chloride and pH effects. *Surf. Sci.*, **473**, 133–142.

84. Sanchez-Cortes, S. and Garcia-Ramos, J.V. (2000) Surface-enhanced Raman of 1,5-dimethylcytosine adsorbed on a silver electrode and different metal colloids: effect of charge transfer mechanism. *Langmuir*, **16**, 764–770.

85. Liang, E.J., Ye, X.L., Kiefer, W., Liang, E.J., Ye, X.L. and Kiefer, W. (1997) Interaction of halide and halate ions with colloidal silver and their influence on surface-enhanced Raman scattering of pyridine with near-infrared excitation. *Vib. Spectrosc.*, **15**, 69–78.

86. Liang, E.J. and Kiefer, W. (1996) Chemical effect of SERS with near-infrared excitation. *J. Raman Spectrosc.*, **27**, 879–885.

87. Dines, T.J. and Wu, H.J. (1995) Surface-enhanced resonance Raman spectroscopic studies of the PbII complex of 1-(2-pyridylazo)-2-naphthol adsorbed on Ag Sol. *J. Chem. Soc. Faraday Trans.*, **91**, 463–468.

88. Grochala, W., Kudelski, A. and Bukowska, J. (1998) Anion-induced charge-transfer enhancement in SERS and SERRS spectra of Rhodamine 6G on a silver electrode: how important is it? *J. Raman Spectrosc.*, **29**, 681.

89. Su, X., Zhang, J.W., Sun, L., Koo, T.W., Chan, S., Sundararajan, N., Yamakawa, M. and Berlin, A.A. (2005) Composite organic–inorganic nanoparticles (COINs) with chemically encoded optical signatures. *Nano Lett.*, **5**, 49.

90. Campbell, M., Lecomte, S. and Smith, W.E. (1999) Effect of different mechanisms of surface binding of dyes on the surface-enhanced resonance Raman scattering obtained from aggregated colloid. *J. Raman Spectrosc.*, **30**, 37.

91. Basu, S., Pande, S., Jana, S., Bolisetty, S. and Pal, T. (2008) Controlled interparticle spacing for surface-modified gold nanoparticle aggregates. *Langmuir*, **24**, 5562–5568.

92. Ruan, C.M., Wang, W. and Gu, B.-H. (2007) Single-molecule detection of thionine on aggregated gold nanoparticles by surface enhanced Raman scattering. *J. Raman. Spectrosc.*, **38**, 568–573.

93. Park, S.Y., Lytton-Jean, A.K.R., Lee, B., Weigand, S., Schatz, G.C. and Mirkin, C.A. (2008) DNA-programmable nanoparticle crystallization. *Nature*, **451**, 553–556.

94. Novak, J.P. and Feldheim, D.L. (2000) Assembly of phenylacetylene-bridged silver and gold nanoparticle arrays. *J. Am. Chem. Soc.*, **122**, 3979–3980.

95. Sardar, R., Heap, T.B. and Shumaker-Parry, J.S. (2007) Versatile solid phase synthesis of gold nanoparticle dimers using an asymmetric functionalization approach. *J. Am. Chem. Soc.*, **129**, 5356–5357.

96. Wang, X., Li, G., Chen, T., Yang, M., Zhang, Z., Wu, T. and Chen, H. (2008) Polymer-encapsulated gold-nanoparticle dimers: facile preparation and catalytical application in guided growth of dimeric ZnO-nanowires. *Nano Lett.*, **8**, 2643–2647.

97. Li, W., Camargo, P.H.C., Lu, X. and Xia, Y. (2009) Dimers of silver nanospheres: facile synthesis and their use as hot spots for surface-enhanced Raman scattering. *Nano Lett.*, **9**, 485–490.

98. Talley, C.E., Jackson, J.B., Oubre, C., Grady, N.K., Hollars, C.W., Lane, S.M., Huser, T.R., Nordlander, P. and Halas, N.J. (2005) Surface-enhanced Raman scattering from individual Au nanoparticles and nanoparticle dimer substrates. *Nano Lett.*, **5**, 1569–1574.

99. McMahon, J.M., Henry, A.I., Wustholz, K.L., Natan, M.J., Freeman, R.G., Van Duyne, R.P. and Schatz, G.C. (2009) Gold nanoparticle dimer plasmonics: finite element method calculations of the electromagnetic enhancement to surface-enhanced Raman spectroscopy. *Anal. Bioanal. Chem.*, **394**, 1819–1825.

100. Mandal, M., Jana, N.R., Kundu, S., Ghosh, S.K., Panigrahi, M. and Pal, T.

(2004) Synthesis of Au-core-Ag-shell type bimetallic nanoparticles for single molecule detection in solution by SERS method. *J. Nanopart. Res.*, **6**, 53–61.

101. Fang, J.-H., Huang, Y.-X., Li, X. and Dou, X.-M. (2004) Aggregation and surface-enhanced Raman activity study of dye-coated mixed silver-gold colloids. *J. Raman. Spectrosc.*, **35**, 914–920.

102. Freeman, R.G., Hommer, M.B., Grabar, K.C., Jackson, M.A. and Natan, M.J. (1996) Ag-clad Au nanoparticles: novel aggregation, optical, and surface-enhanced Raman scattering properties. *J. Phys. Chem.*, **100**, 718–724.

103. Rivas, L., Sanchez-Cortes, S., Garcia-Ramos, J.V. and Morcillo, G. (2000) Mixed silver/gold colloids: a study of their formation, morphology, and surface-enhanced Raman activity. *Langmuir*, **16**, 9722–9728.

104. Jana, N.R. (2003) Silver coated gold nanoparticles as new surface enhanced Raman substrate at low analyte concentration. *Analyst*, **128**, 954–956.

105. Kim, K., Kim, K.L. and Lee, S.J. (2005) Surface enrichment of Ag atoms in Au/Ag alloy nanoparticles revealed by surface enhanced Raman scattering spectroscopy. *Chem. Phys. Lett.*, **403**, 77–82.

106. Wang, Y.-L., Chen, H.-J., Dong, S.-J. and Wang, E.-K. (2006) Surface-enhanced Raman scattering of silver-gold bimetallic nanostructures with hollow interiors. *J. Chem. Phys.*, **125**, 044710.

107. Cui, Y. and Gu, R.-A. (2005) Preparation of AgcoreAushell composite bimetallic nanoparticles and its surface-enhanced Raman spectroscopy. *Chem. J. Chin. U.*, **26**, 2090–2092.

108. Pande, S., Ghosh, S.K., Praharaj, S., Panigrahi, S., Basu, S., Jana, S., Pal, A., Tsukuda, T. and Pal, T. (2007) Synthesis of normal and inverted gold-silver core-shell architectures in β-Cyclodextrin and their applications in SERS. *J. Phys. Chem. C*, **111**, 10806–10813.

109. Lu, L.-H., Zhang, H.-J., Sun, G.-Y., Xi, S.-Q., Wang, H.-S., Li, X.-L., Wang, X. and Zhao, B. (2003) Aggregation-based fabrication and assembly of roughened composite metallic nanoshells: application in surface-enhanced Raman scattering. *Langmuir*, **19**, 9490–9493.

110. Tian, Z.-Q., Ren, B., Li, J. and Yang, Z. (2007) Expanding generality of surface-enhanced Raman spectroscopy with borrowing SERS activity strategy. *Chem. Commun.*, 3514–3534.

111. Guo, S.-J., Dong, S.-J. and Wang, E.-K. (2009) Rectangular silver nanorods: controlled preparation, liquid-liquid interface assembly, and application in Surface-enhanced Raman scattering. *Cryst. Growth Des.*, **9**, 372–377.

112. Gole, A. and Murphy, C.J. (2004) Seed-mediated synthesis of gold nanorods: role of the size and nature of the seed. *Chem. Mater.*, **16**, 3633–3640.

113. Orendorff, C.J., Gole, A., Sau, T.K. and Murphy, C.J. (2005) Surface-enhanced Raman spectroscopy of self-assembled monolayers: sandwich architecture and nanoparticle shape dependence. *Anal. Chem.*, **77**, 3261–3266.

114. Jena, B.K. and Raj, C.R. (2008) Seedless, surfactantless room temperature synthesis of single crystalline fluorescent gold nanoflowers with pronounced SERS and electrocatalytic activity. *Chem. Mater.*, **20**, 3546–3548.

115. Orendorff, C.J., Gearheart, L.A., Jana, N.R. and Murphy, C.J. (2006) Aspect ratio dependence on surface enhanced Raman scattering using silver and gold nanorod substrates. *Phys. Chem. Chem. Phys.*, **8**, 165–170.

116. Zou, X.-Q., Ying, E.-B. and Dong, S.-J. (2006) Seed-mediated synthesis of branched gold nanoparticles with the assistance of citrate and their surface-enhanced Raman scattering properties. *Nanotechnology*, **17**, 4758–4764.

117. Jeong, G.H., Lee, Y.W., Kim, M. and Han, S.W. (2009) High-yield synthesis of multi-branched gold nanoparticles and their surface-enhanced Raman scattering properties. *J. Colloid Interface Sci.*, **329**, 97–102.

118. Lu, L.-H., Kobayashi, A., Tawa, K. and Ozaki, Y. (2006) Silver nanoplates with special shapes: controlled synthesis

and their surface plasmon resonance and surface-enhanced Raman scattering properties. *Chem. Mater.*, **18**, 4894–4901.
119. Lu, L.-H., Ai, K.-L. and Ozaki, Y. (2008) Environmentally friendly synthesis of highly monodisperse biocompatible gold nanoparticles with urchin-like shape. *Langmuir*, **24**, 1058–1063.
120. Bakr, O.M., Wunsch, B.H. and Stellacci, F. (2006) High-yield synthesis of multi-branched urchin-like gold nanoparticles. *Chem. Mater*, **18**, 3297–3301.
121. Guo, S.-J., Wang, Y.-L. and Wang, E.-K. (2007) Large-scale, rapid synthesis and application in surface-enhanced Raman spectroscopy of sub-micrometer polyhedral gold nanocrystals. *Nanotechnology*, **18**, 405602.
122. Wang, D.-W., Liu, Y., Zhou, X.-G., Sun, J.-Y. and You, T.-Y. (2007) EDTA-controlled one-pot preparation of novel shaped gold microcrystals and their application in surface-enhanced Raman scattering. *Chem. Lett.*, **36**, 924–925.
123. Tang, X.-L., Jiang, P., Ge, G.-L., Tsuji, M., Xie, S.-S. and Guo, Y.-J. (2008) Poly(N-vinyl-2-pyrrolidone) (PVP)-capped dendritic gold nanoparticles by a one-step hydrothermal route and their high sers effect. *Langmuir*, **24**, 1763–1768.
124. Zhou, J., An, J., Tang, B., Xu, S., Cao, Y., Zhao, B., Xu, W., Chang, J. and Lombardi, J.R. (2008) Growth of tetrahedral silver nanocrystals in aqueous solution and their SERS enhancement. *Langmuir*, **24**, 10407–10413.
125. Kneipp, J., Li, X., Sherwood, M., Panne, U., Kneipp, H., Stockman, M.I. and Kneipp, K. (2008) Gold nanolenses generated by laser ablation-efficient enhancing structure for surface enhanced Raman scattering analytics and sensing. *Anal. Chem.*, **80**, 4247–4251.
126. Ha, H.-Y., Xu, W.-Q., An, J., Li, D.-M. and Zhao, B. (2006) A simple method to synthesize triangular silver nanoparticles by light irradiation. *Spectrochim. Acta A*, **64**, 956–960.
127. Wiley, B.J., Im, S.H., Li, Z.Y., McLellan, J., Siekkinen, A. and Xia, Y. (2006) Maneuvering the surface Plasmon resonance of silver nanostructures through shape-controlled synthesis. *J. Phys. Chem. B*, **110**, 15666–15675.
128. Sun, Y. and Xia, Y. (2003) Gold and silver nanoparticles: a class of chromophores with colors tunable in the range from 400 to 750 nm. *Analyst*, **128**, 686–691.
129. Tao, A.R., Habas, S. and Yang, P. (2008) Shape control of colloidal metal nanocrystals. *Small*, **4**, 310–325.
130. McLellan, J., Siekkinen, A., Chen, J. and Xia, Y. (2006) Comparison of the surface-enhanced Raman scattering on sharp and truncated silver nanocubes. *Chem. Phys. Lett.*, **427**, 122–126.
131. Rycenga, M., Kim, M.H., Camargo, P.H.C., Cobley, C., Li, Z.-Y. and Xia, Y. (2009) Surface-enhanced Raman scattering: comparison of three different molecules on single-crystal nanocubes and nanospheres of silver. *J. Phys. Chem. A*, **113**, 3932–3939.
132. Hrelescu, C., Sau, T.K., Rogach, A.L., Jäckel, F. and Feldmann, J. (2009) Single gold nanostars enhance Raman scattering. *Appl. Phys. Lett.*, **94**, 153113.
133. Khoury, C.G. and Vo-Dinh, T. (2008) Gold nanostars for surface-enhanced Raman scattering: synthesis, characterization and optimization. *J. Phys. Chem. C*, **112**, 18849–18859.
134. Philip, D., Gopchandran, K.G., Unni, C. and Nissamudeen, K.M. (2008) Synthesis, characterization and SERS activity of Au–Ag nanorods. *Spectrochim. Acta A*, **70**, 780–784.
135. Guo, S.-J., Wang, L., Wang, Y.-L., Fang, Y.-X. and Wang, E.-K. (2007) Bifunctional Au@Pt hybrid nanorods. *J. Colloid Interface Sci.*, **315**, 363–368.
136. Hunyadi, S.E. and Murphy, C.J. (2006) Bimetallic silver-gold nanowires: fabrication and use in surface-enhanced Raman scattering. *J. Mater. Chem.*, **16**, 3929–3935.
137. Gunawidjaja, R., Peleshanko, S., Ko, H. and Tsukruk, V.V. (2008) Bimetallic nanocobs: decorating silver nanowires with gold nanoparticles. *Adv. Mater.*, **20**, 1544–1549.

138. Wu, Q.-S., Zhang, C.-B. and Li, F. (2005) Preparation of spindle-shape silver core-shell particles. *Mater. Lett.*, **59**, 3672–3677.
139. Zou, X.-Q. and Dong, S.-J. (2006) Surface-enhanced Raman scattering studies on aggregated silver nanoplates in aqueous solution. *J. Phys. Chem. B*, **110**, 21545–21550.
140. Jana, N.R. and Pal, T. (2007) Anisotropic metal nanoparticles for use as surface-enhanced Raman substrates. *Adv. Mater.*, **19**, 1761–1765.
141. Sant'Ana, A.C., Rocha, T.C.R., Santos, P.S., Zanchetb, D. and Temperinia, M.L.A. (2009) Size-dependent SERS enhancement of colloidal silver nanoplates: the case of 2-amino-5-nitropyridine. *J. Raman Spectrosc.*, **40**, 183–190.
142. Sabur, A., Havel, M. and Gogotsi, Y. (2008) SERS intensity optimization by controlling the size and shape of faceted gold nanoparticles. *J. Raman Spectrosc.*, **39**, 61–67.
143. Tiwarib, V.S., Olega, T., Darbhaa, G.K., Hardya, W., Singhb, J.P. and Ray, P.C. (2007) Non-resonance SERS effects of silver colloids with different shapes. *Chem. Phys. Lett.*, **446**, 77–82.
144. Jackson, J.B., Westcott, S.L., Hirsch, L.R., West, J.L. and Halas, N.J. (2003) Controlling the surface enhanced Raman effect via the nanoshell geometry. *Appl. Phys. Lett.*, **82**, 257–259.
145. Zhang, J.-T., Li, X.-L., Sun, X.-M. and Li, Y.-D. (2005) Surface enhanced Raman scattering effects of silver colloids with different shapes. *J. Phys. Chem. B*, **109**, 12544–12548.
146. Nikoobakht, B. and El-Sayed, M.A. (2003) Surface-enhanced Raman scattering studies on aggregated gold nanorods. *J. Phys. Chem. A*, **107**, 3372–3378.
147. Alvarez-Puebla, R.A. and Aroca, R.F. (2009) Synthesis of silver nanoparticles with controllable surface charge and their application to surface-enhanced Raman scattering. *Anal. Chem.*, **81**, 2280–2285.
148. Wang, Z.-L. (2001) *Characterization of Nanophase Materials*, John Wiley & Sons, Ltd, Weinheim.
149. Olson, T.Y. and Zhang, J.-Z. (2008) Structural and Optical Properties and emerging applications of metal nanomaterials. *J. Mater. Sci. Technol.*, **24**, 433–446.
150. Zhang, D.-F., Niu, L.-Y., Jiang, L., Yin, P.-G., Sun, L.-D., Zhang, H., Zhang, R., Guo, L. and Yan, C.-H. (2008) Branched gold nanochains facilitated by polyvinylpyrrolidone and their SERS effects on p-aminothiophenol. *J. Phys. Chem. C*, **112**, 16011–16016.
151. Zhao, N., Wei, Y., Sun, N., Chen, Q., Bai, J., Zhou, L., Qin, Y., Li, M. and Qi, L. (2008) Controlled synthesis of gold nanobelts and nanocombs in aqueous mixed surfactant solutions. *Langmuir*, **24**, 991–998.
152. Liu, X.-H., Huang, R. and Zhu, J. (2008) Functional faceted silver nano-hexapods: synthesis, structure characterizations, and optical properties. *Chem. Mater.*, **20**, 192–197.
153. Sherry, L.J., Chang, S.-Hui, Schatz, G.C., Van Duyne, R.P., Wiley, B.J. and Xia, Y. (2005) Localized surface plasmon resonance spectroscopy of single silver nanocubes. *Nano Lett.*, **5**, 2034–2038.
154. Kho, K.W., Shen, Z.X., Zeng, H.C., Soo, K.C. and Olivo, M. (2005) Deposition method for preparing SERS-active gold nanoparticle substrates. *Anal. Chem.*, **77**, 7462–7471.
155. Hu, J.-W., Zhao, B., Xu, W.-Q., Fan, Y.-G., Li, B.-F. and Ozaki, Y. (2002) Simple method for preparing controllably aggregated silver particle films used as surface-enhanced Raman scattering active substrates. *Langmuir*, **18**, 6839–6844.
156. Hu, J.-W., Zhao, B., Xu, W.-Q., Fan, Y.-G., Li, B.-F. and Ozaki, Y. (2002) Aggregation of silver particles trapped at an air-water interface for preparing new SERS active substrates. *J. Phys. Chem. B*, **106**, 6500–6506.
157. Wu, D.-Y., Li, J.-F., Ren, B. and Tian, Z.-Q. (2008) Electrochemical surface-enhanced Raman spectroscopy of nanostructures. *Chem. Soc. Rev.*, **37**, 1025–1041.
158. Ulha, M.C., Kahraman, M., Tokman, N. and Türkoğlu, G. (2008)

Surface-enhanced Raman scattering on aggregates of silver nanoparticles with definite size. *J. Phys. Chem. C*, **112**, 10338–10343.

159. Suzuki, M., Niidome, Y., Kuwahara, Y., Terasaki, N., Inoue, K. and Yamada, S. (2004) Surface-enhanced nonresonance Raman scattering from size- and morphology-controlled gold nanoparticle films. *J. Phys. Chem. B*, **108**, 11660–11665.

160. Hossain, M.K., Kitahama, Y., Huang, G.G., Kaneko, T. and Ozaki, Y. (2008) SPR and SERS characteristics of gold nanoaggregates with different morphologies. *Appl. Phys. B*, **93**, 165–170.

161. Sztainbucha, I.W. (2006) The effects of Au aggregate morphology on surface-enhanced Raman scattering enhancement. *J. Chem. Phys.*, **125**, 124707.

162. Doering, W.E. and Nie, S.-M. (2003) Spectroscopic tags using dye-embedded nanoparticles and surface-enhanced Raman scattering. *Anal. Chem.*, **75**, 6171–6176.

163. Doering, W.E., Piotti, M.E., Natan, M.J. and Freeman, R.G. (2007) SERS as a foundation for nanoscale, optically detected biological labels. *Adv. Mater.*, **19**, 3100–3108.

3
Quantitative SERS Methods

Steven E. J. Bell and Alan Stewart

3.1
Introduction

The goal of transforming SERS from an interesting novelty into a viable quantitative analytical technique has been pursued for many years, but it is only recently that the first indications that quantitative SERS could be generally achievable have appeared. Up to this point, the challenges of preparing sensitive and reproducible enhancing media and the need to develop a deep understanding of the fundamental mechanisms of the effect (or at least to have a working knowledge of the relationship between the microstructure of the enhancing materials, their surface chemistry and the enhancements that they provide; see also Chapter 1) have been sufficient to occupy the attention of researchers who were interested in using SERS for quantitative measurements. However, many of these issues have now either been fully resolved or at least brought to the stage where they are no longer impediments. For example, new generations of enhancing media have been developed; the links between microstructure, plasmon resonances and enhancements have been established; and the role of the media's surface chemistry is being addressed. This chapter discusses these new developments and also considers the extent to which the separate strands can be combined to create standard, generally applicable methods for quantitative SERS. In addition, it highlights the need to begin the process of judging the success of new SERS methods against competing technologies, rather than against previous SERS results.

3.2
SERS Media

The choice of enhancing media for SERS studies has grown to an extraordinary extent, although they can still be broadly divided into solution-phase suspensions of nanoparticles and micro-textured solid substrates. This division dates back to the earliest work in the area, which used either colloidal suspensions of silver or gold nanoparticles [1] or the surfaces of roughened noble-metal electrodes [2]. Since this

Surface Enhanced Raman Spectroscopy: Analytical, Biophysical and Life Science Applications. Edited by Sebastian Schlücker
Copyright © 2011 WILEY-VCH Verlag GmbH & Co. KGaA, Weinheim
ISBN: 978-3-527-32567-2

early work, a huge amount of effort and ingenuity has been put into finding novel methods of preparing both particles and solid substrates. For example, although citrate-reduced silver and gold nanoparticles prepared by methods developed many years ago are still widely used [3], the increased understanding of how to direct particle growth in solution has given us methods to prepare colloids in various different sizes and shapes (see also Chapter 2). Now, in addition to quasi-spherical particles, there are prisms [4, 5], rods [6], cubes [7] and nanostars [8, 9], as well as bumpy gold nanoparticles [10], hollow gold nanoparticles [11] and nanoshells [12], all of which have been used to enhance Raman signals. In some cases, this control over size and shape has allowed optimization of the colloids. For example, Figure 3.1 shows that changing the particle diameter has a large effect on the enhancement given by aggregated monodisperse gold nanospheres and that 46 nm particles give significantly larger enhancements than either 21 or 146 nm particles [13].

Similarly, work on preparing solid materials has, if anything, given an even more diverse range of enhancing substrates. It is useful to divide these into materials that show significant disorder on the micro/nanoscale and those that do not. Disordered materials include roughened electrodes, evaporated metal island films and chemically reduced metal films (e.g. films prepared using Tollens' reagent where glucose is the reducing agent) [14], as well as simple extensions of the earlier work where nanoparticles are first prepared and then deposited in a layer or layers on the substrate [15]. These materials can be contrasted with 'plasmonic' materials, which are fabricated using methods that give highly regular structures even on the nano- and microscale. In these, one of the goals is to control the structure

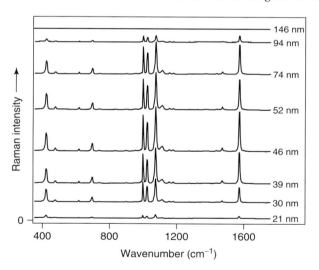

Figure 3.1 SERS spectra of thiophenol obtained using aggregated Au colloids with the diameters marked on the figure. Data have been vertically offset for clarity but not re-scaled. In these raw data, the 46 nm particles give significantly larger enhancements than either 21 or 146 nm particles. (Reproduced with permission from the PCCP owner societies [13].)

to allow the optical properties of the materials to be optimized for particular applications, for example, to provide maximum enhancement at specific excitation wavelengths [16]. Moreover, it is hoped that the fine control exercised over the structure should help to increase the reproducibility of the enhancements they provide. Significant progress has been made in the preparation of these plasmonic materials, and methods for preparing substrates with pyramidal holes,[1] silver films deposited on regular arrays of nanospheres [17] and pyramidal arrays [18] are now also well known.

The choice between colloids and solid substrates is not straightforward. Although, in principle, either type of medium could be used for analysis of most analyte types, some of their inherent features make one preferable over the other. For example, if the target analytes are dissolved in an aqueous solution, mixing aqueous colloidal nanoparticles or immersing a solid substrate will both result in adsorption of the analyte on the metal surface. In this case, if a routine test is required, the low cost per sample for colloidal SERS (due to the fact that the materials can be manufactured in bulk and then divided into small aliquots) will be a distinct advantage. Not only is a low material cost advantageous for general running costs, but it is also important since it allows the medium to be used once and then thrown away, eliminating cross-contamination between samples. However, solid substrates have the advantage that they can be immersed in a sample and then withdrawn, or that samples that are dissolved in non-aqueous solvents can be analysed by depositing them onto the surfaces and allowing the solvent to evaporate.

3.3
Stability and Shelf Life

Stability of the enhancing media will always be a significant issue. In principle, solid substrates should present fewer problems because they can be packed in an inert environment and stored ready for use. With colloids, there is a larger problem because of the potential of the particles to chemically react with other compounds present in the aqueous environment in which they are stored. An approach that circumvents storage issues is to generate enhancing surfaces *in situ*. The most obvious way to do this is through microfluidic techniques. For example, Kier *et al.* have shown a system in which the rapid borohydride reduction of silver is used to create SERS-active nanoparticles within a microfluidic system [19], and the area has recently been reviewed [20]. However, such methods are not without their own difficulties, particularly blocking of the microchannels. An interesting alternative has been developed by Schneider *et al.*, who used the probe laser in the experiment to reduce a silver salt *in situ* thereby creating the enhancing medium in exactly the region of interest within the sample, typically within a gel matrix [21]. Of course, if roughened electrodes are used, they are typically converted to the active form immediately before use by carrying out oxidation/reduction cycles,

---

1) Klarite, a commercially available solid SERS substrate, is available from D3 Technologies Ltd, 5, Nova Technology Park, Glasgow G33 1AP, UK.

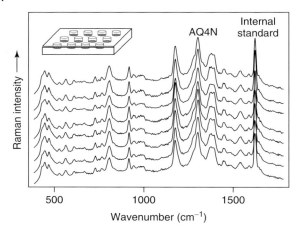

Figure 3.2 Replicate SERRS spectra of an anthraquinone-derived drug (AQ4N) enhanced using Ag colloid protected within a swellable gel-forming polymer. Spectra were recorded in a 96-well microtitre plate and normalized against a SERRS-enhanced internal standard. (Reproduced with permission from the Royal Society of Chemistry [23].)

so the active textured surface is generated freshly before use. This circumvents any storage problems but any strongly binding materials that are then added are typically not removed in subsequent oxidation/reduction cycles. An interesting new approach to removing impurities from electrode surfaces has been to immerse the sample in a solution containing I^-, which is strongly binding and therefore capable of displacing other adventitious impurities from the surface. The I^- is then electrochemically oxidized to I_2, which is weakly binding and therefore released from the surface leaving a fresh surface [22].

A useful method of preserving colloidal particles in storage, which still allows them sufficient freedom to interact with analytes when required, is to suspend the particles in gel-forming hydrophilic polymers, such as a polycarbophil. These 'gel-colls' not only increase the shelf life of the particles but can also be the basis for a strategy that eliminates batch-to-batch reproducibility problems, since a single large batch can be prepared and then subdivided into numerous individual portions, which are identical because they all have a common source. Figure 3.2 shows the spectra obtained from several wells in a 96-well plate which were treated with the same concentration of an anthraquinone-derived drug, which was used as a simple test analyte [23].

3.4
Reproducibility and Internal Standards

Even within conventional Raman analysis, it is normal to use an internal standard to correct for the numerous potential sources of variation in the magnitude of the detected Raman signal that are not associated with changes in concentration. It is relatively straightforward to monitor and automatically correct for changes in the laser power [24]. However, other more subtle changes such as drift in the optical alignment and variation in the positioning of the sample cannot be corrected in this way. Even if they could, it would still be necessary to try to adjust the laser intensity to compensate for self-absorption within coloured samples, or scattering effects within turbid and opaque samples. In contrast, measuring the ratio of the bands of the analyte to those of a standard material in the sample corrects for all these effects in a single step because they affect the signals of analyte and standard to the same extent. This is obviously preferable to directly measuring the absolute signal intensity of the analyte and combining this with numerous checks for sources of variation. Moreover, with many types of samples, such as tableted pharmaceuticals (see also Chapter 6) or solutions of analytes in organic solvents, the sample matrix will typically give strong bands that can be used directly as the internal standard.

In SERS, all the experimental variables that give uncertainty in the absolute signal in normal Raman experiments are still present, but they are joined by another variable that can increase uncertainty: the enhancement factor of the media. Since the enhancement factor can be very large, the potential for extreme excursions from a normal value is particularly high. In early work on enhancing materials, the enhancement factors could vary enormously between different colloid preparations and on different roughened electrodes. More recently, increased understanding of the preparation of nanoscale materials in general has meant that sample-to-sample variations in enhancing media of all types have significantly improved. We are currently at the stage where, between samples, standard deviations of just a few percent are common. At this point, it is pertinent to ask whether spending huge effort pursuing yet higher substrate reproducibility is worthwhile, particularly since the use of internal standards will readily compensate for such variations in enhancement factors of these levels and it may be desirable to use internal standards to correct for laser power drift or alignment variations in any case.

In SERS experiments, it is occasionally possible to use an unenhanced component that is present in the sample as an internal standard. Figure 3.3 shows an example where SERS is used to detect the signal from a particularly dangerous 'ecstasy' variant 2,5,-dimethoxy-4-bromoamphetamine (DOB) [25]. This compound is occasionally seized in the same locations where typical ecstasy-type 'designer' amphetamines such as 3,4-methylenedioxy-N-methylamphetamine (MDMA) are distributed and consumed. The compound is particularly hazardous to users because its larger than usual induction time can lead to accidental overdose, which with this particular drug is associated with instances of serious self-harm. In addition, the average mass of the drug in the tablets is much lower than with other designer amphetamines, making it impossible to detect the drug using conventional Raman methods. However, Figure 3.3 shows that treatment of model lactose/DOB tablets with a colloid yields spectra in which the DOB bands are clearly visible, even at a drug level of 15 μg/tablet. Indeed, at the drug levels typically encountered in seized tablets, the excipient bands are completely concealed under the very large preferentially enhanced Raman signal of DOB. For this screening type of application, the fact that the signal saturates is relatively unimportant, since the test

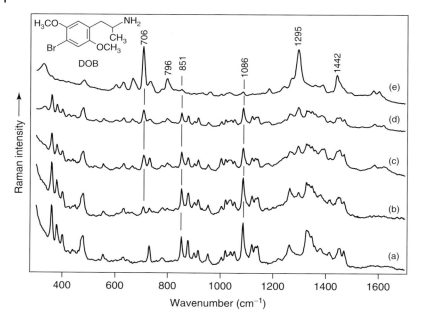

Figure 3.3 SERS spectra of a series of model tablets prepared from lactose with increasing amounts of DOB. Tablets are 400 mg lactose plus (a) 0, (b) 15 µg, (c) 60 µg, (d) 250 µg and (e) 1 mg DOB. With 1 mg per 400 mg lactose, the spectra are entirely dominated by the DOB signal. A similar effect is observed for authentic, seized samples. At 15 µg, the 706 cm^{-1} DOB band is barely visible among the much larger lactose peaks. (Reproduced with permission from Wiley Interscience [25].)

would still need to be confirmed by conventional HPLC methods for the purposes of criminal prosecution. However, more generally, SERS methods that do not include some way of correcting for variations in the enhancement level should be regarded as semi-quantitative at best.

A more widely applicable approach is to use an internal standard that is SERS enhanced. The best internal standards will be materials that respond in the same way as the analyte to changes in the experimental conditions, such as a change in the enhancing medium which alters either the signal that is given by an adsorbed target or the probability of absorption. The first factor is easier to track. For example, it has been suggested that the strong Ag–Cl band reported to lie at 230 cm^{-1} can be used to normalize the SERS spectra of rhodamine 6G on immobilized silver nanoparticles on Si [26]. However, even this approach is problematic in situations where the expected variation gives changes in the binding properties of the analyte, possibly through something as simple as a change in the surface potential. Unless the target and the standard respond in the same way to these perturbations, such changes will alter the ratio of the target : standard bands and, therefore, the apparent concentration of the analyte. This requirement to have similar responses to perturbations of the enhancing medium is most readily met if the target and standards are chemically similar. For example, Figure 3.4 shows

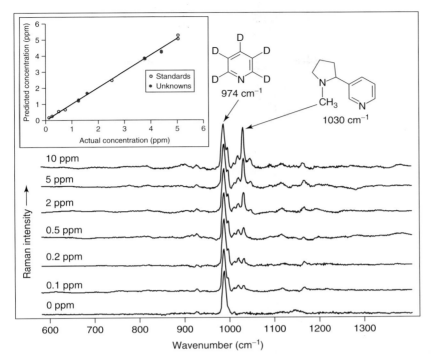

Figure 3.4 Data from a SERS assay for nicotine (structure shown at top right) using polymer-stabilized Ag colloid ('gel-coll') as the enhancing medium. Spectra are normalized to the intensity of the 974 cm^{-1} band of the d_5-pyridine internal standard. Linearized calibration data from the spectra is shown in the inset. (Reproduced with permission from The Royal Society of Chemistry [3].)

the Raman spectrum of nicotine, which was obtained using Ag nanoparticles that were suspended in a transparent polymer gel [27]. In this case, pyridine was an obvious choice as the internal standard since it is a near-analogue of nicotine and has a very similar aromatic nitrogen binding site. It was not possible to use normal pyridine as the internal standard because both nicotine and pyridine have bands near 1030 cm^{-1} and so the internal standard signal interfered with that of the analyte. A simple solution to this problem was to use isotopically substituted pyridine (in this case the deuterated d_5 isotopomer), in which the pyridine signal is shifted to 974 cm^{-1}. Under these conditions, the root mean square (RMS) error of prediction in the concentration of nicotine in the range 1–5 ppm was found to be 0.10 ppm. Isotopomers of the target molecule are ideal internal standards since they are chemically identical to the target but give distinct spectra [28]. It has been demonstrated very convincingly for rhodamine 6G and its d_4 isotopomer that even over very wide concentration ranges analytical accuracy is preserved when the isotopomer is used as an internal standard. Conversely, when a different (but still SERS-enhanced) internal standard was used, the analytical performance degraded significantly because the standard did not fully track the changes of the analyte under different experimental conditions. The only disadvantage of this 'isotope-edited' method is that an isotopically substituted form of the analyte must be available.

3.5
Selectivity

After the enhancing medium has been prepared, the next essential step is to bring the analyte in close proximity to the enhancing surface. This was traditionally achieved either through mixing the aqueous analyte with the colloid or by immersing an electrode into the test solution. In some cases, the required adsorption occurs readily, and most of the earliest SERS studies centred on those compounds that were attracted to the enhancing metal surfaces used.

There are numerous mechanisms of adsorption but the most direct is that the analyte coordinates to the surface, such as was originally observed with pyridine and is now exploited in several classes of surface-seeking molecules, such as triazoles and thiols. Alternatively, charged analytes may be attracted by electrostatic forces. With metal particles, spontaneous adsorption occurs if the analyte carries a charge opposite to that of the particles. Most colloidal metal particles prepared by reduction of metal salts carry a negative charge as a result of an adsorbed layer of stabilizing anions (e.g. citrate in citrate-reduced colloids). Therefore, positively charged analytes, particularly those that can also give resonance effects, are particularly easy to study using surface-enhanced resonance Raman scattering (SERRS) (see also Chapter 11). The anthraquinone derivative whose spectrum is shown in Figure 3.2 is an example; dyes such as rhodamine 6G and crystal violet are also widely studied because they also easily yield very large signals.

If the target carries the same charge as the existing surface species, it will not be electrostatically attracted to the surface, but it may be able to bind by displacing an existing ion from a site on the surface. For example, Figure 3.5 shows a series of spectra where the layer of citrate ions on an as-prepared colloid is displaced by the addition of dipicolinic acid (DPA) [29]. DPA may in turn be displaced by Cl^- which can be displaced by the more strongly binding Br^- (not shown). This understanding that the relative affinities of various species determines what compounds can be detected on a substrate with a given surface chemistry allows rational design of detection systems. For example, negatively charged mononucleotides such as adenosine monophosphate (AMP) cannot be detected on citrate-reduced colloids that have been aggregated with Cl^--containing salts, since they have a surface Cl^- layer that AMP cannot displace. However, aggregation with salts that allow the surface to retain the citrate layer does allow AMP detection, since AMP binds to the surface more strongly than citrate and is thus able to displace it [30].

Unfortunately, there are many analytes whose chemistry means that they do not spontaneously adsorb to the enhancing medium (either by displacement or by electrostatic attraction). One obvious solution in these cases is to force the target analyte to come into physical contact with the enhancing material by drying droplets

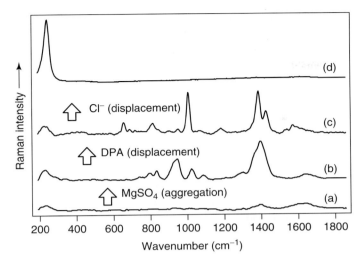

Figure 3.5 SERS spectra taken after sequential addition of sulfate, DPA and Cl⁻ to a citrate-reduced silver colloid. (a) Untreated colloid with featureless spectrum. (b) Colloid aggregated with 5×10^{-4} mol dm^{-3} MgSO$_4$ which shows only citrate bands. (c) DPA added to the colloid, which displaces citrate and gives DPA bands. (d) NaCl added to solution (c) to give a Cl⁻ concentration of 5×10^{-3} mol dm^{-3} in the final mixture, and the DPA signal has now been replaced with the strong AgCl band at 245 cm^{-1}. (Adapted with permission from the American Chemical Society [29].)

of the sample solution onto the substrate. However, droplet drying can result in the formation of 'coffee rings', where the solute is preferentially deposited at the perimeter of the evaporating droplet. This gives a narrow circular deposit where the solute is in much higher concentration than would result from uniform evaporation. This preconcentration effect has been shown to be excellent for protein studies, where the differential deposition of protein and interfering fluorescent impurities gives a combination of higher analyte signals and lower fluorescence background levels [31]. However, outside this specialist application, inhomogeneous drying will clearly cause significant problems for quantitative analytical methods since it means that the absolute size of the signal will vary when different areas of the dried sample are probed.

There are numerous alternative methods of modifying substrates to promote adsorption of the analyte. For example, anions can be attracted to electrodes simply by application of a positive potential. This is an obvious approach when electrochemically roughened electrodes are used as the enhancing medium. More recently, the same approach has been used for Au and Ag nanoparticles distributed on the surface of indium tin oxide (ITO) electrodes [32]. In the case of colloidal suspensions, direct application of potentials is not possible but essentially equivalent effects can be obtained by decorating their surfaces with modifying compounds that carry a charge. One of the simplest and most effective ways of doing this is to use ω-functionalized alkanethiols with various head groups. These readily form strong covalent Ag–S or Au–S bonds at room temperature, which attach them to the particles' surfaces. For example, attachment of amino-terminated alkane thiols can switch the ζ potential of citrate-reduced Au particles from strongly negative to strongly positive values [33]. Alternatively, it has been shown that it is possible to tune the ζ potential of Ag and Au particles by changing pH [34, 35].

Although the ability to attract the target analyte to the surface is useful, and will be important in situations where the objective is simply to quantify a target analyte that is dissolved in a pure solvent, it is not the only factor that would be desirable in an ideal SERS medium. Specifically, many of the potential applications of SERS, such as environmental monitoring (see also Chapter 5) or screening for chemical or biological weapons, will require analysis of samples that contain numerous other, potentially interfering, chemical species. Under these conditions, a 'universal' enhancing medium that gives the same enhancements for all compounds would not be useful because of the huge complexity in the spectra that would be generated. Indeed, pursuit of raw enhancing power without a degree of selectivity is likely to lead to significant problems when fully validated analytical procedures based on these materials need to be developed. While there have been some demonstrations of successful analyses carried out using unmodified media in complex matrices such as saliva [36], sea water [37] and fermentation broths [38], these particular methods work because potentially interfering compounds fortuitously give low SERS signals. This can be either because they are in low concentration and do not adsorb to the surface or they give bands which lie in a spectral region that does not interfere with the analyte signal (e.g. thiocyanide, which is found in saliva but whose strongest peak is at 2095 cm^{-1}) [36].

In general, an approach where the enhancing media are deliberately tuned so as to promote adsorption and therefore detection of the required analyte while discriminating against potential interfering materials is likely to give significant benefits. The most promising approach is to use chemical modification of the surface (see also Chapter 5). The most selective surface modifiers will be anti-bodies but they typically show only changes in their spectra on target binding [39]. Attention to this point has focused on simpler chemical modifiers that can introduce a degree of selectivity or promote binding as well as signal attachment through direct observation of additional bands characteristic of the analyte itself. Figure 3.6 shows a selection of the surface-binding modifiers (predominantly thiols) [40–48] that have already been used to promote adsorption of specific target analytes for SERS.

A potentially very useful extension of chemical modification with a single compound is the use of mixtures of compounds to modify the surface. For example, mixed monolayers of differently substituted alkane thiols have been extensively investigated as models for biomembranes [49] and have been used specifically to modify an SERS-active surface for glucose adsorption [17, 18]. In this latter case, the mixed monolayer served a dual function of promoting adsorption, and therefore detection, along with suppression of signals from the biological matrix. More generally, within mixed monolayers, there is the possibility of varying the properties of the surface between the extreme values found for full monolayer coverage of either component and thus of designing and building surfaces with continuously tuneable properties. Alternatively, the properties of surfaces treated

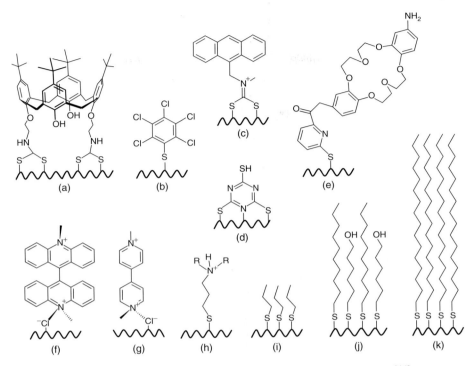

Figure 3.6 Surface-modifying agents that have been used for SERS: (a) dithiocarbamate calix[4]arene derivative [40], (b) pentachlorobenzenethiol [41], (c) N-(9methylanthracene)N'-methyl-dithiocarbamate [40], (d) trimercaptotriazine [42], (e) thiol derivatized dibenzo-18 crown-6 [43], (f) lucigenin [44], (g) paraquat [44], (h) cysteamine derivatives including R = H or CH_3 [45], (i) propane-1-thiol [46], (j) mixed decane-1-thiol and 6-mercaptohexan-1-ol [47] and (k) hexadecane-1-thiol [48].

with mixed monolayers may be different from those created with either constituent separately.

Figure 3.7 shows an example where the properties can be switched between extreme values by altering the ratio of the two modifiers that have been used, in this case mercaptopropanesulfonic acid (MPS) and pentanethiol (PT) [50]. Figure 3.7a shows the SERS spectra of colloids modified with solutions, which were 100 : 0, 75 : 25, 50 : 50, 25 : 75 and 0 : 100 MPS : PT. These SERS spectra reflect the composition of the mixed surface monolayer; more detailed analysis shows that the relative surface coverage of the two species does approximately follow the composition of the modifying feedstock. This is not universally the case, since differences in the free energy of binding can lead to a single component dominating the surface even when it is present as a minority species in the coating feedstock. Figure 3.7b shows the data where this series of modified colloids were each tested for their ability to detect a positively charged analyte, tetra-4-*n*-methylpyridyl porphyrin (TMPyP). It is clear from the data that at 100% PT (0% MPS, bottom

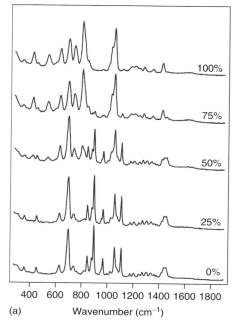

Figure 3.7 (a) SERS spectra of Ag colloids modified with mixed solutions of mercaptopropanesulfonic acid (MPS) and pentanethiol (PT). The percentage of MPS in the feedstock is marked on the figure. These SERS spectra reflect the composition of the mixed surface-modifying thiol monolayer. (b) The same colloids as in (a) after addition of a positively charged analyte, tetra-4-n-methylpyridyl porphyrin (TMPyP). Some of the strongest TMPyP bands are marked by arrows and are obvious even down to 25% MPS/75% PT. However, at 100% PT (0% MPS, bottom spectrum) addition of the analyte solution to the colloid does not give any additional bands, presumably because the neutral hydrocarbon coating prevents binding of the analyte [50].

spectrum) addition of the analyte solution to the colloid does not give any additional bands, presumably because the neutral hydrocarbon coating prevents binding of the TMPyP analyte. However, addition of as little as 25% MPS to the feedstock gives a colloid that can absorb the analyte. Indeed, all colloids where the feedstock was between 25 and 100% MPS give TMPyP signals with similar absolute magnitude. This implies that, even if the MPS is a minor species on the surface, it is able to promote adsorption of TMPyP almost as well as a full MPS monolayer.

3.6
Conclusion

There are now numerous examples in the literature showing that quantitative SERS measurements are possible. One notable observation is how wide a range of substrates and procedures can provide useful analytical precision. In part, this is because developments in nanotechnology have made it much easier to prepare and characterize reproducible enhancing substrates using a broad range of approaches. The second factor that has had a significant effect on the robustness of the calibrations is the increasingly widespread use of internal standards, particularly those that are chemically similar to the target species. One major challenge that remains is the need to build selectivity into the enhancing materials. This was not a significant issue when SERS was being developed using clean samples under laboratory conditions but will become increasingly significant as the methods are applied to real-world situations where interference from other compounds in the samples will be more problematic.

Finally, if SERS is to be widely adopted, it is important that the methods that are developed are shown not only to be improvements over previous SERS methods but are also competitive with alternative analytical techniques. Technical advances are driving down the cost and size of not only Raman spectrometers but also of competing instruments, such as GCMS systems. There will clearly be areas where the advantages of SERS make it the method of choice, but it is important to identify which applications will benefit the most from quantitative SERS methods.

References

1. Creighton, J.A., Blatchford, C.G. and Albrecht, M.G. (1979) Plasma resonance enhancement of Raman-scattering by pyridine adsorbed on silver or gold sol particles of size comparable to the excitation wavelength. *J. Chem. Soc. Farad. Trans. 2*, **75**, 790–798.
2. Fleischman, M.P., Hendra, P.J. and McQuillan, A.J. (1974) Raman-spectra of pyridine adsorbed at a silver electrode. *Chem. Phys. Lett.*, **26**, 163–166.
3. Bell, S.E.J. and Sirimuthu, N.M.S. (2008) Quantitative surface-enhanced Raman spectroscopy. *Chem. Soc. Rev.*, **37**, 1012–1024.
4. Aherne, D., Ledwith, D.M., Gara, M. and Kelly, J.M. (2008) Optical properties and growth aspects of silver nanoprisms produced by a highly reproducible and rapid synthesis at room temperature. *Adv. Funct. Mater.*, **18**, 2005–2016.
5. Yang, Y., Matsubara, S., Xiong, L.M., Hayakawa, T. and Nogami, M. (2007) Solvothermal synthesis of multiple shapes of silver nanoparticles and their SERS properties. *J. Phys. Chem. C*, **111**, 9095–9104.
6. Nikoobakht, B. and El-Sayed, M.A. (2003) Surface-enhanced Raman scattering studies on aggregated gold nanorods. *J. Phys. Chem. A*, **107**, 3372–3378.
7. McLellan, J.M., Siekkinen, A., Chen, J.Y. and Xia, Y.N. (2006) Comparison of the surface-enhanced Raman scattering on sharp and truncated silver nanocubes. *Chem. Phys. Lett.*, **427**, 122–126.
8. Rodriguez-Lorenzo, L., Alvarez-Puebla, R.A., Pastoriza-Santos, I., Mazzucco, S., Stephan, O., Kociak, M., Liz-Marzan, L.M. and de Abajo, F.J.G. (2009) Zeptomol detection through controlled ultrasensitive surface-enhanced Raman scattering. *J. Am. Chem. Soc.*, **131**, 4616–4618.
9. Esenturk, E.N. and Walker, A.R.H. (2009) Surface-enhanced Raman scattering spectroscopy via gold nanostars. *J. Raman Spectrosc.*, **40**, 86–91.
10. Yu, K.F., Kelly, K.L., Sakai, N. and Tatsuma, T. (2008) Morphologies and surface plasmon resonance properties of monodisperse bumpy gold nanoparticles. *Langmuir*, **24**, 5849–5854.

11. Schwartzberg, A.M., Oshiro, T.Y., Zhang, J.Z., Huser, T. and Talley, C.E. (2006) Improving nanoprobes using surface-enhanced Raman scattering from 30-nm hollow gold particles. *Anal. Chem.*, **78**, 4732–4736.
12. Gellner, M., Kustner, B. and Schlucker, S. (2009) Optical properties and SERS efficiency of tunable gold/silver nanoshells. *Vib. Spectro.*, **50**, 43–47.
13. Bell, S.E.J. and McCourt, M.R. (2009) SERS enhancement by aggregated au colloids: Effect of particle size. *Phys. Chem. Chem. Phys.*, **11**, 7455–7462.
14. Aroca, R. (2006) *Surface-Enhanced Vibrational Spectroscopy*, John Wiley & Sons, Ltd, Chichester.
15. Grabar, K.C., Freeman, R.G., Hommer, M.B. and Natan, M.J. (1995) Preparation and characterization of Au colloid monolayers. *Anal. Chem.*, **67**, 735–743.
16. Lal, S., Grady, N.K., Kundu, J., Levin, C.S., Lassiter, J.B. and Halas, N.J. (2008) Tailoring plasmonic substrates for surface enhanced spectroscopies. *Chem. Soc. Rev.*, **37**, 898–911.
17. Stuart, D.A., Yuen, J.M., Lyandres, N.S.O., Yonzon, C.R., Glucksberg, M.R., Walsh, J.T. and Van Duyne, R.P. (2006) In vivo glucose measurement by surface-enhanced Raman spectroscopy. *Anal. Chem.*, **78**, 7211–7215.
18. Jensen, T.R., Malinsky, M.D., Haynes, C.L. and Van Duyne, R.P. (2000) Nanosphere lithography: Tunable localized surface plasmon resonance spectra of silver nanoparticles. *J. Phys. Chem. B*, **104**, 10549–10556.
19. Keir, R., Igata, E., Arundell, M., Smith, W.E., Graham, D., McHugh, C. and Cooper, J.M. (2002) SERRS. In situ substrate formation and improved detection using microfluidics. *Anal. Chem.*, **74**, 1503–1508.
20. Chen, L.X. and Choo, J.B. (2008) Recent advances in surface-enhanced Raman scattering detection technology for microfluidic chips. *Electrophoresis*, **29**, 1815–1828.
21. Trachta, G., Schwarze, B., Sagmuller, B., Brehm, G. and Schneider, S. (2004) Combination of high-performance liquid chromatography and SERS detection applied to the analysis of drugs in human blood and urine. *J. Mol. Struct.*, **693**, 175–185.
22. Li, M.D., Cui, Y., Gao, M.X., Luo, J., Ren, B. and Tian, Z.Q. (2008) Clean substrates prepared by chemical adsorption of iodide followed by electrochemical oxidation for surface-enhanced Raman spectroscopic study of cell membrane. *Anal. Chem.*, **80**, 5118–5125.
23. Bell, S.E.J. and Spence, S.J. (2001) Disposable, stable media for reproducible surface-enhanced Raman spectroscopy. *Analyst*, **126**, 1–3.
24. Bell, S.E.J. (2008) in *Pharmaceutical Applications of Raman Spectroscopy*, 1st edn (ed. S. Sasic), John Wiley and Sons, Inc., pp. 29–64.
25. Bell, S.E.J., Fido, L.A., Sirimuthu, N.M.S., Speers, S.J., Peters, K.L. and Cosbey, S.H. (2007) Screening tablets for DOB using surface-enhanced Raman spectroscopy. *J. Forensic. Sci.*, **52**, 1063–1067.
26. Caro, C., Lopez-Cartes, C., Zaderenko, P. and Mejias, J.A. (2008) Thiol-immobilized silver nanoparticle aggregate films for surface enhanced Raman scattering. *J. Raman Spectrosc.*, **39**, 1162–1169.
27. Bell, S.E.J. and Sirimuthu, N.M.S. (2004) Rapid, quantitative analysis of ppm/ppb nicotine using surface-enhanced Raman scattering from polymer-encapsulated Ag nanoparticles (gel-colls). *Analyst*, **129**, 1032–1036.
28. Zhang, D.M., Xie, Y., Deb, S.K., Davison, V.J. and Ben-Amotz, D. (2005) Isotope edited internal standard method for quantitative surface-enhanced Raman spectroscopy. *Anal. Chem.*, **77**, 3563–3569.
29. Bell, S.E.J. and Sirimuthu, N.M.S. (2005) Surface-enhanced Raman spectroscopy as a probe of competitive binding by anions to citrate-reduced silver colloids. *J. Phys. Chem. A*, **109**, 7405–7410.
30. Bell, S.E.J. and Sirimuthu, N.M.S. (2006) Surface-enhanced Raman spectroscopy (SERS) for sub-micromolar detection of DNA/RNA mononucleotides. *J. Am. Chem. Soc.*, **128**, 15580–15581.

31. Ortiz, C., Zhang, D.M., Xie, Y., Ribbe, A.E. and Ben-Amotz, D. (2006) Validation of the drop coating deposition Raman method for protein analysis. *Anal. Biochem.*, **353**, 157–166.
32. Lacharmoise, P.D., Le Ru, E.C. and Etchegoin, P.G. (2009) Guiding molecules with electrostatic forces in surface enhanced Raman spectroscopy. *ACS Nano*, **3**, 66–72.
33. Lin, Y.C., Yu, B.Y., Lin, W.C., Lee, S.H., Kuo, C.H. and Shyue, J.J. (2009) Tailoring the surface potential of gold nanoparticles with self-assembled monolayers with mixed functional groups. *J. Colloid Interface Sci.*, **340**, 126–130.
34. Alvarez-Puebla, R.A. and Aroca, R.F. (2009) Synthesis of silver nanoparticles with controllable surface charge and their application to surface-enhanced Raman scattering. *Anal. Chem.*, **81**, 2280–2285.
35. Alvarez-Puebla, R.A., Arceo, E., Goulet, P.J.G., Garrido, J.J. and Aroca, R.F. (2005) Role of nanoparticle surface charge in surface-enhanced Raman scattering. *J. Phys. Chem. B*, **109**, 3787–3792.
36. Farquharson, S., Gift, A.D., Shende, C., Maksymiuk, P., Inscore, F.E. and Murran, J. (2005) Detection of 5-fluorouracil in saliva using surface-enhanced Raman spectroscopy. *Vib. Spectrosc.*, **38**, 79–84.
37. Murphy, T., Schmidt, H. and Kronfeldt, H.D. (1999) Use of sol-gel techniques in the development of surface-enhanced Raman scattering (SERS) substrates suitable for in situ detection of chemicals in sea-water. *Appl. Phys. B-Lasers O.*, **69**, 147–150.
38. Clarke, S.J., Littleford, R.E., Smith, W.E. and Goodacre, R. (2005) Rapid monitoring of antibiotics using Raman and surface enhanced Raman spectroscopy. *Analyst*, **130**, 1019–1026.
39. Chen, J.W., Jiang, J.H., Gao, X., Liu, G.K., Shen, G.L. and Yu, R.Q. (2008) A new aptameric biosensor for cocaine based on surface-enhanced Raman scattering spectroscopy. *Chem. Eur. J.*, **14**, 8374–8382.
40. Guerrini, L., Garcia-Ramos, J.V., Domingo, C. and Sanchez-Cortes, S. (2009) Self-assembly of a dithiocarbamate calix 4 arene on ag nanoparticles and its application in the fabrication of surface-enhanced Raman scattering based nanosensors. *Phys. Chem. Chem. Phys.*, **11**, 1787–1793.
41. Kim, K., Jang, H.J. and Shin, K.S. (2009) Ag nanostructures assembled on magnetic particles for ready SERS-based detection of dissolved chemical species. *Analyst*, **134**, 308–313.
42. Zamarion, V.M., Timm, R.A., Araki, K. and Toma, H.E. (2008) Ultrasensitive SERS nanoprobes for hazardous metal ions based on trimercaptotriazine-modified gold nanoparticles. *Inorg. Chem.*, **47**, 2934–2936.
43. Heyns, J.B., Sears, L.M., Corcoran, R.C. and Carron, K.T. (1994) SERS study of the interaction of alkali-metal ions with a thiol-derivatized dibenzo-18-crown-6. *Anal. Chem.*, **66**, 1572–1574.
44. Guerrini, L., Garcia-Ramos, J.V., Domingo, C. and Sanchez-Cortes, S. (2009) Nanosensors based on viologen functionalized silver nanoparticles: Few molecules surface-enhanced Raman spectroscopy detection of polycyclic aromatic hydrocarbons in interparticle hot spots. *Anal. Chem.*, **81**, 1418–1425.
45. Ruan, C.M., Wang, W. and Gu, A.H. (2006) Surface-enhanced Raman scattering for perchlorate detection using cystamine-modified gold nanoparticles. *Anal. Chim. Acta*, **567**, 114–120.
46. Costa, J.C.S., Sant'Ana, A.C., Corio, P. and Temperini, M.L.A. (2006) Chemical analysis of polycyclic aromatic hydrocarbons by surface-enhanced Raman spectroscopy. *Talanta*, **70**, 1011–1016.
47. Dieringer, J.A., McFarland, A.D., Shah, N.C., Stuart, D.A., Whitney, A.V., Yonzon, C.R., Young, M.A., Zhang, X.Y. and Van Duyne, R.P. (2006) Surface enhanced Raman spectroscopy: new materials, concepts, characterization tools, and applications. *Faraday Discuss.*, **132**, 9–26.
48. Shafer-Peltier, K.E., Haynes, C.L., Glucksberg, M.R. and Van Duyne, R.P. (2003) Toward a glucose biosensor based on surface-enhanced Raman scattering. *J. Am. Chem. Soc.*, **125**, 588–593.

49. Donten, M.L., Krolikowska, A. and Bukowska, J. (2009) Structure and composition of binary monolayers self-assembled from sodium 2-mercaptoetanosulfonate and mercaptoundecanol mixed solutions on silver and gold supports. *Phys. Chem. Chem. Phys.*, **11**, 3390–3400.
50. Stewart, A., McCourt, M.R. and Bell, S.E.J. (submitted for publication).

4
Single-Molecule- and Trace Detection by SERS
Nicholas P.W. Pieczonka, Golam Moula, Adam R. Skarbek, and Ricardo F. Aroca

4.1
Introduction

4.1.1
SERS

Surface-enhance Raman scattering (SERS) and surface-enhanced resonance Raman scattering (SERRS) are two of the most commonly used plasmon-enhanced optical techniques [1, 2]. The rich and unique vibrational information contained within Raman spectra makes SERS/SERRS one of the most powerful analytical techniques for trace detection and the characterization of molecular moieties down to single-molecule detection (SMD) [3–7]. It is through the coupling of the Raman scattering of a molecular system to a localized surface plasmon resonance (LSPR) of a silver or gold nanostructure (see also Chapters 1 and 2) that it is possible to capture the vibrational fingerprint of a single molecule. Since the first reports of single-molecule Surface-enhanced Raman scattering (SM-SERS) [6, 7], this ultimate sensitivity has been successfully demonstrated in a host of studies [3, 4, 8]. The validity of the claim to single-molecule spectroscopy has been further confirmed through recent investigations showing the ability to distinguish individual spectra in bi-analyte systems and identifying isotopic differences in both enriched systems [9, 10] and systems of natural origin [11]. Notwithstanding the advances in the field of SM-SERS, there are challenging theoretical and experimental questions waiting to be fully addressed. At the bottom of the pyramid is the need for controlled fabrication of reproducible nanostructures with a known quantity of spatial locations of high electric field enhancement or 'hot spots'. This need has been expressed by many practitioners in many different forms. For instance: "..., *however, of SM-SERS requires synthetic or fabrication methods capable of routinely delivering substrates with single-molecule activity. Hitherto, pursuit of this goal has been hindered by a lack of knowledge about the specific nanostructures that give rise to SM-SERS*" [5]. To break open the potential for understanding new phenomena and the applications that will surge from the information that can be extracted from single-molecule spectra, it is necessary to determine the experimental conditions that minimize the 'instrumental factors' which will allow the acquisition of reliable SM-SERS spectra reflecting a molecule and its interactions. A brief review of the current literature reveals that many groups are tackling these problems and are providing new references for SM-SERS. The approach selected in our group to study single-molecule surface-enhanced resonance Raman scattering (SM-SERRS) exploits the advantages of the fabrication of monomolecular layers (mixed dye–fatty acid Langmuir–Blodgett (LB) monolayers) that can be deposited on nanostructured Ag, Au or Ag/Au films. The ratio of the two components in the LB films can be varied quantitatively, and the effects of dye concentration can be followed, down to the single-molecule event. The target analytes employed in our work are dyes dispersed in monolayers of fatty acid, and all our spectra correspond to SM-SERRS. The results are spectral maps that highlight the points of the most efficient coupling of the probe molecule with electromagnetic hot spots present on the nanostructured film [12].

4.1.2
The Two Regimes: Ensemble and Trace/SM

The statistical average and the statistical breakdown of the SERS/SERRS spectra clearly define two regimes of the experimental phenomena and show the way for potential applications in 'average SERS' and also a tool for SMD. In practice, one may find that the behaviour of the ensemble SERS/SERRS experiment is very similar to that of normal Raman measurements. While there are additional perturbations to the characteristic Raman features in the ensemble SERS spectrum, such as modifications that arise because of interaction with a nanoparticles surface, the measured parameters such as frequency, relative peak intensity and bandwidth are generally reproducible (see also Chapter 3) [13]. This is the regime of the average or ensemble SERS. What differentiates the SERS experiments from that of a normal Raman measurement is that the absolute intensity, that is, the measured photons per second, is greatly increased, upwards to 10^6 to that recorded without the plasmonic enhancement provided by a special class of nanoparticles. A defining characteristic of this regime is that a large number of the molecules in a probed volume or area experience a similar signal enhancement. Hence, the signal is stable and reproducible in time with dynamic systems such as SERS-active colloidal solutions and from spatially different areas from static substrates such as nanoparticle arrays.

The second regime, which is where the focus of this chapter is on, is often referred to as the *trace-level* or the *single-molecule regime* (see also Chapter 5). Here, the above measurables tend to show a great degree of variability. A key feature is that the measured Raman signal is dominated by a small percentage of the probed molecules, in particular those that experience the largest enhancement, which is a consequence of non-uniform field enhancement (see also Section 1.3.2). These highly enhanced points are generated in spots where the coupling between the excitation frequency and a mode of the complex plasmonic field is most efficient. These zones of large field enhancement are an area of ongoing interest, and a more in-depth exploration of this topic can be found in several chapters in this edition (for example, see Chapters 1 and 14) as well as a vast amount of literature that has been dedicated to this topic.

4.1.3
Requirements for SM-SERS

From the first reports of SM detection with SERS, there has been great strides in understanding the underlying foundations of the effect. Unlike other SM techniques, the SERS signal is not just a property of the molecule but is a combination of the response of a molecular system and of a localized plasmon-supporting nanostructure (Figure 4.1). These two components contribute to the success of SM detection. For a given instrumentation, there will be a minimum amount of signal required for detection. The degree of enhancement needed for trace and SM detection is strongly dependent on the intrinsic Raman scattering cross section of the molecule. This is the reason that most of the reported SM-SERS are under resonance Raman conditions, that is, SERRS. The difference in cross section that can be experienced under resonance conditions can be 10^6 times greater then under non-resonance excitation (Figure 4.2). This greatly lowers the needed enhancement for SM detection accordingly, and the extreme field enhancements associated with hot spots may not be necessary. A relatively large enhancement is still required and the evidence to date suggests this is attained from only a small number of nanostructures. Though a multitude of nanoparticles of all shapes and sizes have been fabricated and used for SERS (see also Chapter 2), there are only a few classes of silver and gold nanostructures that have shown enhancement factors large enough for SMD. These are colloid aggregates, metal evaporated films and, most recently, modified scanning tunnelling microscope (STM) tips [14, 15].

This complex topic of the SERS enhancement factor has been discussed in great detail in a recent work of Etchegoin *et al.* [16]. The importance of the intrinsic Raman cross section is exemplified by the fact that SM detection has only been

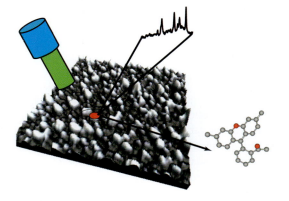

Figure 4.1 Basic elements needed for SM-SERRS.

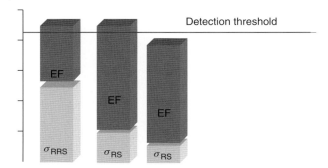

Figure 4.2 A graphical representation of interplay between the intrinsic Raman cross section of a molecule and the required enhancement factor needed for single-molecule detection.

achieved with a small set of molecules. All the reported systems are 'soft', highly polarizable molecules that have relatively large intrinsic Raman cross sections.

The 'long tail' interpretation [17] for the enhancement distribution from complex plasmonic systems (see also Section 1.3.2) reveals an important facet of the SM-SER(R)S experiment, that even at relatively large concentrations of the target analyte, SM events can be recorded. This has been demonstrated in our work with evaporated metal island and LB films [12] and Etchegoin group's work with colloidal systems [17]. This has implications for future experimental designs for the collection of SM-SERRS data. In practice then, we can accumulate SM spectra with relative ease for a handful of systems. What remains now are the many open questions surrounding the interpretation of these spectra.

In the rest of this chapter we discuss a set of systems that provide not only insight into SM-SERRS but also lay the groundwork for possible new biological applications.

4.2
Experiments and Results

4.2.1
The Langmuir–Blodgett Method for SM-SERRS

Our approach to SM studies works on the strengths of the LB method, a technique that gives precise control over the molecular concentration in a unit area. Langmuir films are monomolecular films that can be fabricated from amphiphiles at the air–water interface. The resulting film can then be transferred to a solid substrate such as a glass slide. The LB technique adapts well to the nature of many SERS/SERRS substrates, which can be easily coated with a monomolecular thick film. For our SERS/SERRS studies, the target molecule is treated as a dopant in a monolayer matrix of a spectrally silent fatty acid (fatty acids have relatively low Raman scattering cross sections compared to the molecules studied in LB films). The concentration of target analytes can be easily varied by adjusting the ratio of fatty acid to the target molecules. With this approach, the number of analyte molecules in a given probe area can be estimated with a high amount of certainty. An LB film of pure arachidic acid (AA) will contain approximately 4×10^6 molecules in the field of view of our micro-Raman spectrometer with a 50× objective (NA 0.75) which is about 1 μm^2. Hence, for different doping ratios, we can have a very good estimate of the upper limit of target molecules in our probe area. For example, a 1 : 10 dopant to matrix ratio film for AA will have about 3.6×10^5 target molecules in the probed area.

In the past, our group has used this technique to detect the SM-SERRS signal from a variety of chromophores such as several perylene derivatives [18, 19] and rhodamines [20]. The LB approach has allowed us to explore many interesting aspect of SM-SERRS including systematically demonstrating the loss of ensemble averaging and exploring the differences of SM spectra to that of the ensemble, as well as the first observation of overtone and combination bands from a single molecule [21]. Most recently, the LB techniques have been used for bi-analyte studies, which has further solidified the validity of the SM claim [20].

4.2.2
LB SM-SERRS to Biologically Relevant Systems

Recently, our research efforts have advanced towards applying this technique to biologically relevant systems. Langmuir films have long been used as mimetic systems for biological membranes [22–25]. To monitor properties such as diffusion and membrane interactions, larger molecules such as lipids and proteins are often tagged with fluorescent moieties. These investigations frequently involve the use of single-molecule fluorescence spectroscopy for tracking the behaviour of individual chromophores [26, 30]. While single-molecule fluorescence (SMF) has been extensively used, it does have some inherent limitations. The information content in a broad fluorescence signal can be difficult to differentiate as a result of spectral overlapping. It has been recognized that the SERRS cross sections for common fluorophores are similar to fluorescence and has the additional advantages of possessing much higher information content: the spectrum contains many characteristic vibrational peaks [31]. These possible advantages for SERRS have been identified, and the exploration of the potential of SERRS for multiplexing applications is an active area of research [32, 33].

The next section is an overview of our first forays into assessing the feasibility of SM-SERRS with some of the same materials used routinely in SMF. Here, we have applied our experience with LB and SERRS to systems of tagged phospholipids and lipophilic probe molecules and have incorporated them into our films for the purpose of SM detection (Figures 4.3 and 4.4). This has not only given us a new set of systems to explore the basic questions of SMD but is also a step towards developing possible new applications for SM-SERRS. The following section highlights these recent efforts with several examples.

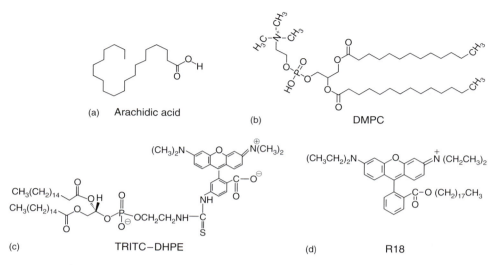

Figure 4.3 The components of an LB SERRS film. The matrix elements of (a) arachidic acid or (b) phospholipids, and the target chromophores (c) TRITC tagged phospholipids and (d) R18.

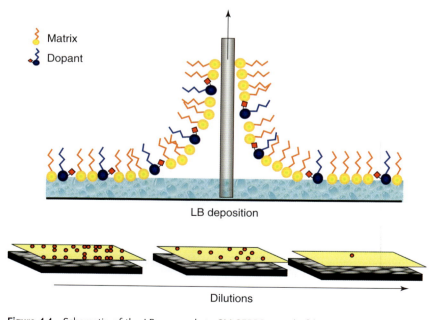

Figure 4.4 Schematic of the LB approach to SM-SERRS sample fabrication.

4.2.3
Experimental Details

For all the following examples, the same general approach was used. Silver island films (SIFs) of 9 nm mass thickness were vacuum evaporated onto Corning glass microscope slides at a pressure of $\sim 10^{-6}$ Torr at a temperature of 150 °C (maintained for 1 h after evaporation). These films are composed of a disordered collection of nanoparticles between 40 and 80 nm as determined by AFM and have a broad plasmon absorption centred at about 490 nm (Figure 4.5).

LB monolayers doped with the target analyte at various concentrations were prepared at the air–water interface and deposited onto the Ag nanostructured films. Spreading solutions were prepared so as to achieve the desired surface concentration, which could be varied from $\sim 10^5$ to 1 molecule(s) within 1 µm², which was the approximate scattering area being probed by our instrumental setup. Film transfers were carried out using Z-deposition with an electronically controlled dipping device at a constant surface pressure corresponding to the condensed phase of the Langmuir monolayers, which consistently resulted in transfer ratios near unity. Ultra-pure water (18.2 MΩ cm), maintained at a constant temperature of 20 °C, containing small amounts of $CdCl_2$ (2.5×10^{-4} M) was employed as the subphase for the preparation of all Langmuir monolayers.

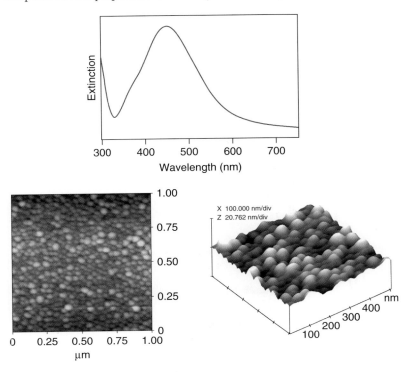

Figure 4.5 Extinction spectrum and AFM image of a 9 nm Ag SIF.

All micro-Raman scattering experiments were conducted using a Renishaw InVia system, with laser excitation at 514.5 nm, with powers of 10–20 µW at the sample. All measurements were made in a backscattering geometry, using a 50× microscope objective with an NA value of 0.75, providing collection areas of about 1 µm². Spectra were recorded with 4 cm^{-1} resolution with 1–10 s accumulation times, while 2D mapping results were collected through the rastering of a computer-controlled two-axis encoded (XY) motorized stage.

4.2.4
Single-Molecule Examples

4.2.4.1 Tagged Phospholipid

As an example of the type of systems that can be used, we present here a commercially available phospholipid that has a chromophore attached to the head group. In this instance, the targeted chromophore is an isothiocyanate derivative of a tetramethylrhodamine dye, TRITC, and the TRITC–DHPE system is one that is routinely used in SMF studies [29, 34]. This system was incorporated into two

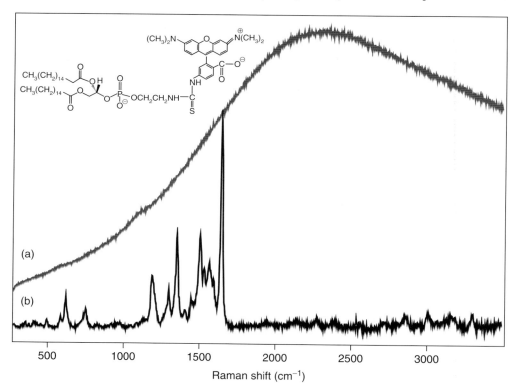

Figure 4.6 Comparison of fluorescence and SERRS taken from a 10 : 1 TRITC–DHPE : AA on a 9 nm Ag SIF.

types of matrices: a simple fatty acid (AA) used in our previous SM works, and a phospholipid matrix (DMPC).

The TRITC has an electronic absorption at about 540 nm and resonance Raman can be achieved with 514.5 nm excitation. The RR and hence the SERRS spectrum is dominated by several bands corresponding to the central xanthene. Ensemble SERRS measurements were attained from a deposited film with a 10 : 1 ratio of AA to TRITC-DMPC. Excitation with 514 nm is resonant with both TRITC and the main LSPR of the SIF. For a comparison of the spectral features of SERRS to that of fluorescence, spectra were taken from the same 10 : 1 film deposited on glass and SIF. As can be seen in Figure 4.6, the fluorescence is quite broad while the SERRS spectrum is composed of several easily identifiable narrow Raman bands. It would be prudent to point out that a small baseline correction has been applied to the SERRS spectrum for ease of comparison but, while there is a small background, most of the fluorescence has been quenched as a result of proximity to the metal surface.

For trace detection levels, low concentration samples are fabricated from a dilute spreading solution that contains a ratio of the matrix material to target analyte

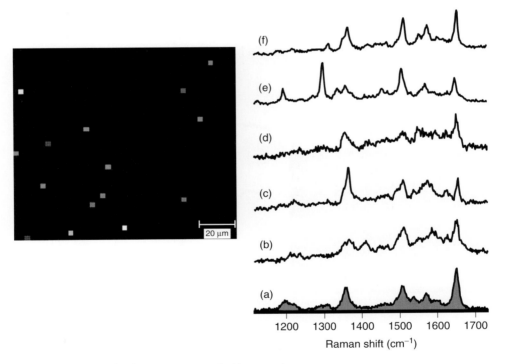

Figure 4.7 Single-molecule map constructed with a sample of 10 TRITC–DHPE molecules per square micrometre excited with 514.5 nm. Step sizes were 3 × 3 μm and the map was constructed from 1156 collected spectra. Intensity map was generated from correlation to reference ensemble spectrum.

which results in a film that has about 10 molecules/μm². Previous work has shown that a handful of spectra collected from this surface coverage are indeed SM spectra, and there is generally a higher chance of recording SM events compared to samples containing 1 molecule/μm².

Our approach to acquiring SM spectra has been to raster-scan our samples under low laser power to avoid photodegradation and collect spectra every 2 or 3 μm so as to avoid any contribution from overlapping areas. The data set is analysed with the native Wire 3.0 software [35]. Correlation maps are generated on the basis of spectral matching to a representative spectrum for the target system. Only data points of highest coincidence based on the degree of correlation to the reference spectrum are distinguished. These correspond to the light spots or pixels seen in the data maps shown in the following figures. After the data maps are generated, each highlighted spot is examined and its spectral character verified, to eliminate false positives. With this method, we can attain a number of spectra that are very likely SM spectra for a given system. For example, the map shown in Figure 4.7 was generated from a total of 1156 spectra and only spectra that have a good correlation with the reference spectra appear as highlighted spots, which, as can be seen from the map, is a small percentage of the total spectra collected. A sampling of the SM spectra recorded from the TRITC–DHPE system is shown in the figure. The variations in the Raman spectra that have become one of the key indicators of SM

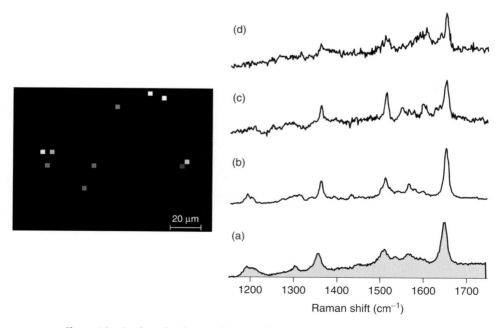

Figure 4.8 Single-molecule map fabricated from TRITC–DHPE in a DMPE matrix. For this map, 1505 spectra were collected: (a) TRITC reference from 10 : 1 sample and (b–d) SM spectra.

detection with SERRS are present. Deviations in the relative intensities of bands, narrowing of band widths and occasional appearances of new bands, can all be seen. One factor common to all is the absence of any fluorescence in the collected signal. This could indicate that the molecules reporting the SM signal are in direct contact with the nanoparticle surface and experiencing the best enhancement.

The same experiment was conducted with a commercially available phospholipid, DMPC, as the matrix. While there are additional considerations for Langmuir films fabricated with DMPC, the underlying approach is the same. The data set presented in Figure 4.8 is for a 10 molecules/μm^2 sample of TRITC–DHPE in a DMPC matrix overlaying an SIF. The resulting data map was generated from 1505 collected spectra. Again, the number of SM events is a small percentage of the whole and the tell tale signs of SM spectra are present. With these studies, it is demonstrated that SM-SERRS is at least feasible with these types of systems.

4.2.4.2 R18, Octadecyl Rhodamine B

We have also worked with the lipophile, octadecyl rhodamine B (R18). The system is an 18 carbon alkyl chain attached with a rhodamine chromophore which exhibits a strong RRS with 514.5 nm excitation. Similar in structure to the SERRS standard reference molecule R6G, R18 has proven to be a good test system in previous bi-analyte studies [20].

The R18 samples were fabricated in a similar fashion as previously discussed and proved to be a good example of the breakdown of ensemble averaging and used to demonstrate in several ways the importance of the coupling of the probe molecule with the electromagnetic 'hot spots' present in these nanostructured films. The rarity of the SM events in our maps is an indication of the very few matching coincidences of 'molecule–hot spot' in the surface area rastered. In Figure 4.9, two data maps collected with the 514.5 nm line are presented with the first map corresponding to a 10 : 1, DMPC : R18 sample and the second to a diluted sample

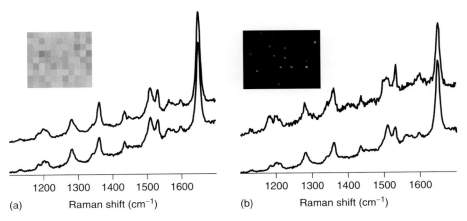

Figure 4.9 R18 in DMPC. A comparison of ensemble SERRS to trace: (a) 10 : 1 sample and (b) from about 2200 R18 square micrometre map composed 1116 data points.

that has about 2200 R18 molecules/μm² in a DMPC matrix. From the 10 : 1 sample, the characteristic SERRS profile of R18 can be recorded from any spot with little variation in the spectral profile, as indicated by the fairly uniform data map, and is good representation of 'average' or ensemble SERRS. In contrast, looking at the data map taken from the dilute sample, only a very few of the 1116 points probed correlate to the R18 spectrum. While it is a challenge in any SM-SERRS measurement to designate a spectrum SM, these spectra possess many of the features seen in SM samples. For this particular system then, R18 in a DMPC matrix, the breakdown of ensemble averaging and the possibility for SM spectra seem to be apparent at a relatively high surface coverage of 2200 R18/μm².

SM-SERRS of R18 was also recorded from samples that were prepared to have 10 molecules/μm² both in an AA and a DMPC matrix and these are presented in Figure 4.10.

Variations in the parameters were again seen in the SM spectra. An example of the changes measured is shown in Table 4.1. Here, values for relative intensities

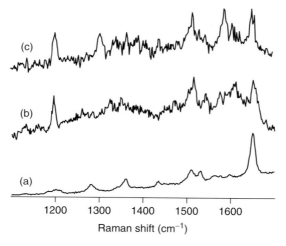

Figure 4.10 Single molecule of R18: (a) reference SERRS spectrum; (b) single molecule of R18 in DMPC and (c) R18 in AA.

Table 4.1 Comparison of relative intensities and bandwidths for peaks of R18.

	Relative intensities			Bandwidth (cm⁻¹)		
	1650	1510	1195	1650	1510	1195
1 : 10	1.00	0.46	0.17	16	28	25
1 : 1000	1.00	0.30	0.20	12	26	14
10 mol/μm²	1.00	1.00	1.18	11	11	8

and bandwidths for the three peaks measured at three different surface coverages have been tabulated. Comparing the values for an SM spectrum to that of the ensemble, the large difference in the relative intensities of bands is noteworthy, as is the narrowing of the bandwidths. These trends are representative of all the SM spectra collected.

4.3
Conclusions

What we hoped to have demonstrated here is not only the possibility of SM-SERRS for these biologically relevant systems but also the flexibility of the LB approach to acquiring SM spectra from targeted systems. As was laid out in the beginning of this chapter, the major challenges to SM-SERS is to create a framework to understand the variations in spectral parameters so as to tap into the wealth of information that may be present. The examples presented here are some of the first steps towards developing application for SM-SERRS for biological systems. And while SM spectra can be routinely attained, the relative rarity of the SM event attest to the need for the development of new silver or gold nanostructures for LB applications, which is one of the most pressing challenges for further developments.

References

1. Aroca, R. (2006) *Surface-Enhanced Vibrational Spectroscopy*, John Wiley & Sons, Ltd, Chichester.
2. Le Ru, E.C. and Etchegoin, P.G. (2009) *Principles of Surface Enhanced Raman Spectroscopy (and Related Plasmonic Effects)*, Elsevier, Amsterdam.
3. Pieczonka, N.P.W. and Aroca, R.F. (2008) Single molecule analysis by surfaced-enhanced Raman scattering. *Chem. Soc. Rev.*, **37**, 946–954.
4. Etchegoin, P.G. and Le Ru, E.C. (2008) A perspective on single molecule SERS: current status and future challenges. *Phys. Chem. Chem. Phys.*, **10**, 6079–6089.
5. Camden, J.P., Dieringer, J.A., Wang, Y., Masiello, D.J., Marks, L.D., Schatz, G.C. and Van Duyne, R.P. (2008) Probing the structure of single-molecule surface-enhanced raman scattering hot spots. *J. Am. Chem. Soc.*, **130**, 12616–12617.
6. Nie, S. and Emory, S.R. (1997) Probing single molecules and single nanoparticles by surface-enhanced Raman scattering. *Science*, **275**, 1102–1106.
7. Kneipp, K., Wang, Y., Kneipp, H., Perelman, L.T., Itzkan, I., Dasari, R.R. and Feld, M.S. (1997) Single molecule detection using surface-enhanced Raman scattering (SERS). *Phys. Rev. Lett.*, **78**, 1667–1670.
8. Camden, J.P., Dieringer, J.A., Zhao, J. and Van Duyne, R.P. (2008) Controlled plasmonic nanostructures for surface-enhanced spectroscopy and sensing. *Acc. Chem. Res.*, **41**, 1653–1661.
9. Blackie, E., Le Ru, E.C., Meyer, M., Timmer, M., Burkett, B., Northcote, P. and Etchegoin, P.G. (2008) Bi-analyte SERS with isotopically edited dyes. *Phys. Chem. Chem. Phys.*, **10**, 4147–4153.
10. Dieringer, J.A., Lettan, R.B., Scheidt, K.A. and Van Duyne, R.P. (2007) A frequency domain existence proof of single-molecule surface-enhanced Raman spectroscopy. *J. Am. Chem. Soc.*, **129**, 16249–16256.

11. Etchegoin, P.G., Le Ru, E.C. and Meyer, M. (2009) Evidence of natural isotopic distribution from single-molecule SERS. *J. Am. Chem. Soc.*, **131**, 2713–2716.
12. Goulet, P.J.G., Pieczonka, N.P.W. and Aroca, R.F. (2005) Mapping single-molecule SERRS from Langmuir-Blodgett monolayers on nanostructured silver island films. *J. Raman Spectrosc.*, **36**, 574–580.
13. Pieczonka, N.P.W. and Aroca, R.F. (2005) Inherent complexities of trace detection by surface-enhanced Raman scattering. *Chem. Phys. Chem.*, **6**, 2473–2484.
14. Domke, K.F., Zhang, D. and Pettinger, B. (2006) Toward Raman fingerprints of single dye molecules at atomically smooth Au(111). *J. Am. Chem. Soc.*, **128**, 14721–14727.
15. Zhang, W., Yeo, B.S., Schmid, T. and Zenobi, R. (2007) Single molecule tip-enhanced Raman spectroscopy with silver tips. *J. Phys. Chem. C*, **111**, 1733–1738.
16. Ru, E.C.L., Blackie, E., Meyer, M. and Etchegoin, P.G. (2007) Surface enhanced Raman scattering enhancement factors: a comprehensive study. *J. Phys. Chem. C*, **111**, 13794–13803.
17. Le Ru, E.C., Etchegoin, P.G. and Meyer, M. (2006) Enhancement factor distribution around a single surface-enhanced Raman scattering hot spot and its relation to single molecule detection. *J. Chem. Phys.*, **125**, 204701.
18. Constantino, C.J.L., Lemma, T., Antunes, P.A. and Aroca, R. (2002) Single molecular detection of a perylene dye dispersed in a Langmuir-Blodgett fatty acid monolayer using surface-enhanced resonance Raman scattering. *Spectrochim. Acta, Part A*, **58A**, 403–409.
19. Goulet, P., Pieczonka, N. and Aroca, R. (2003) Single molecule SERRS of mixed perylene Langmuir-Blodgett monolayers on novel metal island substrates. *Can. J. Anal. Sci. Spectros.*, **48**, 146–152.
20. Goulet, P.J.G. and Aroca, R.F. (2007) Distinguishing individual vibrational fingerprints: Single-molecule surface-enhanced resonance Raman scattering from one-to-one binary mixtures in Langmuir-Blodgett monolayers. *Anal. Chem.*, **79**, 2728–2734.
21. Goulet, P.J.G., Pieczonka, N.P.W. and Aroca, R.F. (2003) Overtones and combinations in single-molecule surfaced-enhance resonance Raman scattering spectra. *Anal. Chem.*, **75**, 1918–1923.
22. Leblanc, R.M. (2006) Molecular recognition at Langmuir monolayers. *Curr. Opin. Chem. Biol.*, **10**, 529–536.
23. Korchowiec, B., Paluch, M., Corvis, Y. and Rogalska, E. (2006) A Langmuir ?lm approach to elucidating interactions in lipid membranes: 1,2-dipalmitoyl-sn-glycero-3-phosphoethanolamine/cholesterol/metal cation systems. *Chem. Phys. Lipids*, **144**, 127–136.
24. Mohwald, H. (1990) Phospholipid and phospholipid-protein monolayers at the air/water interface. *Annu. Rev. Phys. Chem.*, **41**, 441–476.
25. Hwang, J., Tamm, L.K., Bohm, C., Ramalingam, T.S., Betzig, E. and Edidin, M. (1995) Nanoscale complexity of phospholipid monolayers investigated by near-field scanning optical microscopy. *Science*, **270**, 610–614.
26. Naumann, C.A. and Ke, P.C. (2001) Lateral diffusion of non-tethered phospholipids in polymer-tethered phospholipid membranes studied at the single molecule level. *Abstr. Pap. Am. Chem. Soc.*, **221**, 147-PHYS.
27. Benda, A., Fagul'ova, V., Deyneka, A., Enderlein, J. and Hof, M. (2006) Fluorescence lifetime correlation spectroscopy combined with lifetime tuning: new perspectives in supported phospholipid bilayer research. *Langmuir*, **22**, 9580–9585.
28. Sharonov, A., Bandichhor, R., Burgess, K., Petrescu, A.D., Schroeder, F., Kier, A.B. and Hochstrasser, R.M. (2008) Lipid diffusion from single molecules of a labeled protein undergoing dynamic association with giant unilamellar vesicles and supported bilayers. *Langmuir*, **24**, 844–850.

29. Ke, P.C. and Naumann, C.A. (2001) Single molecule fluorescence imaging of phospholipid monolayers at the air-water interface. *Langmuir*, **17**, 3727–3733.
30. Schmidt, T., Schütz, G.J., Baumgartner, W., Gruber, H.J. and Schindler, H. (1996) Imaging of single molecule diffusion. *Proc. Natl. Acad. Sci. U.S.A*, **93**, 2926–2929.
31. Sabatte, G., Keir, R., Lawlor, M., Black, M., Graham, D. and Smith, W.E. (2008) Comparison of surface-enhanced resonance Raman scattering and fluorescence for detection of a labeled antibody. *Anal. Chem.*, **80**, 2351–2356.
32. McKenzie, F., Ingram, A., Stokes, R. and Graham, D. (2009) SERRS coded nanoparticles for biomolecular labelling with wavelength-tunable discrimination. *Analyst*, **134**, 549–556.
33. Faulds, K., McKenzie, F., Smith, W.E. and Graham, D. (2007) Quantitative simultaneous multianalyte detection of DNA by dual-wavelength surface-enhanced resonance Raman scattering. *Angew. Chem.Int. Ed. Engl.*, **46**, 1829–1831.
34. Slaughter, B.D., Unruh, J.R., Price, E.S., Huynh, J.L., Urbauer, R.J.B. and Johnson, C.K. (2005) Sampling unfolding intermediates in calmodulin by single-molecule spectroscopy. *J. Am. Chem. Soc.*, **127**, 12107–12114.
35. Renishaw plc (2008).

5
Detection of Persistent Organic Pollutants by Using SERS Sensors Based on Organically Functionalized Ag Nanoparticles

Luca Guerrini, Patricio Leyton, Marcelo Campos-Vallette, Concepción Domingo, José V. Garcia-Ramos, and Santiago Sanchez-Cortes

5.1
Introduction

Surface-enhanced Raman spectroscopy (SERS) is an extremely sensitive analytical technique based mainly on the giant electromagnetic enhancement induced by metal nanoparticles (NPs) [1–3]. This enhancement (see also Chapter 1) is attributed to the large absorption of light by NPs resulting from localized surface plasmon resonance (LSPR) [2, 4]. Most currently employed SERS substrates (see also Chapter 2) are metal NPs in suspension (colloids) or immobilized and distributed on a solid surface (island films). The enormous amplification of the optical signal is largely related to the electromagnetic properties of the nanostructures. In addition, the so-called chemical mechanism has to be considered to account for other processes, as for instance molecular resonances and charge-transfer transitions, which affect often drastically the SERS spectra of molecules directly adsorbed on the metal surface [1]. This technique has been extensively used in the identification and orientation of adsorbates on a surface. Great efforts have been devoted to adapt SERS as a useful tool for biological research because of the detection of single molecules [5] or small numbers of molecules [6] reported in recent years. These applications were successful in the case of resonant molecules leading to surface-enhanced resonance Raman scattering (SERRS; see also Chapters 10 and 11). In general, the molecules that are active in SERS show some affinity for the metal, resulting in the necessary approach to the surface. Nevertheless, many other molecules of great interest lack this affinity and their SERS detection is not possible. Among the last ones there is a large list of pollutants that are extremely inactive in SERS due to their poor affinity to the metal surface. This is the case of many persistent organic pollutants (POPs), such as polycyclic aromatic hydrocarbons (PAHs) and chlorinated pesticides. SERS-based sensors have been applied in different modalities to detect POPs at low concentrations. For instance, optical fibres modified with metal NPs were employed in the 1990s for environmental monitoring [7–9]. Likewise, continuous SERS devices for the

Surface Enhanced Raman Spectroscopy: Analytical, Biophysical and Life Science Applications. Edited by Sebastian Schlücker
Copyright © 2011 WILEY-VCH Verlag GmbH & Co. KGaA, Weinheim
ISBN: 978-3-527-32567-2

detection of trace organic pollutants in aqueous systems have also been employed [10]. From the very beginning, SERS has demonstrated its ability in the trace detection of environmental pollutants [11, 12] and drugs [13, 14].

The affinity of the analyte towards the metal surface can be increased by modifying or functionalizing the interface with organic molecules. Since thiol groups strongly interact with both Ag and Au (the most active metals in SERS), alkyl thiols have been extensively employed in the organic functionalization of these metals in the fabrication of SERS-based sensors [15]. In the past few years, the group of Van Duyne has developed an interesting method for the preparation of nanoscale biosensors based on the thiol functionalization of nanostructured metals obtained by nanolithography [16, 17]. However, although thiols linkers are suitable organic molecules in some cases, they are not convenient in the detection of POPs or small aromatic compounds since they form very tight self-assembled monolayers (SAMs) because of the strong intermolecular interactions between aliphatic chains [18]. In these SAMs, there are no cavities or intermolecular spaces where the ligand could be located. In fact, only an on-top adsorption was observed in these systems [18].

In this context, we have developed in our laboratory several strategies aimed at increasing the applicability of SERS techniques in the sensitive and selective detection of trace pollutants by using host cavitands. In a first approach, we have shown that by properly modifying the chemical properties of the metal surface it is possible to increase drastically the adsorption of environmental pollutants on the substrate [19, 20]. This allowed both the sensitive and selective detection of POPs, in particular PAHs and polychlorinated pesticides (PCPs). These pollutants have been reported to be strong carcinogens [21], and therefore their trace detection is nowadays a very important goal. In addition, these molecules show very low affinity for adsorption on a metallic surface, thus limiting the use of surface-enhanced techniques in their detection. Identification and quantification of POPs in aqueous solutions is usually carried out by HPLC with UV–visible absorption, fluorimetric or amperometric detection or by means of GC–MS or GC–FID (flame ionization detection) and most of them include a pre-concentration step [22]. These steps are time consuming and require a great deal of effort, thus making the analysis unsuitable for routine control analysis. The detection of POPs by means of SERS is of interest because of the ready detection of pollutants directly in multicomponent samples. The detection of PAHs with SERS has also been carried out by other groups by using aliphatic organic molecules [23, 24]. In contrast, the work developed in our group consisted in the metal surface functionalization by specific cavitands aiming at increasing the sensitivity and selectivity of the fabricated nanosensors towards pollutants of different nature. This functionalization was mainly done by using Ag NPs as substrates for SERS spectroscopy because of the highest intensification achieved with these substrates. The metal NPs employed in the work presented here were prepared by chemical reduction of Ag^+ using citrate or hydroxylamine as chemical reducers, both in aqueous suspensions or immobilized on a substrate, depending on the conditions specified in each reference.

In a second approach, we have used bifunctional linkers to further increase the sensitivity of the technique by stimulating the creation of interparticle junctions acting as electromagnetic hot spots. In fact, it is generally accepted that the possibility of single molecule or few molecules detection depends on the existence of interparticle gaps where the main part of the electromagnetic field intensification occurs [25].

Aggregated colloids are the main source for single-molecule detection [26] but the fabrication of such hot spots become uncontrollable in experiments under macro conditions, their existence being a matter of luck provided by the specific aggregation pattern and the morphology induced by the aggregates adsorbed on NPs. Thus, the molecules adsorbed on the metal surface may effectively play a crucial role in the formation of these hot spots [27].

Previous experience in our laboratory has allowed us to perform a classification of host molecules employed for functionalization of metal surfaces according to the mechanism of interaction with the ligand. Host molecules can then be classified into three main groups: *inclusion*, *occlusion* and *contact* hosts (Figure 5.1). The conditions that these molecular receptors must fulfil are (i) good adherence to the metal surface, (ii) the existence or formation of intra- or intermolecular cavities able to locate the ligand, (iii) high affinity towards the analyte of interest and (iv) no overlapping spectroscopic signals. In this overview, we will show examples of metal surface functionalization with organic molecules corresponding to the above-mentioned host groups applied to the detection of pollutants with different nature.

Functionalized metal NPs must be characterized by surface-enhanced optical spectroscopy prior to their application in the detection. To accomplish this, we have used SERS, but also surface-enhanced infrared absorption (SEIRA) and surface-enhanced fluorescence (SEF). Finally, plasmon resonance was used to follow the formation of hot spots and multimers in NP suspensions, while electron microscopies (TEM and SEM) and AFM (atomic force microscopy) were used to characterize the morphology and aggregation of NPs. Functionalized metal NPs were employed in the detection of a large number of POP pollutants of different nature, including PAHs and chlorinated pesticides. From our studies, a good relationship between the structure and detection ability could be deduced, as is summarized in this overview.

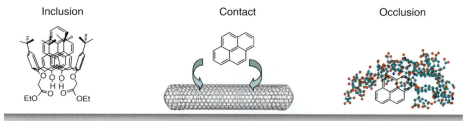

Figure 5.1 Host–guest interaction mechanisms.

5.2
Inclusion Hosts

5.2.1
Calixarenes

Calixarenes are synthetic cyclooligomers that are formed via a phenol–formaldehyde condensation. They exist in a 'cup'-like shape with a defined upper and lower rim and a central annulus (Figure 5.2). Calixarenes have interesting applications as host molecules as a result of their pre-formed cavities [28, 29]. By changing the chemical groups of the upper and/or lower rim, it is possible to prepare various derivatives with differing selectivities for various guest ions and molecules. Recently, we have reported the application of SERS and SEIRA techniques to a series of calix[4]arene derivatives on Ag NPs and their complexes with PAHs [30–33]. Calixarene bearing four t-butyl groups in the upper rim have been demonstrated to have a high selectivity in the detection of four-ring PAH molecules. Among all the analysed functionalizations of the lower rim, the ester one was demonstrated to be the most effective in the interaction with metals. Furthermore, SEIRA spectra revealed the existence of remarkable differences between the adsorption behaviour of ester-functionalized calix[4]arenes depending on the substitution pattern of the lower rim. DCEC, with a 1,3-dicarboethoxy substitution (Figure 5.2a), is anchored on both Ag and Au films through a bidentate interaction. The lower rim aperture is induced to better accommodate the ester groups towards the surface. The SEIRA technique also demonstrated that DCEC interacts more strongly with Au than Ag.

The adsorbed DCEC displays an appropriate conformation to interact with PAH pollutants. However, the ester groups in TCEC, which bears a 1,2,3,4-tricarboethoxy substitution, are linked to the surface through a monodentate interaction due to the more open conformation in its lower rim, constrained by the steric hindrance of the carboethoxy groups (Figure 5.2a). This hinders seriously its possible use in the detection of the PAH pyrene (PYR) and other PAHs. On the other hand, the interaction of DCEC with PYR leads to significant structural changes mainly affecting the carboethoxy union with the metal. Indeed, this accounts for the lower activity of TCEC calixarene in the PAH detection, which is due to a shutdown of the t-butyl groups in the upper calixarene rim [18].

The activity of DCEC calixarene, measured in terms of SERS intensity, was tested with a group of PAHs with different structures (Figure 5.3). The SERS spectra of PYR are shown in Figure 5.2b. As can be seen, the SERS spectra of all the four-ringed PAHs assayed (PYR, triphenylene (TP) and benzo[c]phenanthrene (BcP)) are selectively more enhanced. In particular, PYR and TP are the PAH molecules with the highest specificity to bind DCEC since they lead to the maximum SERS intensification. In contrast, the biggest PAHs (coronene (COR), rubicene (RUB) and dibenzoanthracene (DBA)) displayed the lowest SERS intensities. The size specificity is probably determined by the cavity, in terms of size and accessibility, of the calixarene and the t-butyl groups placed in the upper rim. It must be noted that chrisene (CHR) led to a much lower SERS intensity than BcP, although both PAHs

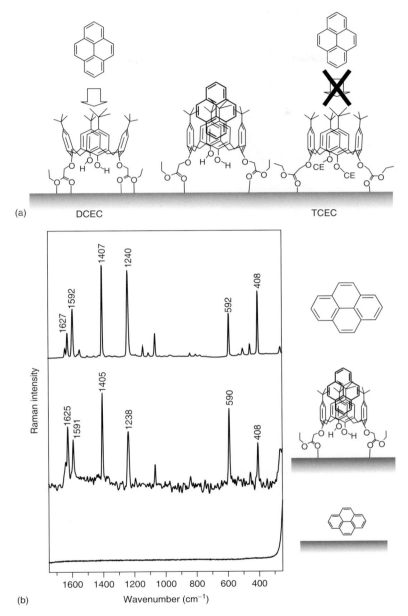

Figure 5.2 (a) Selective recognition of PAHs by DCEC due to the optimum conformation of the cavity in contrast to the inactivity of TCEC. (b) SERS spectra of pyrene obtained on citrate Ag NPs (bottom), showing no bands in absence of calixarene, and DCEC/pyrene complex (middle) displaying intense pyrene features, with slight intensity and wavenumber changes in relation to the Raman spectrum of solid pyrene (top). Excitation at 1064 nm.

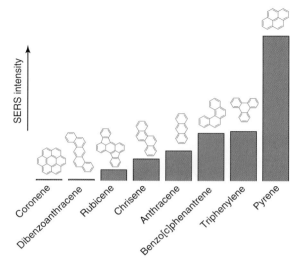

Figure 5.3 SERS intensity of different PAH pollutants on DTCE-functionalized Ag NPs.

are phenanthrene derivatives. This indicates that the binding properties of each PAH must change depending on their chemical structure, revealing the existence of a size and shape selectivity of DCEC to interact with PAH molecules.

In a recent work, we reported that a modification of the lower rim of calixarene with dithiocarbamate (DT) enhances the affinity of this kind of host molecules and also the sensitivity in the detection of PAHs [34]. This is due to the ability of DT to strongly interact with the surface of metals [35, 36]. Thus, the combination of the good host properties of calixarenes and the high affinity of the DT group in the same molecule has shown to be a good strategy to design new sensitive and selective functionalized SERS substrates for PAH detection.

A key issue in the application of calixarenes in general, and dithiocarbamate calixarene (DTCX) in particular, is the study of the self-assembly of the host molecule on the metal surface. In a recent work, we have demonstrated that this assembly depends on the surface coverage as expected from an adsorbate of the size of DTCX [37]. The most important DTCX structural marker bands can be deduced from the SERS bands because of the rich structural information provided by this vibrational spectroscopy. Such marker bands are related to the interaction geometry with the metal [the $\upsilon C=N^+$ thioureide band and the $I(\upsilon C=S)/I(\upsilon C-S)$ ratio), the conformational state of the aromatic intramolecular cavity (the $I(702)/I(665)$ ratio) and the orientation of aromatic benzene rings with respect to the surface (the $I(702)/I(571)$ ratio)]. The two first parameters are related to the lower rim of the calixarene molecule, and the two latter parameters are associated to the cavity conformation, that is, with the opening degree of this cavity, which has a great impact on the host ability of this molecule. Figure 5.4 shows the variation of some of these parameters with the DTCX concentration, together with the proposed structure of DTCX.

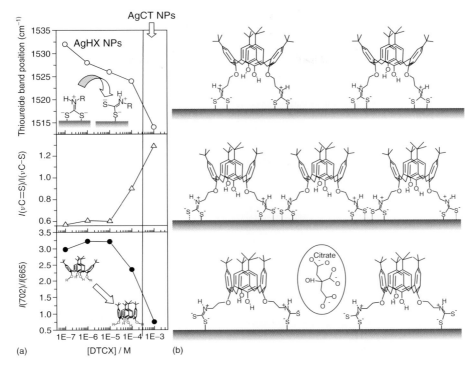

Figure 5.4 (a) Variation of structural DTCX marker parameters measured on Ag NPs prepared with hydroxylamine at different calixarene concentrations: (top) position of thioureide $\nu C=N^+$ band; (middle) $I(\nu C=S)/I(\nu C-S)$ ratio and (bottom) $I(702)/I(665)$ ratio. Data plotted at concentration 10^{-3} M correspond to the value of these parameters obtained on Ag NPs prepared by citrate by using a DTCX concentration of 5×10^{-4} M, thus corroborating the high coverage character on this surface. (b) Conformation of DTCX and proposed interaction geometry adopted by the calixarene when adsorbed on hydroxylamine Ag NPs at (top) submonolayer coverage (10^{-7} M); (middle) full surface coverage (5×10^{-4} M) and (bottom) on citrate Ag NPs (5.10^{-4} M).

Once the adsorption of DTCX on Ag was known, the trace detection of PAHs was accomplished using DTCX-functionalized Ag NPs as SERS substrates [38]. As a molecular spectroscopic technique, SERS not only reports the existence of pollutants but also reveals important structural information from both the host and the analyte, which is crucial to deduce the host–guest interaction mechanism. The effectiveness of this system was checked for a group of PAHs: PYR, BcP, TP and COR. The analysis of the spectral data allowed us to estimate the affinity constants and the limit of detection (LOD) for each pollutant (Table 5.1). From the table, it can be seen that the affinity of the binding to DTCX increases in the following order: TP < PYR < BcP < COR. The LOD ranges from 10^{-8} M in the case of PYR and TP to 10^{-10} M for COR. These LODs are similar to those reported by employing chromatographic and fluorimetric methods [39–42], with the additional advantage that no pre-concentration step was necessary. COR represents a special case. The

Table 5.1 Values of the affinity constant (log K) and LOD values deduced from the graphs in Figure 5.6 for the analysed PAHs.

	log K	LOD
BcP	4.09 ± 0.10	10^{-9} M (204 ppt)
PYR	3.93 ± 0.35	10^{-8} M (2.02 ppb)
TP	3.19 ± 0.30	10^{-8} M (2.04 ppb)
COR	7.13 ± 2.01	10^{-10} M (30 ppt)

higher affinity of this analyte to DTCX, the lower influence on the calixarene structure and the lowest LOD compared to the other PAHs suggest a different interaction mechanism with the host, involving two calixarene molecules and the formation of interparticle hot spots, where the sensitivity is highly enhanced (Figure 5.5). This supramolecular assembly brings about interparticle junctions or

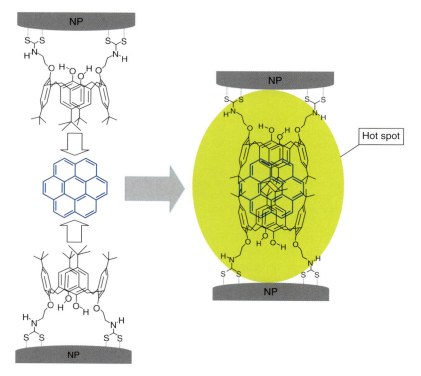

Figure 5.5 (a) Interaction model of COR (middle) with two DTCX molecules (top and bottom) in different Ag NPs leading to the formation of (b) a highly sensitive hot spot interparticle junction.

gaps, where the electromagnetic field is significantly enhanced (it is estimated to be about 10^3 times greater compared to the rest of the surface, for interparticle distances of 2–3 nm) [25].

The sensitivity of PAH detection by SERS can be further increased, and very low limits of detection reached, if micro-SERS analysis is employed [38]. For instance, the LOD of PYR in SERS macro conditions is 10^{-8} M, while 10^{-10} M was the lowest concentration that could be detected when measured with a micro-Raman device (spot size about on 1 µm^2). This corresponds to 20 ppt in the case of PYR. A calculation based on the overall available metal surface existing in the above area (spot size) at the mentioned concentration reveals that the SERS signal obtained corresponds to that from about 50 molecules.

One of the most important advantages of DTCX-functionalized Ag NPs is its application in the detection of PAH mixtures or multicomponent systems without any pre-treatment. Figure 5.6 shows strong SERS features measured for a sample containing a mixture of PAHs at different final concentrations. The comparison of the partial intensities corresponding to each PAH molecule to the *absolute* ones revealed that, while COR underwent a marked SERS intensity increase in the presence of the other PAHs, as a result of its increased solubility (its SERS intensity is approximately twofold larger than that observed in the absence of other PAHs at a partial concentration of 10^{-6} M), while marked SERS intensity decreases were observed in the case of PYR, BcP and TP.

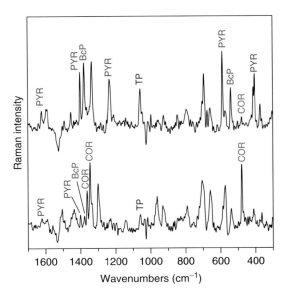

Figure 5.6 Multicomponent analysis: SERS difference spectra DTCX/total PAHs – DTCX measured at the following concentration ratios: (a) 10^{-4} M/10^{-6} M and (b) 10^{-4} M/10^{-5} M, displaying Raman features corresponding to all the PAHs contained in the mixture. The SERS spectra were obtained by using hydroxylamine Ag NPs and laser excitation at 785 nm.

5.2.2
α,ω-Aliphatic Diamines

The structure of linear α,ω-aliphatic diamines (ADs) is characterized by the existence of a linear aliphatic chain between two amino groups. SAMs of ADs on metal surfaces are interesting materials with powerful applications in many fields, such as the detection of pollutants. The molecular orientation and self-assembly of ADs on metal surfaces can be deduced from the relative intensity of bands in SERS spectra. For instance, a relative enhancement of the C–C skeletal stretching modes, the CH_2 wagging vibration and the NH_3^+ stretching vibration are observed. However, an intensity decrease for the CH_2 twisting band and the CH_2 bending modes is seen. According to the selection rules of the SERS effect [43], the selective enhancement of the Raman modes possessing a strong perpendicular component with respect to the metal surface indicates a preferably perpendicular orientation of the adsorbed diamines on the NPs, as outlined in Figure 5.7. This perpendicular arrangement implies the interaction of both amino head groups with different Ag NPs, favouring NP aggregation thanks to the bifunctional nature of ADs and leading to the formation of an NP dimer, where interparticle junctions acting as hot spots are readily formed. Additionally, these bifunctional diamines are completely protonated in the colloidal suspension, presenting two positively charged nitrogen atoms at the side ends of the alkyl chain, which are able to strongly

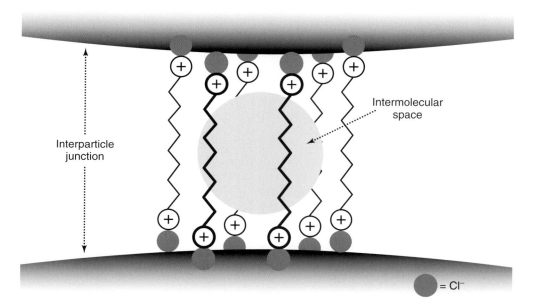

Figure 5.7 Hexagonal adsorption model for AD_8 on Ag NPs and formation of interparticle hot spot junctions and intermolecular cavities where the analyte can be hosted.

attach the NPs through the formation of ion pairs with chloride anions previously adsorbed on the metal surface. The electrical repulsion between these positive charges leads to the formation of intermolecular spaces, which are necessary for the sensor activity of these molecules as mentioned in the introductory part. Therefore, the functionalization of Ag NPs with ADs induces the creation of selective molecular recognizing cavities just in hot spots, where one expects the highest SERS sensitivity. These cavities act as actual inclusion pockets that are able to interact with the ligand, as depicted in Figure 5.7 for a hexagonal inclusion pocket model of AD in NP–NP gaps.

The use of ADs as NP linkers and host molecules in the design of nanosensors based on LSPR is also advantageous because of the relative weakness of their Raman features and their structural simplicity. This is also a condition to be fulfilled by host molecules to facilitate the SERS detection of analytes. The former property is essential to avoid interferences in the SERS spectra of the analytes, while the latter is crucial for making an appropriate interpretation of their adsorption onto the metal surface.

A deeper study on the self-assembly of ADs on Ag NPs revealed that the length of the aliphatic chain and the surface coverage of ADs are key parameters that determine the dimensions of the intermolecular AD–AD and NP–NP interparticle distances. These distances can be followed by the position of the plasmon resonance of NP aggregates and the specific molecular parameters (*trans/gauche* isomerization along each chain and the interchain lateral interactions) [44] (Figure 5.8). These structural considerations are of great importance in the application of combined AD–metal NP systems in molecular sensing [45].

Ag NPs functionalized with ADs are highly selective since they are able to detect PCPs while not being capable of detecting aromatic PAHs compounds. This is attributed to the ability of ADs to form intermolecular spaces of aliphatic nature, which tend to host non-aromatic POPs. Figure 5.9 shows the SERS spectrum of the polychlorinated insecticide aldrin (ALD) by Ag NPs functionalized with the diamine 1,8-octyldiamine (AD_8). This pesticide was demonstrated to interact with the plasmatic membrane of neuronal cells, modifying the nervous current toxicity [46]. Since this membrane is integrated by the alkylic chains of phospholipids, the SAM formed by dicationic ADs on metal NPs could mimic the biological target of this pollutant. As a result, the self-assembly of diamines on NPs modifies the chemical properties of the surface in order to favour the approaching of PCPs.

In the absence of diamines, no SERS spectrum from ALD could be obtained because of the low affinity towards the metal surface, whereas the functionalization of the NP surface with ADs (in this case AD_8) allowed its SERS detection due to the formation of the host/guest complex (inset schemes Figure 5.9). The characteristic bands of ALD can be better seen in the difference spectrum (Figure 5.9b). The low wavenumber region shows the Raman bands mainly assigned to C–Cl stretching motions and constituted an actual fingerprint of ALD. The comparison with the Raman spectrum of solid ALD (Figure 5.9c) reveals some differences due to the interaction with AD_8 which involve the Raman bands of both the

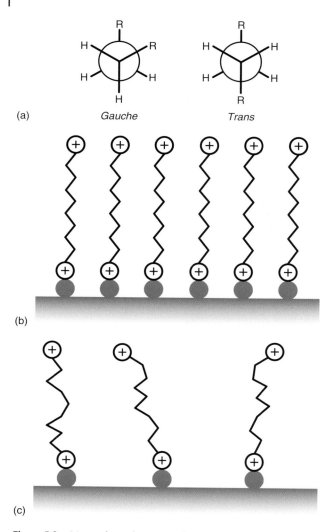

Figure 5.8 (a) *gauche* and *trans* conformations in C–C alkyl chains. (b) Predominant all-*trans* conformations in AD_8 adsorbed on Ag NPs at high surface coverages. (c) Disordering of the aliphatic chains and low lateral package at low surface coverage with diamines.

diamine and the pesticide. A more detailed study of the analytical application of diamine-functionalized NPs is currently in progress in our laboratory [47]. The adsorption of the ALD near or at the interparticle junctions formed by the diamine linkers, allows the detection of the pollutant down to ∼5 ppb.

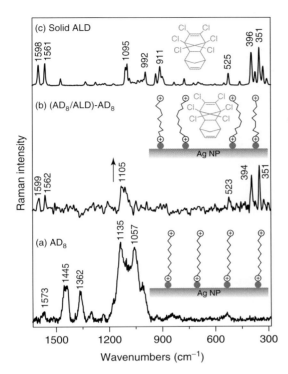

Figure 5.9 SERS spectra of (a) AD$_8$ (10^{-3} M) and (b) AD$_8$/ALD (10^{-3} M/10^{-4} M) difference spectrum. (c) Raman spectrum of ALD in the solid state. All the spectra were recorded at $\lambda_{exc} = 514.5$ nm by using hydroxylamine Ag NPs. (Insets) Schemes of the interaction of the pesticide aldrin with the interparticle cavities induced by the self-assembly of AD$_8$.

5.3
Contact Hosts

5.3.1
Viologens

Viologen dication (VGD) species have different properties of interest: they are electron acceptors [48, 49], and thus are able to interact strongly with the metal surface; besides, they are able to form charger-transfer (CT) complexes with electron donor species like PAHs [48] or chlorinated pesticides. Furthermore, the bifunctional nature of some of the VGD species, such as lucigenin (LG), makes them able to induce the formation of hot spots, thus acting as molecular hosts in the detection of analytes just in the highly sensitive region of the so-formed hot spots (Figure 5.10a) as in the case of dicationic ADs.

According to previous works [49–52], LG is able to interact with the surface of Ag through the formation of a CT complex with Cl$^-$ anions already adsorbed onto the metal. This interaction is stronger than the ionic one occurring in AD–Ag NP systems (Section 5.2.2). Besides, the aromatic nature of LG allows the interaction with aromatic pollutants such as PAHs and, to a lower extent, with chlorinated pesticides [53]. Additionally, LG is able to form intermolecular cavities (Figure 5.10b), whose dimensions are modulated by the positive residual charge

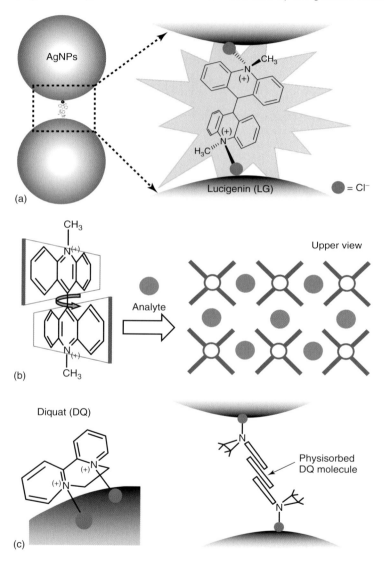

Figure 5.10 Self assembly of Lucigenin (LG) on AgNPs: formation of (a) interparticle hot spot and (b) intermolecular cavities for analyte detection. (c) Self-assembly of Diquat (DQ) of AgNPs: formation of three-unit interparticle bridge.

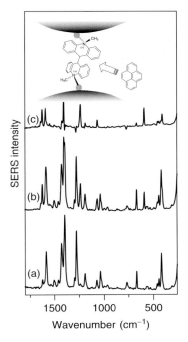

Figure 5.11 SERS spectra of (a) LG (10^{-6} M) and (b) LG/PYR complex (10^{-6} M/10^{-6} M). (c) Difference spectrum showing strong bands of PYR. The spectra were obtained by using hydroxylamine Ag NPs and laser excitation at 514.5 nm.

of LG and the rotation of the two acridine moieties to better locate the analyte. All these structural parameters can be followed by the SERS and plasmon resonance spectra [49].

The hosting ability of LG was checked with a group of different pollutants [49, 54]. From these studies, a correlation between the pollutant structure and the detection ability of LG was deduced. In the case of PYR, strong SERS features were observed in the presence of LG on the metal surface (Figure 5.11). These bands are not seen in the absence of the viologen. This result indicates the existence of an effective interaction of PYR with LG, which seems to take place through ring-stacking leading to a CT complex. This interaction induces a rotation of the acridine rings, as deduced from the shift of the corresponding inter-ring C–C stretching band at 1280 cm^{-1}, and also of the aromatic C–H bending of the acridine moiety at 1190 cm^{-1}. The formation of intermolecular cavities between LG dications due to the repulsion between the positive charges existing in LG may be an important factor to allow the interaction of the pollutant with VGD through CT complexes. Thus, as in the case of AD diamines, the main advantage of LG is that its adsorption onto Ag NPs induces the creation of intermolecular cavities prevalently localized in interparticle junctions. By using this linker, the detection of PYR and BcP at concentrations as low as 10^{-8} M (about 2 ppb) was possible. The sensitivity towards BcP was lower, likely due to the non-planar structure of this molecule, which hinders the ring-stacking between the host and the analyte rings. Other VGD linkers were also tested for the detection of pollutants: diquat (DQ) and paraquat (PQ). These two viologens exhibited a lower ability in the detection of

PYR due to their lower aromaticity and the different structure of these molecules [49, 55]. For instance, DQ has the two N atoms directed to the same side, which limits seriously its ability to form interparticle junctions where hot spots could be active, since they interact simultaneously with the metal surface (Figure 5.10c, left). However, in the latter case the activity is markedly increased when increasing the linker concentration, which can be attributed to the need of a more than a single viologen to create hot spot interparticle junctions (Figure 5.10c, right).

Intense SERS features of the pesticide endosulfan at concentrations as low as 10^{-6} M can be also seen when the Ag NPs are previously functionalized by the viologen LG [53]. The LOD deduced from this technique is 400 ppb, which is much higher than that observed for PAHs. This is attributed to the lower affinity of chlorinated pesticides to LG. In addition to the trace detection of endosulfan, the SERS technique afforded valuable structural information about the interaction mechanism. For instance, the isomerization of this pesticide upon interaction with LG was deduced, as well as a structural change of LG in the presence of the analyte.

5.3.2
Carbon Nanotubes

Single wall carbon nanotube (SWCNT) composites represent a potentially significant advance in the field of pre-concentration and analytical separations due to their high surface area and high thermal conductivity. These two characteristics suggest in particular that carbon nanotube composites could be used as pre-concentrators to facilitate trace contaminant detection. In general, aromatic compounds such as anthracene, pyrene and phthalocyanine derivatives interact with the graphitic walls of nanotubes [56, 57] by $\pi - \pi$ stacking of aromatic rings. The immobilization of SWCNT on metal surfaces has already been characterized by Corio et al. [58] by using different excitation laser lines and metal surfaces. From these studies, it was evident that the resonance Raman effect of nanotubes can be applied to investigate their electronic characteristics and their interaction with the metal surface. Therefore, since SWCNT are able to strongly interact with the surface and, on the other hand, to link polyaromatic compounds, we have employed these materials in the functionalization of metal NPs to be applied in the detection of PAHs by SERS [59].

The SERS spectrum of the assembled metallic single-walled nanotube and that of the SWCNT/PYR system deposited on the silver surface, both obtained by confocal micro-Raman analysis, are shown in Figure 5.12. The SERS spectra of the PYR adduct display all the typical pollutant bands at concentrations as low as 10^{-9} M, that is, 200 ppt (Figure 5.12b), corresponding to approximately 500 PYR molecules. This is probably still far from the single-molecule detection referred previously for very active SERS molecules, but it is a very satisfactory result when one considers that PYR is not active from the point of view of the SERS effect.

A detailed analysis of the SERS of PYR (Figure 5.12b) shows marked changes in both the intensity of bands and the peak positions in comparison to the normal Raman spectrum of the solid pollutant (Figure 5.12a). These changes are more evident as the PYR concentration is lowered. This is attributed to the higher analyte

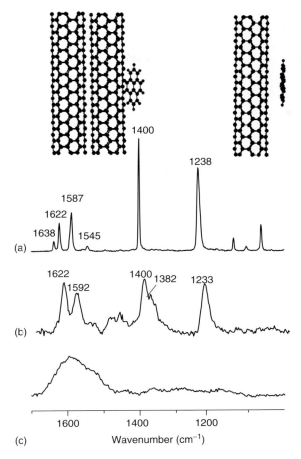

Figure 5.12 (a) Micro-Raman spectrum of solid PYR; (b) micro-SERRS spectra of PYR adsorbed onto a nanotube/silver surface and (c) micro-SERRS spectrum of the nanotube. SERS spectra were obtained using citrate dispersion Ag NPs and laser excitation at 514.5 nm. Top: proposed structural models for the SWCNT/PYR interaction mechanism (single and bundles).

dispersion on the exposed surface which induces the formation of a monolayer on the nanotube surface were the molecules adopt a particular orientation.

The band at 1622 cm^{-1} is markedly enhanced, most probably due to a resonance induced by the CT complex formed with the carbon nanotube. In addition, the 1592/1622 cm^{-1} ratio is inverted. These changes also occur in the SERS spectrum of calixarene/PYR complexes, which were also attributed to such an interaction. However, in contrast to what happens in the presence of calixarene, the main bands of PYR almost do not change their positions. Thus, we have deduced that the interaction strength is lower in the case of the complex with the nanotube.

The in-plane C–H bending bands in the 1200–1000 cm^{-1} region almost disappear as a result of the surface effect. This is attributed to the parallel orientation of

PYR on the nanotube surface. Furthermore, theoretical calculations carried out for the SWCNT/PYR systems indicate that the pollutant is interacting with SWCNT bundles adopting a parallel orientation with respect to the nanotube surface [59] (Figure 5.12, top panel).

5.4
Occlusion Hosts

5.4.1
Humic Substances

Humic acids (HAs) are also able to interact with POPs, in particular PAHs, due to their particular structural characteristics with the existence of hydrophobic cavities that display a high tendency in trapping these environmental contaminants by an occlusion mechanism [60–62]. The detection ability of HA highly depends on the three-dimensional structure of these complex macromolecules, which can be monitored by SERS spectroscopy. Recently, we have demonstrated that the SERS intensity of HA depends on parameters such as the pH and the degree of aromaticity [63, 64]. In particular, the SERS intensity of HAs decreases very much at pH between 5.0 and 7.0. These facts encouraged us to employ these natural compounds in the detection of pollutants involving the combination of advanced materials based on the HA-functionalized metal nanostructures and the enormous analytical power of surface-enhanced optical spectroscopies [65]. HAs have been already employed in layer-by-layer functionalization systems for pollutant detection [66]. HA from lignite can be used to build HA-functionalized systems capable of detecting PAHs. These compounds have a high aromaticity, which makes them ideal in the detection of highly aromatic substances such as PAH pollutants, as illustrated in Figure 5.13 for the interaction with PYR. In addition, lignite HA compounds display a rather weak SERS spectrum, and then a low overlapping with the bands of the analyte.

Figure 5.13 Interaction and detection of PAHs by HA-functionalized metal NPs through an occlusion mechanism.

The SERS spectrum of HAs from lignite has previously been reported [67]. The SERS spectra display two typical strong and large bands at about 1375 and 1600 cm^{-1}, which correspond to polycondensed aromatic moieties existing in the sample. In general, the SERS spectrum of humic substances changes very much when varying the pH [63]. This spectral behaviour by pH effect is interpreted in terms of conformational changes adopted by the HA macromolecules. At neutral pH, the molecule adopts a nearly globular structure and the SERS signal is lower, while a pH increase causes a decurling of the HA [68] which may lower their affinity towards PAHs. Thus, SERS experiments on HA-functionalized NPs were carried out at neutral conditions, in order to both avoid the interference from the host molecule and to preserve the hydrophobic cavities naturally existing in the native HA conformation.

The SERS spectrum of HA/PYR complex is shown in Figure 5.14. The appearance of strong PYR bands indicates that HA is a good receptor molecule to trap the pollutant approaching its surface. Most of the SERS bands in the spectrum of PYR in Figure 5.14 display spectral modifications related to both frequency shifts and relative intensity variations. In general terms, the SERS bands of the

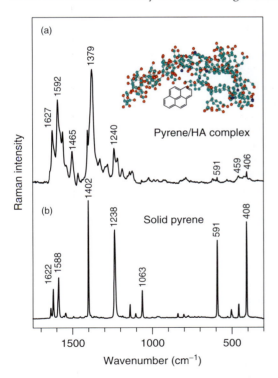

Figure 5.14 (a) SERS spectrum of HA (10 mg/ml)/PYR (10^{-6} M) complex as compared to (b) the Raman of pyrene in the solid state. The SERS spectra were obtained on citrate Ag NP with 514.5 nm laser excitation.

pollutant undergo a broadening associated with the specific interaction with the aromatic residues in HA. The νCC aromatic bands above 1450 cm^{-1} shift to higher wavenumbers while keeping a similar intensity compared to the normal Raman spectrum of the solid. The strongest νCC band at 1405 cm^{-1} in the solid shifts to 1379 cm^{-1} because of the interaction, while the νCC/δCH band at 1240 cm^{-1} does not shift but drastically decreases in relative intensity as a result of the same effect. Spectral modifications in the region of the δCH bands and an intensity decrease of the out-of-plane bending ρCH vibrations below 1000 cm^{-1}, appearing at 591, 456 and 408 cm^{-1}, are observed. These changes can be attributed to a specific molecular orientation or a resonant Raman effect upon interaction with the HA. The above changes showing frequency shifts to lower and higher wavenumbers is unambiguously associated with a short-range HA/PYR interaction which causes a π electronic redistribution of the pollutant molecule.

The SERS spectrum of PYR in the presence of HA displays larger changes compared to the same pollutant in the host molecules studied in the previous sections. We have attributed this effect to a stronger interaction due to the formation of an occlusion complex in this case. The existence of many binding points in HA and the relative flexibility of the HA structure, capable of suiting the size and shape of possible ligands, are the structural factors playing a key role in the interaction affinity with PAHs.

5.5
Conclusions

The sensitivity and selectivity of metal NPs can be remarkably enhanced by changing the interfacial properties of the metal surfaces. One of the possible modifications can be the functionalization with molecules that are able to activate the formation of intra- or intermolecular cavities, thus acting as a host to accommodate the analytes. The main properties that must be fulfilled by the host are (i) presence of cavities with an appropriate size, (ii) strong affinity to the metal and to the analyte, (iii) good self-assembly on the metal–liquid interface, (iv) no self-association or formation of multilayers, (v) high selectivity and (vi) minimal spectral band overlap with the Raman bands of the probe molecule. In this chapter, we have reported the use of molecular hosts of different nature for the surface functionalization of metal NPs with the aim of detecting POPs. The mechanism of host–guest interaction followed by these molecules can be inclusion, $\pi - \pi$ stacking or contact and occlusion. Table 5.2 summarizes the sensitivity and selectivity of the host molecules employed for the functionalization of Ag NPs in the detection of PAHs and PCPs.

Inclusion assemblers, such as calixarenes, are good molecules to fulfil the above conditions, as they are able to adsorb onto the surface by forming SAMs and display a high specificity regarding the analysed pollutants. However, these molecules are specially indicated in the detection of aromatic pollutants such as PAHs. ADs selectively detect non-aromatic PCPs because of the aliphatic nature

Table 5.2 Selectivity and sensitivity of Ag NPs functionalized with different host molecules in relation to aromatic (PAHs) and non-aromatic (PCPs) pollutants, when using SERS as analytical detection technique.

Host molecule	Selectivity and LOD detected by SERS	
	PAHs	PCPs
Calixarenes	Detected LOD: 2 ppb (PYR) LOD: 30 ppt (COR)	Not detected
Aliphatic diamines	Not detected	Detected LOD: 5 ppb (Aldrin)
Viologen dications	Detected LOD: 2 ppb (PYR)	Detected LOD: 400 ppb (Endosulfan)
Carbon nanotubes	Detected LOD: 200 ppt (PYR)	Not detected
Humic acids	Detected LOD: 20 ppb (PYR)	No results

of these compounds. Likewise, these compounds are able to form intermolecular cavities at interparticle junctions, displaying a large affinity towards highly aliphatic analytes such as PCPs. ADs and viologens are a new class of host molecules that combine the host properties with the modification of the NPs architecture leading to the formation of highly sensitive points (hot spots) and lowering the LOD. Viologens display a selective capacity to detect PAHs, but they are also able to detect non-aromatic molecules such as PCPs, although with a lower effectiveness. Carbon nanotubes are contact assemblers that produce mixed materials (metal/nanotube) with interesting optoelectronic properties to act as optical transducers. These materials are able to detect PAHs at very low concentrations. Finally, HAs are occlusion materials with promising applications in the detection of a large number of pollutants. HAs are able to interact strongly with PAHs, modifying strongly their structure. This is attributed to the existence of many binding points in these macromolecules.

The main advantage of SERS applications in the detection of pollutants is that this technique is not only able to detect the existence of the analyte but also provides information on the structural state of the molecule. The choice of the most appropriate host for surface functionalization in SERS nanosensors will obviously depend on the nature and size of the molecule to be detected. The work presented here aims to elaborate a correlation between the structure of possible hosts and guests in order to facilitate the design of more efficient, localized, surface-plasmon-resonant nanosensor systems.

Acknowledgements

The authors acknowledge financial support from projects FIS2007-63065 from Ministerio de Ciencia e Innovación (Spain), S-0505/TIC/0191 MICROSERES from Comunidad de Madrid, Fondecyt 1040640 and 1070078 from CONICYT (Chile) and Proyecto de Cooperación Internacional CONICYT/CSIC 2007CL029.

References

1. Moskovits, M. (1985) Surface-enhanced spectroscopy. *Rev. Mod. Phys.*, **57**, 783–826.
2. Aroca, R. (2006) *Surface-Enhanced Vibrational Spectroscopy*, John Wiley & Sons, Ltd, Chichester.
3. Moskovits, M. (2005) Surface-enhanced spectroscopy. *J. Raman Spectrosc.*, **36**, 485–496.
4. Schwartzberg, A.M. and Zhang, J.Z. (2008) Novel optical properties and emerging applications of metal nanostructures. *J. Phys. Chem. C*, **112**, 10323–10337.
5. (a) Kneipp, K., Kneipp, H., Itzkan, I., Dasari, R.R. and Feld, M.S. (1999) Ultrasensitive chemical analysis by Raman spectroscopy. *Chem. Rev.*, **99**, 2957; (b) Constantino, C.J.L., Lemma, T., Antunes, P.A. and Aroca, R. (2001) Single-molecule detection using surface-enhanced resonance Raman scattering and Langmuir-Blodgett monolayers. *Anal. Chem.*, **73**, 3674–3678.
6. Xu, H.X., Bjerneld, E.J., Kall, M. and Borjesson, L. (1999) Spectroscopy of single hemoglobin molecules by surface enhanced Raman scattering. *Phys. Rev. Let.*, **83**, 4357–4360.
7. Lieberman, R.A. (1993) Recent progress in intrinsic fiberoptic chemical sensing-II. *Sens. Act. B-Chem.*, **11**, 43–55.
8. Alarie, J.P., Stokes, D.L., Sutherland, W.S., Edwards, A.C. and Vodinh, T. (1992) Intensified charge coupled device-based fiberoptic monitor for rapid remote surface-enhanced raman-scattering sensing. *Appl. Spectrosc.*, **46**, 1608–1612.
9. Crane, L.G., Wang, D.X., Sears, L.M., Heyns, B. and Carron, K. (1995) Sers surfaces modified with a 4-(2-Pyridylazo)resorcinol disulfide derivative – detection of copper, lead, and cadmium. *Anal. Chem.*, **67**, 360–364.
10. Weissenbacher, N., Lendl, B., Frank, J., Wanzenbock, H.D. and Kellner, R. (1998) Surface enhanced Raman spectroscopy as a molecular specific detection system in aqueous flow-through systems. *Analyst*, **123**, 1057–1060.
11. Campiglia, A.D. and VoDinh, T. (1996) Fiber optic sensor for laser-induced room-temperature phosphorescence detection of polycyclic aromatic compounds. *Talanta*, **43**, 1805–1814.
12. Helmenstine, A., Uziel, M. and Vodinh, T. (1993) Measurement of DNA-adducts using surface-enhanced Raman-spectroscopy. *J. Toxicol. Environ. Health*, **40**, 195–202.
13. Perez, R., Ruperez, A. and Laserna, J.J. (1998) Evaluation of silver substrates for surface-enhanced Raman detection of drugs banned in sport practices. *Anal. Chim. Acta*, **376**, 255–263.
14. Ruperez, A. and Laserna, J.J. (1996) Surface-enhanced Raman spectrometry of chiral beta-blocker drugs on colloidal silver. *Anal. Chim. Acta*, **335**, 87–94.
15. Hill, W., Wehling, B. and Klockow, D. (1994) Chemo-optical sensing by SERS at organically modified surfaces. *Sens. Act. B-Chem.*, **18–19**, 188–191.
16. Yonzon, C.R., Stuart, D.A., Zhang, X.Y., McFarland, A.D., Haynes, C.L. and Van Duyne, R.P. (2005) Towards advanced chemical and biological nanosensors – An overview. *Talanta*, **67**, 438–448.
17. Anker, J.N., Hall, W.P., Lyandres, O., Shah, N.C., Zhao, J. and Van Duyne, R.P. (2008) Biosensing with plasmonic nanosensors. *Nat. Mater.*, **7**, 442–453.

18. Domingo, C., Garcia-Ramos, J.V., Sanchez-Cortes, S. and Aznarez, J.A. (2003) Surface-enhanced infrared absorption of DMIP on gold-germanium substrates coated by self-assembled monolayers. *J. Mol. Struct.*, **661**, 419–427.
19. Domingo, C., Guerrini, L., Leyton, P., Campos-Vallette, M., Garcia-Ramos, J.V. and Sanchez-Cortes, S. (2007) in *New Approaches in Biomedical Spectroscopy*, ACS Symposium Series, Vol. 963 (eds K. Kneipp, R. Aroca, H. Kneipp and E. Wentrup-Byrne), pp. 138–151.
20. Sanchez-Cortes, S., Guerrini, L., Garcia-Ramos, J.V. and Domingo, C. (2007) Funcionalización de nanopartículas metálicas con receptores sintéticos y naturales para la detección de hidrocarburos policíclicos aromáticos mediante espectroscopías intensificadas por superficies. *Opt. Pura Apl.*, **40**, 235–242.
21. Harvey, R.G. (1997) *Polycyclic Aromatic Hydrocarbons*, John Wiley & Sons, Ltd, Chichester.
22. Barro, R., Regueiro, J., Llompart, M. and Garcia-Jares, C. (2009) Analysis of industrial contaminants in indoor air: Part 1. Volatile organic compounds, carbonyl compounds, polycyclic aromatic hydrocarbons and polychlorinated biphenyls. *J. Chromat. A*, **1216**, 540–566.
23. Olson, L.G., Uibel, R.H. and Harris, J.M. (2004) C18-modified metal-colloid substrates for surface-enhanced raman detection of trace-level polycyclic aromatic hydrocarbons in aqueous solution. *Appl. Spectrosc.*, **58**, 1394–1400.
24. Schmidt, H., Bich Ha, N., Pfannkuche, J., Amann, H., Kronfeldt, H.-D. and Kowalewska, G. (2004) Detection of PAHs in seawater using surface-enhanced Raman scattering (SERS). *Mar. Poll. Bull.*, **49**, 229–234.
25. Le Ru, E.C., Etchegoin, P.G. and Meyer, M. (2006) Enhancement factor distribution around a single surface-enhanced Raman scattering hot spot and its relation to single molecule detection. *J. Chem. Phys.*, **125**, 204701.
26. Aroca, R.F., Alvarez-Puebla, R., Pieczonka, N., Sanchez-Cortes, S. and Garcia-Ramos, J.V. (2005) Surface-enhanced Raman scattering on colloidal nanostructures. *Adv. Colloid Interface Sci.*, **116**, 45.
27. Anderson, D.J. and Moskovits, M. (2006) A SERS-active system based on silver nanoparticles tethered to a deposited silver film. *J. Phys. Chem. B*, **110**, 13722–13727.
28. Gutsche, C.D. (1992) Calixarenes, in *Monographs in Supramolecular Chemistry* (ed. J.F. Stoddart), Royal Society of Chemistry, Cambridge.
29. Kim, J.H., Kim, Y.G., Lee, K.H., Kang, S.W. and Koh, K.N. (2001) Size selective molecular recognition of calix[4]arenes in Langmuir-Blodgett monolayers. *Synth. Mater.*, **117**, 145–148.
30. Leyton, P., Sanchez-Cortes, S., Campos-Vallette, M., Domingo, C., Garcia-Ramos, J.V. and Saitz, C. (2005) Surface-enhanced micro-Raman detection and characterization of calix[4]arene-polycyclic aromatic hydrocarbon host-guest complexes. *Appl. Spectrosc.*, **59**, 1009–1015.
31. Leyton, P., Sanchez-Cortes, S., Garcia-Ramos, J.V., Domingo, C., Campos-Vallette, M., Saitz, C. and Clavijo, R.E. (2004) Selective molecular recognition of polycyclic aromatic hydrocarbons (PAHs) on calix[4]arene-functionalized Ag nanoparticles by surface-enhanced Raman scattering. *J. Phys. Chem. B*, **108**, 17484–17490.
32. Leyton, P., Domingo, C., Sanchez-Cortes, S., Campos-Vallette, M. and Garcia-Ramos, J.V. (2005) Surface enhanced vibrational (IR and Raman) spectroscopy in the design of chemosensors based on ester functionalized p-tert-butylcalix[4]arene hosts. *Langmuir*, **21**, 11814–11820.
33. Leyton, P., Domingo, C., Sanchez-Cortes, S., Campos-Vallette, M., Díaz, G.F. and Garcia-Ramos, J.V. (2007) Reflection-absorption IR and surface-enhanced IR spectroscopy of tetracarboethoxy t-butyl-calix[4]arene, as a host molecule with potential applications in sensor devices. *Vib. Spectrosc.*, **43**, 358–365.
34. Guerrini, L., Garcia-Ramos, J.V., Domingo, C. and Sanchez-Cortes, S.

(2006) Functionalization of Ag nanoparticles with dithiocarbamate calix[4]arene as an effective supramolecular host for the surface-enhanced Raman scattering detection of polycyclic aromatic hydrocarbons. *Langmuir*, **22**, 10924–11926.
35. Sanchez-Cortes, S., Domingo, C., Garcia-Ramos, J.V. and Aznarez, J.A. (2001) Surface-enhanced vibrational study (SEIR and SERS) of dithiocarbamate pesticides on gold films. *Langmuir*, **17**, 1157–1162.
36. Sanchez-Cortes, S., Vasina, M., Francioso, O. and Garcia-Ramos, J.V. (1998) Raman and surface-enhanced Raman spectroscopy of dithiocarbamate fungicides. *Vib. Spectrosc.*, **17**, 133–144.
37. Guerrini, L., Garcia-Ramos, J.V., Domingo, C. and Sanchez-Cortes, S. (2009) Self-assembly of a dithiocarbamate calix[4] arene on Ag nanoparticles and its application in the fabrication of surface-enhanced Raman scattering based nanosensors. *Phys. Chem. Chem. Phys.*, **11**, 1787–1793.
38. Guerrini, L., Garcia-Ramos, J.V., Domingo, C. and Sanchez-Cortes, S. (2009) Sensing polycyclic aromatic hydrocarbons with dithiocarbamate-functionalized Ag nanoparticles by surface-enhanced Raman scattering. *Anal. Chem.*, **81**, 953–960.
39. Bourdat-Deschamps, M., Daudin, J.J. and Barriuso, E. (2007) An experimental design approach to optimise the determination of polycyclic aromatic hydrocarbons from rainfall water using stir bar sorptive extraction and high performance liquid chromatography-fluorescence detection. *J. Chromat. A*, **1167**, 143–153.
40. Zuazagoitia, D., Millan, E. and Garcia, R. (2007) A screening method for polycyclic aromatic hydrocarbons determination in water by headspace SPME with GC-FID. *Chromatographia*, **66**, 773–777.
41. Cai, Z.Q., Zhu, Y.X. and Zhang, Y. (2008) Simultaneous determination of dissolved anthracene and pyrene in aqueous solution by synchronous fluorimetry. *Spectrochim. Acta Part A-Mol. Biomol. Spect.*, **69**, 130–133.
42. de Sousa, J.R., Parente, M.M.V., Diogenes, L.C.N., Lopes, L.G.F., Neto, P.D., Temperini, M.L.A., Batista, A.A. and Moreira, I.D. (2004) A correlation study between the conformation of the 1,4-dithiane SAM on gold and its performance to assess the heterogeneous electron-transfer reactions. *J. Electroanal. Chem.*, **566**, 443–449.
43. Moskovits, M. and Suh, J.S. (1984) Surface selection-rules for surface-enhanced Raman-spectroscopy – calculations and application to the surface-enhanced Raman-Spectrum of phthalazine on silver. *J. Phys. Chem.*, **88**, 5526–5530.
44. Guerrini, L., Garcia-Ramos, J.V., Domingo, C. and Sanchez-Cortes, S. (2009) Self-assembly of α,ω-aliphatic diamines on Ag nanoparticles as an effective localized surface plasmon nanosensor based in interparticle hot spots. *Phys. Chem. Chem. Phys.*, **11**, 7363–7371.
45. Marques, M.P.M. and de Carvalho, L. (2007) Vibrational spectroscopy studies on linear polyamines. *Biochem. Soc. Trans.*, **35**, 374–380.
46. Luo, M. and Bodnaryk, R.P. (1988) The effect of insecticides on (Ca-2++Mg-2+)-Atpase and the Atp-dependent calcium-pump in MOTH brain synaptosomes and synaptosome membrane-vesicles from the bertha armyworm, mamestra-configurata Wlk. *Pest. Biochem. Phys.*, **30**, 155–165.
47. Guerrini, L., Izquierdo-Lorenzo, I., Rodriguez-Oliveros, R., Sanchez-Gil, J.A., Sanchez-Cortes, S., Garcia-Ramos, J.V., and Domingo, C. (2010) α,ω-Aliphatic Diamines as Molecular Linkers for Engineering Ag Nanoparticle Clusters: Tuning of the Interparticle Distance and Sensing Application. *Plasmonics*. DOI 10.1007/s11468-010-9143-x
48. Monk, P.M.S. (1998) *The Viologens: Physicochemical Properties, Synthesis and Applications of the Salts of 4,4'-Bipyridine*, John Wiley & Sons, Ltd, Chichester.
49. Guerrini, L., Garcia-Ramos, J.V., Domingo, C. and Sanchez-Cortes, S. (2008) Building highly selective hot spots in Ag nanoparticles using bifunctional viologens: application to the SERS

detection of PAHs. *J. Phys. Chem. C*, **27**, 7527–7530.

50. Guerrini, L., Jurasekova, Z., Domingo, C., Perez-Mendez, M., Leyton, P., Campos-Vallette, M., Garcia-Ramos, J.V. and Sanchez-Cortes, S. (2007) Importance of metal-adsorbate interactions for the surface-enhanced Raman scattering of molecules adsorbed on plasmonic nanoparticles. *Plasmonics*, **2**, 147–156.

51. Millan, J.I., Garcia-Ramos, J.V., Sanchez-Cortes, S. and Rodriguez-Amaro, R. (2003) Adsorption of lucigenin on Ag nanoparticles studied by surface-enhanced Raman spectroscopy: effect of different anions on the intensification of Raman spectra. *J. Raman Spectrosc.*, **34**, 227–233.

52. Millan, J.I., Garcia-Ramos, J.V. and Sanchez-Cortes, S. (2003) Study of the adsorption and electrochemical reduction of lucigenin on Ag electrodes by surface-enhanced Raman spectroscopy. *J. Electroanal. Chem.*, **556**, 83–92.

53. Guerrini, L., Aliaga, A.E., Carcamo, J., Gomez-Jeria, J.S., Sanchez-Cortes, S., Campos-Vallette, M. and Garcia-Ramos, J.V. (2008) Functionalization of Ag nanoparticles with the bis-acridinium lucigenin as a chemical assembler in the detection of persistent organic pollutants by surface-enhanced Raman scattering. *Anal. Chim. Acta*, **624**, 286–293.

54. Guerrini, L., Garcia-Ramos, J.V., Domingo, C. and Sanchez-Cortes, S. (2009) Nanosensors based on viologen functionalized silver nanoparticles: few molecules surface. Enhanced Raman spectroscopy detection of polycyclic aromatic hydrocarbons in interparticle hot spots. *Anal. Chem.*, **81**, 1418–1425.

55. Lopez-Ramirez, M.R., Guerrini, L., Garcia-Ramos, J.V. and Sanchez-Cortes, S. (2008) Vibrational analysis of herbicide diquat: a normal Raman and SERS study on Ag nanoparticles. *Vib. Spectrosc.*, **48**, 58–64.

56. Star, A., Han, T.-R., Gabriel, J.-C.P., Bradley, K. and Grüner, G. (2003) Interaction of aromatic compounds with carbon nanotubes: correlation to the Hammett parameter of the substituent and measured carbon nanotube FET response. *Nano Lett.*, **3**, 1421–1423.

57. Nakashima, N., Tomonari, Y. and Murakami, H. (2002) Water-soluble single-walled carbon nanotubes via non-covalent sidewall-functionalization with a pyrene-carrying ammonium ion. *Chem. Lett.*, **6**, 638–640.

58. Corio, P., Brown, S.D.M., Marucci, A., Pimenta, M.A., Kneipp, K., Dresselhaus, G. and Dresselhaus, M.S. (2000) Surface-enhanced resonant Raman spectroscopy of single-wall carbon nanotubes adsorbed on silver and gold surfaces. *Phys. Rev. B*, **61**, 13202–12211.

59. Leyton, P., Gómez-Jeria, J.S., Sanchez-Cortes, S. and Campos-Vallette, M. (2006) Carbon nanotube bundles as molecular assemblies for the detection of polycyclic aromatic hydrocarbons: surface-enhanced resonance Raman spectroscopy and theoretical studies. *J. Phys. Chem. B*, **110**, 6470–6474.

60. McCain, K.S. and Harris, J.M. (2003) Total internal reflection fluorescence-correlation spectroscopy study of molecular transport in thin sol-gel films. *Anal. Chem.*, **75**, 3616–3624.

61. Larsen, P. and Carlsson, L. (1997) Solubilization of phenanthrene by humic acids. *Chemosphere*, **34**, 817–825.

62. Terashima, M., Tanaka, S. and Fukushima, M. (2003) Distribution behavior of pyrene to adsorbed humic acids on kaolin. *J. Environ. Qual.*, **32**, 591–598.

63. Sanchez-Cortes, S., Francioso, O., Ciavatta, C., Garcia-Ramos, J.V. and Gessa, C. (1998) pH-dependent adsorption of fractionated peat humic substances on different silver colloids studied by surface-enhanced Raman spectroscopy. *J. Colloid Interface Sci.*, **198**, 308–318.

64. Sanchez-Cortes, S., Corrado, G., Trubetskaya, O.E., Trubetskoj, O.A., Hermosin, B. and Saiz-Jimenez, C. (2006) Surface-enhanced Raman spectroscopy of chernozem humic acid and their fractions obtained by coupled size exclusion chromatography-polyacrylamide gel

electrophoresis (SEC-PAGE). *Appl. Spectrosc.*, **60**, 48–53.

65. Leyton, P., Córdova, I., Lizama-Vergara, P.A., Gómez-Jeria, J.S., Aliaga, A.E., Campos-Vallette, M.M., Clavijo, E., García-Ramos, J.V. and Sanchez-Cortes, S. (2008) Humic acids as molecular assemblers in the surface-enhanced Raman scattering detection of polycyclic aromatic hydrocarbons. *Vib. Spectrosc.*, **46**, 77–81.

66. Crespilho, F.N., Zucolotto, V., Siquiera, J.R., Constantino, C.J.L., Nart, F.C. and Oliveira, O.N. Jr. (2005) Immobilization of humic acid in nanostructured layer-by-layer films for sensing applications. *Environ. Sci. Technol.*, **39**, 5385–5389.

67. Francioso, O., Sanchez-Cortes, S., Tugnoli, V., Marzadori, C. and Ciavatta, C. (2001) Spectroscopic study (DRIFT, SERS and H-1 NMR) of peat, leonardite and lignite humic substances. *J. Mol. Struct.*, **565–566**, 481–485.

68. McCarthy, J.F. and Jimenez, B.D. (1985) Interactions between polycyclic aromatic-hydrocarbons and dissolved humic material- binding and dissociation. *Environ. Sci. Technol.*, **19**, 1072–1076.

6
SERS and Pharmaceuticals

Simona Cîntă Pînzaru and Ioana E. Pavel

6.1
Introduction

Developments in the pharmaceutical industry are nowadays driven by the requirement for high efficacy while retaining reduced side effects. For this reason, new medicines are very often of relatively low dose, typically less than 100 mg of drug substance (i.e. less than 10% w/w) in a tablet. This poses major analytical challenges with respect to the rapid chemical identification of active drug substances, their polymorphic forms, possible contaminants, the distribution and interaction between drug substances and excipients, and so on. Drug and excipient particle sizes can vary, but they are generally of micron size, thus resulting in dosage samples that have substance inhomogeneity on a similar scale. Therefore, the analysis of such dosage samples by conventional techniques such as mass spectrometry, elemental analysis, spectrophotometry or nuclear magnetic resonance (NMR) can be very difficult.

With the advent of new instrumentation and intense monochromatic light sources, Raman spectroscopy has become a popular tool for analysing pharmaceutical compounds [1, 2]. However, a major problem with the analysis of pharmaceutical compounds by Raman spectroscopy is the weak intensity of the Raman scattered signal and the fluorescence caused by either the active drug itself or by a small amount of impurity present in the sample. The fluorescence emission of aromatic molecules generally occurs in the near-UV to visible spectral region. Therefore, it can interfere with the Raman signal if it is located within that spectral region. This problem can be circumvented if Raman scattering is excited in the UV or near-infrared (NIR) region of the electromagnetic spectrum, by choosing a laser of appropriate wavelength (e.g. a UV or NIR laser). Moreover, the NIR laser photon does not have enough energy to excite the fluorophore by single-photon excitation. This results in a reduced fluorescence emission. On the other hand, excitation with a UV laser still produces fluorescence, but the Raman spectrum is positioned in the relatively fluorescence-free UV region of the electromagnetic spectrum [3]. However, there are two major disadvantages associated with NIR excitation. The Raman intensity is inversely proportional to the fourth power of

Surface Enhanced Raman Spectroscopy: Analytical, Biophysical and Life Science Applications. Edited by Sebastian Schlücker
Copyright © 2011 WILEY-VCH Verlag GmbH & Co. KGaA, Weinheim
ISBN: 978-3-527-32567-2

the laser wavelength, so the intensity of the Raman signal is weaker with visible excitation. In addition, the efficiency of the typical silicon-based CCD detector falls off in the NIR region.

Surface-enhanced Raman scattering (SERS) spectroscopy can definitely help in such situations. With the enhancement offered by SERS, the normally measured Raman cross sections ($10^{-30}-10^{-25}$ cm^2 per molecule) [4] can, under favourable circumstances, be increased by several orders of magnitude (single-molecule SERS experiments; see also Chapter 4), that is, to a point where it rivals those of intense fluorescence [5]. The advantages of SERS include detection limits at the parts-per-billion level or lower, real-time response, qualitative and quantitative capabilities, a high degree of specificity, simultaneous multiplex detection and trace analysis (see also Chapters 3, 4 and 5). SERS can also be used as a surface investigative tool [1]. It can provide valuable information about the changes in molecular identity of the drug substance and its orientation with respect to the SERS substrate under different physiological conditions (e.g. pH and concentration) [1]. It should be noted that the type and morphology of the SERS-active substrate (see also Chapter 2) are responsible for the chemical nature and geometry of the adsorbed drug species as well as for the enhancement of the SERS signal (see also Chapter 1).

The direct assignment of Raman and SERS bands of pharmaceutical compounds can be often complicated. Theoretical calculations can certainly assist in obtaining a deeper understanding of the vibrational spectra of such molecules. In the last 20 years, quantum chemical methods have so successfully been developed that now they are almost an indispensable complement of experimental studies in chemistry. In particular, recent developments in density functional theory (DFT) have shown that this method is a powerful computational alternative to the conventional quantum chemical methods, since they are much less computationally demanding and take account of the effects of electron correlations. In the present chapter, several successful examples will be presented. Overall, the results will provide a benchmark illustration of the virtues of DFT in aiding the interpretation of rich vibrational spectra as well as in furnishing insight into the relation between the vibrational properties and the nature of the SERS substrate–pharmaceutical interaction.

SERS data on pharmaceutical compounds from the widely used Anatomical Therapeutic Chemical Classification (e.g. antipyretics, analgesics, antimalarial drugs, antibiotics, antiseptics) and others classes (e.g. anticarcinogenic and antimutagenic drugs) are briefly presented in this chapter.

6.2
SERS of Antipyretics and Analgesics

Antipyretics are drugs that reduce body temperature in situations such as fever. However, they will not affect the normal body temperature if one does not have a fever. Such drugs usually lower the thermodetection set point of the hypothalamic heat regulatory centre, with resulting vasodilation and diaphoresis. Most antipyretics are also used for other purposes. For example, the most common antipyretics in the United States are aspirin, paracetamol (i.e. acetaminophen) and non-steroidal anti-inflammatory drugs such as ibuprofen (usually abbreviated NSAIDs). These drugs are primarily used as analgesics (also known as *pain relievers*), and are known to act in various ways on the peripheral and central nervous systems. The analgesic class also includes narcotic drugs such as morphine, synthetic drugs with narcotic properties such as tramadol and various others.

Because water is the preferred medium in SERS experiments, a large number of pharmaceutical compounds investigated by SERS are water soluble. However, some pharmaceuticals (e.g. aspirin, paracetamol) are poorly soluble in water and may not be easily administered and adsorbed by the human body. This makes it hard to obtain their SERS spectra using ordinary aqueous silver colloids. There are several approaches to overcome this problem and some of them are listed below:

1) Organic solvents such as acetone, pyridine, dioxane, ethanol and ethanol/water mixtures can be used to dissolve the samples. The drug solutions are then mixed with the aqueous colloids for SERS detection. However, this method requires that all organic solutions are miscible in water [6–8].

2) The use of non-aqueous silver sols or sol gels is another approach for handling this problem. Several methods for preparing silver hydrosols [9, 10] in different non-aqueous solvents or sol–gels have been already proposed, but to the best of our knowledge only a modest number of spectroscopic studies have been conducted yet to test their SERS activities on pharmaceuticals [11, 12].

3) The sample solution is prepared using an appropriate organic solvent and then transferred onto a solid SERS-active metal substrate by either dipping the substrate into the solution or by dropping the solution onto the surface of the SERS substrate. However, the disadvantage of this method is the possible occurrence of a high background due to the carbonization of the adsorbed samples on the substrate surface exposed to a strong excitation laser [7, 13–15].

4) The sample solution is prepared using an appropriate organic solvent that is immiscible in water. The solution is mixed with an aqueous silver sol and shaken for a few minutes. The mixture is then allowed to separate into two liquid layers: the first one containing the colloidal silver nanoparticles (AgNPs) with adsorbed molecules, and the second one having the chloroform solvent. The aqueous layer is transferred to a cuvette or glass capillary tube for SERS measurements. This extraction method is based on the strong affinity of sample molecules to the metal surface of colloidal NPs [16].

5) The use of a drug–excipient mixture can also overcome the water solubility problem of many drugs such as aspirin and paracetamol. An excipient is an inactive substance used as carrier for the active ingredient (e.g. normal paracetamol) of a medication (e.g. paracetamol sinus tablet). Excipients can be used as antioxidants, coating materials, emulgents, taste and smell improvers, ointment bases, conserving agents, consistency improvers, disintegrating materials, and so on. In particular, disintegrating excipients (e.g. the

starch ingredient in paracetamol sinus medication) can serve as a water uptake facilitator and help prepare the aqueous solutions for SERS investigations. The SERS spectra of the drug–excipient solutions are then simply recorded by using aqueous silver or gold colloids [17, 18].

The last four approaches were recently employed for the detection of the two most common antipyretics and analgesics: aspirin and acetaminophen.

Farquharson et al. [11, 12] used approach 2 and proposed an SERS-based method for the direct detection, identification and quantification of various drugs in human saliva, including aspirin, paracetamol, 5-fluorouracil, 50/50 mixtures of dacarbazine and 5-fluorouracil in water, cocaine and phenobarbital. The collected saliva (less than 0.5 ml) was first treated to effect mutual separation of the drugs of interest and interfering chemicals present in the sample by a chemical treatment, a physical treatment and/or a chromatographic method. The SERS spectra were then recorded using acquisition times of 1 min from silver-doped sol–gel filled glass capillaries with the help of a 785 nm laser. The SERS-active nanomaterial consisted of a chemically synthesized porous structure, that is, the sol–gel prepared utilizing silica-, titania- or zirconia-based alkoxides, and at least one SERS-active metal. Briefly, a silver amine complex, (consisting of a 5 : 1 v/v solution of 1 N $AgNO_3$ and 28% NH_4OH) was mixed with an alkoxide (consisting of a 2 : 1 v/v solution of methanol and tetramethyl orthosilicate (TMOS)) in a 1 : 8 v/v silver amine : alkoxide ratio. A 0.15 µl aliquot of the above mixture was then drawn into a 1 mm diameter glass capillary to fill 15 mm length. After sol–gel formation, the incorporated silver ions were reduced with a dilute sodium borohydride solution, followed by a water wash to remove the residual reducing agent [19–21]. Farquharson et al. suggested that the porous material could also be a silica gel, silica stabilized by zirconia, derivatized silica-based matrix, long-chain alkane particles or derivatized long-chain alkane particles, so that it could produce chemical separations or selective chemical extractions. The presence and concentration of a specific drug and its metabolites [e.g. aspirin (acetylsalicylic acid) and salicylic acid] in the analyte sample were determined by analysing the collected Raman and SERS spectra. A detailed example for specific concentrations is provided for 5-fluorouracil in Section 6.4. Because of the very small volume of saliva required and its facile sampling, the method proved to be well suited for non-invasive and label-free detection of the major metabolites and study of their pharmacokinetics. These SERS-based measurements of drug pharmacokinetics could also be used to provide information regarding patient response based on phenotype and genotype.

Taguenang et al. used approach 3 and recorded the SERS spectra of paracetamol and aspirin with the help of an SERS tip [14]. The tip was a poly(methyl methacrylate) (PMMA) plastic optical fibre of 3 mm core diameter, on which gold nanoparticles (AuNPs) of 30 nm diameter were deposited. SERS spectra were recorded for various concentrations of the analyte. The interfering spectrum of PMMA was then simply subtracted from the overall SERS spectrum to reveal the SERS pattern of the pharmaceutical compound of interest. Using a focussed 633 nm laser, a detection sensitivity of 0.1 pg was established. This method could further simplify the SERS detection by the possibility of connecting the other end of the fibre directly to the spectrometer.

Wang et al. used approach 4 and prepared paracetamol, aspirin and vitamin A solutions in HPLC-grade chloroform (at concentrations of 10^{-4} and 10^{-3} mol l^{-1}), which is immiscible with water [16]. The paracetamol and aspirin solutions were then mixed with an aqueous silver sol at a volume ratio of 1 : 1 and allowed to separate into two layers. The SERS spectra obtained on the AgNPs of the aqueous layer by the extraction method are presented in Figures 6.1 and 6.2. As can be seen, the proposed method works very well for aspirin but is unsatisfactory for trace analysis and structural identification applications in case of paracetamol due to the low signal intensity and poor spectral features. By comparing the SERS spectra with the normal Raman spectra obtained for the solid-state samples, the authors established the interaction sites of the two pharmaceuticals with the metallic surface. Aspirin was found to be adsorbed on the AgNPs via the carboxylate group, as evidenced by the dramatic spectral changes observed for the vibrational modes of this group. The $\nu_a(COO^-)$ asymmetric stretching mode present at 1445 cm^{-1} in the Raman spectrum of aspirin powder could be hardly seen in the SERS spectrum, while the $\nu_s(COO^-)$ symmetric stretching mode appeared as a strong peak at 1346 cm^{-1} in the SERS spectrum. A strong band characteristic of the $\nu(Ag-O)$ stretching mode was detected at 241 cm^{-1}. In case of paracetamol, the adsorption to the AgNP surface occurred through the acetamido nitrogen. This was concluded by the absence of the $\nu(C=O)$ stretching vibration in the SERS spectrum (observed

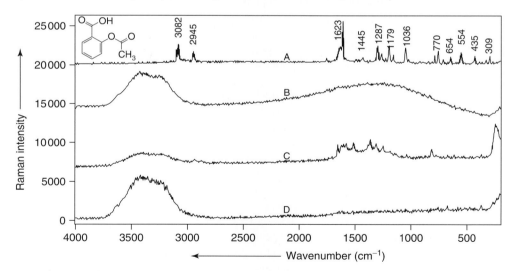

Figure 6.1 Raman spectra of aspirin: (a) powder and (b) saturated aqueous solution. SERS spectra of aspirin adsorbed on the colloidal AgNPs (c) by the extraction method and (d) after mixing with pure CHCl$_3$. Excitation line: 514.5 nm. The inset shows the molecular structure of aspirin. Reproduced with permission.

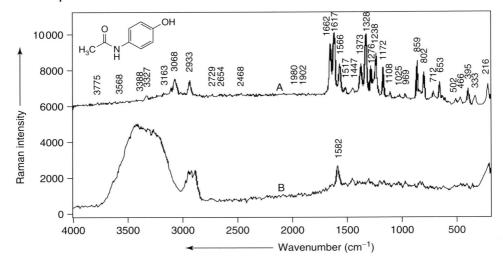

Figure 6.2 (a) Raman spectrum of acetaminophen in solid state; (b) SERS spectrum of acetaminophen adsorbed on the colloidal AgNPs by the extraction method. Excitation line: 514.5 nm. The inset shows the molecular structure of acetaminophen. Reproduced with permission.

in the normal Raman spectrum at 1652 cm^{-1}) and the appearance of an intense peak at 1582 cm^{-1}, which was attributed to the δ(NH) bending mode. To verify the feasibility of this method for quantitative analysis, a series of SERS spectra were then collected only for vitamin A at different concentrations and a calibration curve (i.e. the SERS intensity as a function of concentration) was constructed. The calibration curve was found to be linear, which means that the orientation of the adsorbed molecules remained unchanged in the investigated concentration range. It was concluded that the detection sensitivity of the SERS-based extraction method had comparable values of about 10^{-7} mol l^{-1}, or better than the commonly used spectrophotometric methods [16].

Andronie *et al.* used approach 5 and detected the presence of paracetamol sinus tablet at micromolar concentrations through SERS measurements on an aqueous silver colloid [17]. Beside the active ingredient 4-hydroxyacetanilide, the paracetamol tablet also contains inactive excipients such as lactose powder, maize starch and microcrystalline cellulose. The starch ingredient is known to dissolve readily in water and to help the tablet disperse once swallowed. Small amounts of lubricants (e.g. stearic acid and magnesium stearate) are usually added as well. These lubricants help the tablets, once pressed, to be more easily ejected out of the matrix. Andronie *et al.* recorded the normal Raman spectra of commercially available normal paracetamol and paracetamol sinus (Europharm) using the 514.5 nm excitation line, and noticed that some additional bands appeared in the spectrum of sinus paracetamol due to the presence of the above-mentioned inactive ingredients. However, the Raman signal characteristic of the active ingredient in paracetamol sinus could be easily recognized and had a very similar spectral pattern as that of normal paracetamol. Furthermore, the fast and non-invasive nature of the Raman technique allowed for the molecular probing of the tablet ingredients at various depths while still in plastic packs. This was achieved without sample cross-sectioning using the point-by-point Raman mapping approach with confocal detection. The paracetamol sinus solution was then obtained by simply dissolving one tablet in 10 ml of water at room temperature and recording the normal Raman spectrum of the solution. (Figure 6.3a).

The SERS spectra of the paracetamol sinus solutions were collected on an aqueous sodium citrate-reduced Ag colloid [22] using the 514.5 nm excitation line (Figure 6.3a). The final drug concentration was about 10^{-6} mol l^{-1}. A detailed assignment of all vibrational modes suggested the preponderance of the polymorphic form I of paracetamol [23, 24]. It is well known that there are two major possibilities of molecular adsorption on the metal surface, namely, physisorption and chemisorption. In the case of physisorption, the molecules have an SERS spectrum similar to that of free molecules, due to the relatively large distance between the metal surface and the adsorbed compound. On the contrary, in the case of chemisorption, there is an overlapping of the molecular and metal orbitals, the molecular structure being modified and, consequently, the position and profile of the bands are dramatically changed. In this study, the paracetamol molecule was found to be mainly chemisorbed to the metallic nanosurface in a parallel orientation through the π-electrons of the phenyl ring. The flat orientation of the molecule was established on the basis of the broadening of the bands, changes in relative intensity and the major shifts observed for the vibrational modes characteristic of the aromatic skeletal plane (e.g. the ν(CC) and δ(CCC) ring motions at 1614 cm^{-1}) as well as those its substituents located in the immediate vicinity of the AgNP surface (e.g. the ν(C–O) stretching mode at 1283 cm^{-1}, the ν(C–N) stretching at 1245 cm^{-1}, the δ(C–N) deformation at 778 cm^{-1} and the ν(H$_3$C–C=O) stretching at 1365 cm^{-1}). The proposed orientation was further substantiated by the δ(CH$_3$) deformation at 1073 cm^{-1}, which was found to be one of the most prominent peaks in the SERS spectrum. The interaction mechanism was independent of the SERS substrate used for the drug detection. A similar SERS behaviour was concluded for the normal paracetamol solution (8 × 10^{-2} mol l^{-1}) adsorbed on thin Ag island films of 2.9 and 1.9 Å roughness, using the 514.5 nm excitation line (Figure 6.3b). The SERS bands of the normal paracetamol at 1112, 1081 and 860 cm^{-1} (Figure 6.3b), which were attributed to the out-of-plane δ(C–H) deformation of the phenyl ring, the out-of-plane C–C skeletal and the phenyl–N deformation modes, respectively, were significantly enhanced. This supported the proposed flat orientation with respect to the Ag surface (Figure 6.3a inset).

The drug absorption process at the human body level takes place at different pH values, usually at near-neutral pH values in blood and human plasma [25]. Consequently, the SERS spectra of paracetamol sinus solutions were recorded at different pH values on colloidal AgNPs (Figure 6.4) to determine whether changes in the molecular identity of the drug occur at different physiological conditions. As expected, the nitrogen atom of the molecule was found to undergo deprotonation

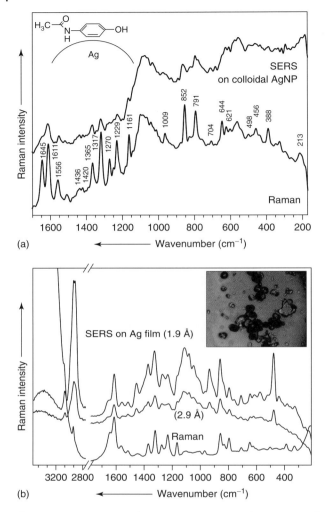

Figure 6.3 (a) Raman and SERS spectra of paracetamol sinus solution on colloidal AgNPs. Excitation line: 514.5 nm. The DFT-optimized geometry of the active ingredient of the drug is inserted; (b) Raman spectrum of 3.3×10^{-1} M normal paracetamol solution (pH 7) (a), SERS spectra of 8×10^{-2} M normal paracetamol on thin films with 2.9 Å (b) and 1.9 Å (c) roughness and its proposed orientation on the silver surface (d). Excitation: 514.5 nm (a–c), 200 mW (a) and 50 mW (b,c). Microscopical images of the normal paracetamol solution (8×10^{-2} M) after drying on the silver films with 2.9 Å (e) and 1.9 Å (f) roughness.

at basic pH values as reflected by the SERS spectra (Figure 6.4a). A possible interaction of the molecule with the AgNP surface via the now available lone pair of the nitrogen atom could not be excluded (a possible v(Ag–N) stretching mode at about 200 cm^{-1}). The clearly defined and sharper SERS signals observed at acidic pH values (Figure 6.4b) suggested a rather tilted orientation to the metallic

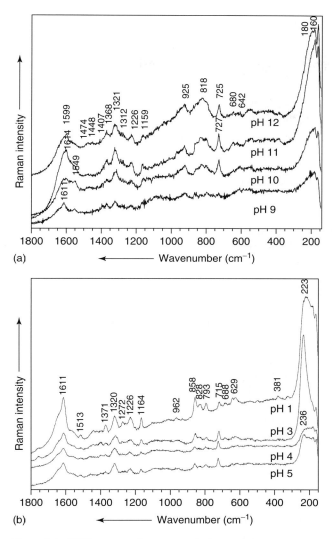

Figure 6.4 SERS spectra of paracetamol sinus solution on colloidal AgNPs at different (a) basic and (b) acid pH values. Excitation 514.5 nm.

surface and a possible chemisorption through the lone pair groups of oxygen (ν(Ag–O) stretching modes at 223 and 236 cm^{-1}). With the decrease in pH, the C=O stretching mode at 1611 cm^{-1} experienced an increase in intensity due to the interaction with the AgNP surface. The δ(O=C–N) deformation mode, which was detected at 925 cm^{-1} in the SERS spectrum at basic pH values, was absent in the SERS spectrum at acidic pH values. Furthermore, the ν(C–N) mode appeared as a distinct peak at 793 cm^{-1} only in acidic conditions. These changes suggested a reorientation of the skeletal plane of the molecule from predominantly planar (at basic and neutral pH values) to tilted on the AgNP surface (at acid pH values), so that the direct interaction through the oxygen atom is facilitated. In conclusion, the paracetamol molecule was found to be mainly chemisorbed through the π-electrons of the aromatic ring (at neutral pH), the oxygen (at acid pH) or the nitrogen (at basic pH) lone pair groups. The parallel to slightly tilted orientation of the molecule to the metallic nanosurface was also found to be pH dependent.

Peica et al. employed approach 5 and recorded the SERS spectra of buffered and unbuffered aspirin solutions on a sodium citrate-reduced Ag colloid [26]. Aspirin in tablet form is fat soluble in stomach acid and diffuses easily through the stomach lining into the cells underneath. Once there, aspirin encounters a much higher pH and ionizes, which prevents it from diffusing back into the stomach and causing bleeding. Adding buffering agents helps to avoid this. It should be noted that the authors were able to detect and differentiate between the two forms of aspirin tablet by simply performing micro-Raman measurements (in spite of the different excipients used in drug formulation). The aspirin tablet solutions (purchased from Europharm) were then prepared by sonication in water at about 50 °C and added to the aqueous colloid for SERS measurements. The SERS detection limit was of 5 and 8 μmol l^{-1} for the unbuffered and buffered species, respectively. The normal Raman spectra of the aspirin tablets in aqueous solutions were collected to help with the interpretation of the SERS spectra. The spectrum of the buffered species was very similar to the one reported by Wang et al. [16]. Both aspirin forms were found to be strongly chemisorbed to the AgNP surface. However, the interaction mechanism was slightly different. The strongest SERS bands of the unbuffered form were observed at 1387, 1635, 1262, 825 and 239 cm^{-1}, whereas for the buffered form the most intense peaks were located at 1614, 1367, 1302, 1239, 1026, 802 and 218 cm^{-1}. The SERS bands at 1387 and at 1367 cm^{-1} were attributed to the ν(COO$^-$) symmetric stretching mode of the unbuffered and buffered compounds, respectively. These signals were more intense in the SERS spectrum than in the normal Raman spectrum, which suggested that the COO$^-$ moiety was located in close proximity to the AgNP surface. A direct interaction through the oxygen atom of the C=O group was concluded on the basis of the significant shift observed in the SERS spectrum for the ν(C=O) stretching mode at 1635 and 1614 cm^{-1}, respectively, in comparison with the normal Raman spectrum. A closer examination of the spectral area characteristic of the ν(Ag–O) stretching modes indicated that both C=O groups (i.e. the ones belonging to the carboxylate and ester moieties) are involved in the interaction with the AgNP surface. Only one stretching mode was observed at 218 cm^{-1} for the buffered species. In conclusion, the buffered and unbuffered forms of aspirin were found to be slightly differently chemisorbed on the AgNP surface. Both pharmaceutical compounds could be detected in tablets using SERS at micromolar concentration level.

6.3
SERS of Antimalarials

Malaria is a vector-borne infectious disease caused by protozoan parasites of the genus *Plasmodium*. Five species of the plasmodium parasite can infect humans. The most serious forms of the disease are caused by *Plasmodium falciparum*. Each year, there are approximately 350–500 million cases of malaria, killing between 1 and 3 million people, the majority of whom are young children in Sub-Saharan Africa [27]. Antimalarial drugs are designed to prevent (i.e. prophylactic drugs) or cure (i.e. therapy drugs). There are many antimalarial drugs currently in the market. Quinine is the oldest antimalarial drug. However, quinine is less effective and more toxic as a blood schizonticidal agent than chloroquine. Chloroquine was until recently the most widely used antimalarial drug. It was the original prototype from which most other methods of treatment were derived. It is also the least expensive, best tested and safest of all available drugs. Chloroquine is also used in the treatment of rheumatoid arthritis and lupus-associated arthritis. Unfortunately, the emergence of drug-resistant parasitic strains is rapidly decreasing its effectiveness. Nevertheless, it is still the first-line drug of choice in most sub-Saharan African countries. Chloroquine has largely been replaced by mefloquine and it is now used in combination with other antimalarial drugs to extend its effective usage. Mefloquine is chemically related to quinine and was developed during the Vietnam War to protect American troops against multi-drug-resistant strains of *P. falciparum*. It is a very potent blood schizonticide with a long half-life, which is thought to contribute to the formation of toxic haem complexes that damage parasitic food vacuoles. Although both chloroquine and mefloquine can be used for prophylaxis and treatment of malaria, a combination therapy has become the standard for many regimes in overcoming drug resistance.

Raman [28, 29] and SERS [30] spectroscopy were recently employed to probe the interaction of antimalarial drugs quinine, chloroquine and mefloquine with haematin. The malarial parasite is known to degrade the haemoglobin of the infected human erythrocyte in a lyosomal food vacuole. The released Fe(II)haem is then rapidly oxidized to Fe(III)haematin and detoxified (i.e. the free haem is turned to an inert micron-sized malaria pigment called *haemozoin*) [31]. Therefore, the malarial parasite detoxifies the haematin by crystallizing it into insoluble and chemically inert β-haematin crystals (i.e. the haemozoin). Haemozoin is an excellent drug target because it is essential for the survival of the malarial parasite and absent from the human host. Crystal structure analysis of synthetic β-haematin suggests the inhibition of further haem aggregation through the adsorption of chloroquine to the haemozoin surface [30]. The interaction of antimalarial drugs (especially quinoline) with haematin has been intensively studied over the years using various techniques (e.g. X-ray diffraction, theoretical calculations and NMR spectroscopy) and different mechanisms have been invoked to account for the primary mode of action of quinoline [30–34]: (i) complexation with haematin in solution, which interferes with the crystallization process of haemozoin; (ii) enzymatic inhibition of a protein that catalyses haemozoin crystallization; and (iii) inhibition of malaria pigment growth by quinoline attachment to the growing crystal.

Polarization-resolved resonance Raman (RR) spectra of haematin and its complex with chloroquine in solution were recorded using Q-band resonance conditions in a rotating cell to avoid sample degradation [28]. A non-covalent interaction of the chloroquine–haem complex was concluded in the electronic ground state on the basis of the very small wavenumber shifts observed in spectrum upon drug addition (i.e. smaller than 2 cm^{-1}). Similar results were obtained for quinine and mefloquine [28]. However, when RR spectroscopy is employed to investigate the chromophore-containing species in the presence of a drug, the available information is selectively related to the structure of the chromophore and its possible perturbation caused by the drug. If the drug targets the nonchromophore part of the macromolecular structure, no information can be extracted using RR spectroscopy.

Cîntă-Pînzaru et al. utilized Raman and SERS spectroscopy to probe the chloroquine–haematin and mefloquine–haematin interaction mechanism. Colloidal AuNPs were prepared by the reduction of gold(III) chloride solution with sodium citrate solution [30] and employed as an SERS substrate based on their narrow size distribution, chemical stability and high biocompatibility. It has been already shown that the AuNPs can have SERS enhancement factors comparable to AgNP when NIR excitation is used [35]. In fact, a new broad band was detected at about 760 nm in the absorption spectrum of the chloroquine–haematin system adsorbed on AuNPs (Figure 6.5). This indicated a clear aggregation of the colloidal AuNPs

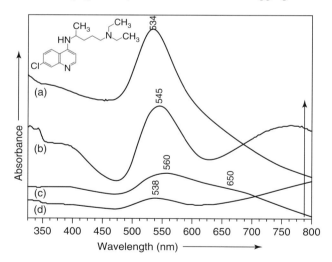

Figure 6.5 Absorption spectra of the (a) gold colloid, (b) hematin-chloroquine-AuNPs, (c) hematin-AuNPs, and (d) chloroquine-AuNPs system. The arrow indicates the NIR-excitation wavelength at 785 nm. The inset shows the molecular structure of chloroquine diphosphate salt.

in the presence of the drug–haematin complex and further supported the use of NIR excitation in the SERS study. In case of the mefloquine–haematin complex adsorbed on AuNPs, this band was small in intensity. As can be seen, the NIR excitation line (i.e. 785 nm from a diode laser) is pre-resonant with the very broad and weak absorption mode of haematin adsorbed on AuNPs (at about 650 nm in Figure 6.5). Therefore, RR scattering from the haematin can be considered negligible, so that the nature of the drug–haematin interaction can be interrogated under non-resonant conditions for haematin. The NIR SERS spectrum of chloroquine–haematin on AuNPs is presented in Figure 6.6 in comparison with those obtained for the pure haematin and chloroquine solutions on AuNPs. Many similarities were observed between the SERS spectrum of the chloroquine–haematin system on AuNPs and the SERS spectrum of the pure chloroquine on AuNPs, which suggested a preferential adsorption of chloroquine to the metallic surface. However, a complete surface coverage with adsorbed chloroquine species was excluded based on the contribution of haematin vibrational modes to the overall SERS pattern (Figure 6.6). The spectral changes observed in the SERS spectrum for several of the vibrational modes of haematin upon the addition of chloroquine solution were explained through an interaction of the two species. The peak found at 1533 cm^{-1} in the SERS spectrum of pure haematin was correlated with the porphyrin core size. This mode was detected at the same position in the SERS spectrum of the chloroquine–haematin complex, which showed that the pyrrole N atoms of the porphyrin macrocycle were not involved in the interaction with the drug. On the other side, the porphyrin skeletal mode located at 941 cm^{-1} presented a large increase in intensity. This

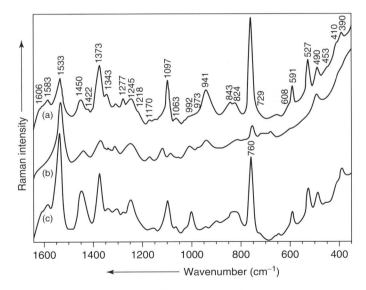

Figure 6.6 NIR-SERS spectra of (a) chloroquine-hematin, (b) hematin, and (c) chloroquine on colloidal AuNPs. Excitation line: 785 nm.

demonstrates the sensitivity of the haem core to the presence of chloroquine upon adsorption onto the AuNPs surface (as an axial π acceptor). No contributions due to the chloroquine aliphatic chain could be detected in the SERS spectrum of the haematin–chloroquine system adsorbed on AuNPs. A similar interaction through the quinoline part was concluded for the mefloquine drug. The pyrimidine moiety of mefloquine was found to be less involved in the interaction with haematin upon adsorption on the AuNPs surface.

Future NIR SERS studies could offer more insight into the *in vivo* mechanisms of malarial drug interactions and may help to design new therapeutic agents.

6.4
SERS of Anticarcinogenics and Antimutagenics

Anticarcinogenics are chemicals that counteract the effect of a cancer-causing agent by reducing its occurrence or severity and/or by acting against it. Antimutagenics are chemicals that reduce or interfere with the mutagenic effects of another substance. As many mutations cause cancer, mutagens are typically also carcinogens. Extensive efforts continue to be directed towards developing better anticancer and antimutagenic drugs that would not harm the healthy normal cells (i.e. targeted therapy). As a result, this drug development research has developed into a multibillion-dollar industry. The most common chemotherapeutic drugs include alkylating agents, antimetabolites, anthracyclines, plant alkaloids, topoisomerase inhibitors, and so on. All of these drugs affect cell division or DNA synthesis and function. The newer targeted agents do not directly interfere with DNA (e.g. monoclonal antibodies and tyrosine kinase inhibitors). There is epidemiological evidence that diets rich in antioxidant vitamins and flavanoids might help prevent or reduce the risk of cancer (e.g. vitamin D and the precursor of vitamin A: β-carotene). In fact, over 80% of the articles published over the past 20 years on anticarcinogens and antimutagens are on plant constituents consumed as foods or used for medical purposes. To this large class of bioactive compounds we can add phenolics, pigments, allylsulfides, glucosinolates, tannins, anthocyans, phytosterols, protease inhibitors and phytoestrogens [36]. In this chapter, we choose to report the SERS studies performed on one well-known substance from each category, namely, 5-fluorouracil (Section 6.4.1) and β-carotene (Section 6.4.2).

6.4.1
5-Fluorouracil

5-Fluorouracil is a fluorinated pyrimidine antimetabolite (Figure 6.7a, R = F) which is structurally similar to uracil (Figure 6.7a, R = H), one of the necessary building blocks in cellular division and growth. The well-known anticarcinogenic drug is mostly employed in the palliation of inoperable malignant neoplasms, especially those of the gastrointestinal tract, breast, liver and pancreas [37, 38]. 5-Fluorouracil drug solutions and creams are also used to treat human skin cancer

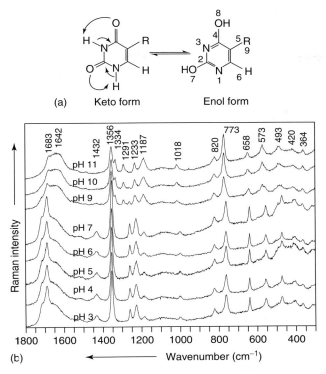

Figure 6.7 (a) The possible tautomeric forms of fluorouracil and the atoms labelling; (b) Raman spectra of 5-fluorouracil at neutral and acid pH values. Excitation line: 514.5 nm.

[38, 39]. All uracil compounds, in which the hydrogen bound to the C5 atom is substituted by halogen atoms (Figure 6.7a, R = F, Cl, Br, I), are presently tested against HIV [38, 40] and used as antitumour [38, 39] and antiviral [40] drugs. Recent Raman and SERS studies on 5-fluorouracil easily detected the presence of the chemotherapy drug and its metabolites at concentrations of 2 µg ml^{-1} in saliva and blood by using metal-doped sol–gels [11, 12]. The detection limit was estimated to be of about 150 ng ml^{-1} for an acquisition time of 5 min, and demonstrated that the method can successfully be used in pharmacokinetic studies to monitor or to regulate patient dosage. It should be noted that genetic-based variations in 5-fluorouracil metabolism can range by as much as fivefold from one patient to another. Although successful, this study was never concerned with the possible appearance of 5-fluorouracil resonance structures or tautomeric forms under various physiological conditions (e.g. change in pH value of the environment).

It is well known that the pK_a value of free 5-fluorouracil is about 7.8. This was proved by titration experiments monitored by the NMR resonance of the N3 proton of 5-fluorouracil [41]. But the possibility that the pK_a value of 5-fluorouracil may change with the environment cannot be ruled out. Conformational dependences of the pK_a value of a base are not uncommon. For example, 5-fluorouracil was found to form only wobble base pairs with guanine even under basic conditions (pH = 9) in the crystal lattice [42]. This is a very surprising result because, at pH 9, a significant portion of the molecule in solution should exist in the ionized form, which can pair with guanine using two H-bonds, in the well-known Watson–Crick geometry [42]. Such changes in the stacking geometry could be connected with the mutagenic action of 5-fluorouracil. Furthermore, the specificity of the hydrogen bonding plays a central role in the transmission of genetic information, in the production of messenger RNA as well as in the polymerization of amino acids to form proteins. Thus, there is considerable interest in possible conformational changes with the pH value or environment and in the way 5-fluorouracil molecules combine.

To examine these aspects, Pavel *et al.* recorded the Raman (10^{-1} mol l^{-1}) and SERS spectra (10^{-3} mol l^{-1}) of 5-fluorouracil solution using an aqueous Ag colloid at different pH values (Figures 6.7–6.9) and discussed them with the help of DFT calculations [43]. The DFT calculation results helped to establish for the first time the most stable resonance structure for each of the tautomeric forms of 5-fluorouracil (i.e. two enol and two enolate forms – Figures 6.7a, 6.8b and 6.9b). It is well known that most carbonyl compounds exist almost exclusively in their keto form at equilibrium and it is usually difficult to isolate the pure enol. Both acids and bases catalyse the keto–enol tautomerism.

The protonation of the carbonyl bond occurred at acidic pH values. The oxygen atom acts as a Lewis base taking the H-atom from the hydrochloric acid and regenerating an intermediate cation that can be represented by two resonance forms. The most stable intermediate cation loses its proton and leads to the enol form (Figure 6.8b). The geometry optimization of the two enol forms performed at the B3LYP/6-311 + G(d) level of theory showed that the species protonated at O7 are more stable than the ones protonated at the O8 atom. This was found to be due to a larger dislocation of the negative charge on the first compound. The presence of the in-plane and out-of-plane δ(N3–H) deformation modes at 1187 and 658 cm^{-1}, respectively, in the normal Raman spectrum of the 5-fluorouracil solution confirmed this conclusion (Figure 6.7b). The first enol form could also be present at these pH values, but in a very low concentration. This is due to the fact that the in-plane and out-of-plane deformation modes of the N1–H group were either absent or weak in the normal Raman spectrum. Moreover, the carbonyl spectral region was dominated by a very strong peak at about 1689 cm^{-1}, which is due to the ν(C4=O8) mode. Because of the protonation at the O7 atom, the ν(C2=O7) stretching is more constrained than the corresponding motion of the C4=O8 bond and, consequently, it should have a much lower intensity. Indeed, the ν(C2=O7) mode appeared as a weak shoulder at higher wavenumbers (acid pH values).

At basic pH values, the base (i.e. the hydroxyl group) deprotonated the N1–H bond, leading to an intermediate anion – an enolate ion – that was again represented by two resonance structures (Figure 6.9b). The Na$^+$ addition to the oxygen atom yields the neutral enolate. The geometry optimization of the two deprotonated forms indicated the N1-deprotonated form as being the most stable

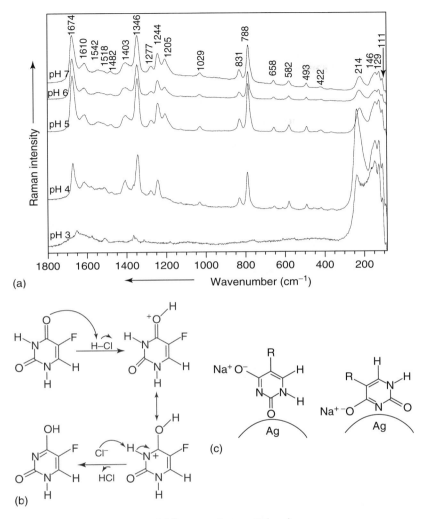

Figure 6.8 (a) SERS spectra of 5-fluorouracil at neutral and acid pH values. Excitation line: 514.5 nm; (b) Protonation reaction of 5-fluorouracil; (c) Possible orientations of the most stable N3-deprotonated form of 5-flurouracil to the AgNP surface at acid pH values.

one by 12.6 kJ mol^{-1}. This energy difference was explained by the fact that the deprotonation at the N3 atom occurs very quickly and leads to a kinetically controlled product, while the deprotonation at the N1 atom occurs slowly leading to a thermodynamically controlled product. The normal Raman spectra allowed experimentally distinguishing between the two enolate forms once more. The ν(C5–F9) stretching and the ring stretching modes at 1233 and 1258 cm^{-1}, which are strongly coupled with the δ(N1–H) deformation mode, decreased significantly in intensity

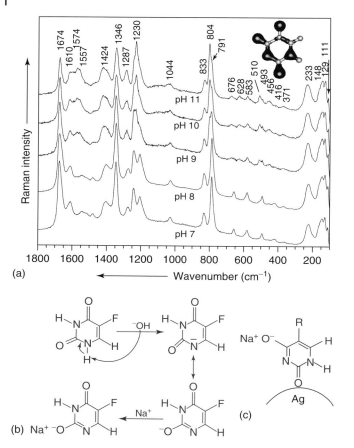

Figure 6.9 (a) SERS spectra of 5-fluorouracil on colloidal AgNPs at neutral and basic pH values. Excitation line: 514.5 nm. The total electron density of the N3-deprotonated form of 5-fluorouracil is inserted (isosurface value: 0.27); (b) Deprotonation reaction of 5-fluorouracil; (c) Orientation of the most stable N3-deprotonated form of 5-fluorouracil to the AgNP surface at neutral and basic pH values.

or disappeared on passing from acid to basic pH values (Figure 6.7b). The δ(N3–H) bending mode at 1187 cm^{-1} presented the opposite behaviour. Moreover, two new signals were detected at 1334 and 1291 cm^{-1}, which were assigned to simultaneous vibrations of the N3–H moiety and the ring. The band at 1432 cm^{-1}, which mainly results from a δ(N1–H) bending mode slightly coupled with a δ(N3–H) deformation and a ring stretching, decreased very much in intensity with the increase in the pH value. Although the ν(CO) stretching modes at these pH values were not well resolved, their shift to smaller wavenumbers and decrease in relative intensity confirmed the Na$^+$ addition at the oxygen atom.

The DFT calculations performed at the B3LYP/LANL2DZ level of theory helped to identify the most stable tautomers and the AgNP-adsorbed species of 5-fluorouracil.

The N1-deprotonated form was determined to be more stable than the N3 form in gaseous phase and solution, but for the AgNP complexes the energy levels were reversed, that is, the second tautomer became more stable. This was explained through a smaller delocalization of the negative charge and a higher dipole moment for the N3-deprotonated form. The SERS spectra supported the DFT results.

A closer examination of the normal Raman (Figure 6.7b) and SERS spectra (Figure 6.9a) at neutral and basic pH showed major differences in band positions and their relative intensities, which indicated the formation of AgNP–5-fluorouracil species. Almost all bands present in the SERS spectrum at neutral pH could also be detected at basic pH values. This fact demonstrated that 5-fluorouracil was adsorbed to the AgNP surface in its deprotonated form even at neutral pH. At basic pH values, some of the peaks that were present at neutral pH disappeared and some new ones emerged. The most likely explanation for the additional bands at high pH is the coexistence of both deprotonated forms. As is shown below, the N3-deprotonated form was found to prevail. The SERS spectra at alkaline pH values were dominated by the peaks at 1674, 1346, 1230 and 804 cm^{-1}. The most significant changes were seen for the bands at 1674 and 1230 cm^{-1}, which presented a significant enhancement in comparison with their relative intensities in the normal Raman spectra at the same pH values (Figures 6.7b and 6.9a). The peak at 1230 cm^{-1} corresponds to the C5–F9 stretching vibration strongly coupled with an N1–H deformation, which indicates the presence of the N3-deprotonated form. This was further confirmed by the broad band at 1424 cm^{-1}, which is mostly due to a δ(N1–H) bending mode slightly coupled with an N3–H deformation and a ring stretching. However, the presence of the second N1-deprotonated form at high pH values could not be excluded because the in-plane deformation mode of the N3–H group also appeared in the SERS spectra as a weak red shoulder to the very strong signal at 1230 cm^{-1}. With the decrease in pH value, the shoulder developed into a clear peak of medium relative intensity (1208 cm^{-1} at pH 7), indicating a considerable increase in the concentration of the second N1-deprotonated form.

The orientation of the 5-fluorouracil molecules on the AgNP surface was then determined with the help of SERS surface selection rules [1–4]. These state that vibrations deriving their intensities from a large value of α_{zz} (the z axis parallel to the surface normal) would be the most intense in the SERS spectrum when the molecule has a perpendicular orientation to the metal surface [44, 45]. The very strong bands at 1674 and 804 cm^{-1} were attributed to the C4=O8 stretching and out-of-plane deformation modes, respectively. Taking into account that at these pH values, the N3-deprotonated form prevailed; a strong adsorption probably through the O7 atom in a perpendicular orientation to the surface or an orientation not significantly tilted from the surface normal would explain the observed enhancement. The orientation was further substantiated by the very strong ring stretching vibration at 1350 cm^{-1} and the C5–F9 stretching mode at 1230 cm^{-1}. According to the SERS surface selection rules, these vibrations should be the most prominent bands in the SERS spectrum when the molecule stands upright on the surface (Figure 6.9c).

In most SERS studies, precise determination of the adsorption site on the AgNP surface is fraught with problems. From the chemical structure, one can observe that the deprotonated forms of 5-fluorouracil have three atoms, namely, the two O atoms and one of the N atoms, capable of forming bonds with the AgNP surface, while in the protonated forms both N atoms and one O atom are available (Figures 6.8b and 6.9b). The authors treated this problem by enumerating the negative charge density on each of these possible active sites. The higher the negative charge density on the atom, the higher is its probability to act as an adsorptive site for the Ag substrate. The calculated plot of the total electron density for the N3-deprotonated form (Figure 6.9a inset) indicated a build-up of charge density on the fluorine (slightly charged), oxygen and nitrogen atoms, as well as nodes at the other atoms. The natural population analysis performed at the B3LYP/6-311 + G(d) level of theory for the dominant N3-deprotonated tautomer showed that the highest negative charge densities are located on the O7 and O8 atoms (i.e. -0.732 and of -0.685 (e), respectively). The negative charge density on the N3 atom was very close to that on the O8 atom. Therefore, it is likely that all three atoms (i.e. O7, O8 and N3) can be active sites for the adsorption process, but the O7 atom has a slightly higher probability. This confirmed the interaction site (i.e. O7 atom) with the AgNP surface. Furthermore, a broad band was detected in the SERS spectra at 233 cm^{-1} and assigned to the ν(Ag–O) stretching mode.

Most of the bands observed at neutral pH were also present at smaller pH, providing evidence that 5-fluorouracil was adsorbed to the AgNP surface in its deprotonated forms even at acidic pH. The pK_a value for the deprotonation at the N atom was changed in the presence of the Ag colloid and generated the enolate forms even at acidic pH values. These enolate forms can replace the citrate ions and interact with the positively charged Ag surface via the negatively charged N atom or the lone pair of the oxygen atoms. The positive charges on the colloidal surface might have lowered the pK_a value of 5-fluorouracil by the coulombic stabilization of the deprotonated anion. As one can notice from the SERS spectra (Figure 6.8a), only a few peaks disappeared or changed their intensity at acidic pH values, which indicated a slight modification in the concentration of one of the two tautomeric forms. More exactly, the band at 1205 cm^{-1}, which corresponds to the δ(N3H) bending mode, disappeared at smaller pH values. Therefore, the concentration of the N3 deprotonated form increased at lower acid pH values. This observation was supported by the change in out-of-plane δ(N3–H) deformation mode at 658 cm^{-1}, whose relative intensity dropped in the SERS spectra in comparison to that in the normal Raman spectrum at the same pH values. Moreover, the broad band at 1403 cm^{-1} (pH 7), which mainly results from a N1–H bending mode slightly coupled with a N3–H deformation and a ring stretching, increased in intensity with the decrease in the pH value. All this indicated the N3-deprotaned tautomer as being the dominant form in these conditions.

At acidic pH values, three very strong bands were observed in the SERS spectra, namely, at 1672, 1344 and 789 cm^{-1}. The ring stretching mode at 1344 cm^{-1} was again a clear indication for the upright orientation of the molecule on the AgNP surface. The other two peaks at 1672 and 789 cm^{-1} correspond to the C4=O8 stretching and out-of-plane deformation modes. According to the SERS surface selection rules, the carbonyl group should be close to the AgNP surface and should have a large Raman polarizability component perpendicular to the surface in order to present such an enhancement. The chemisorption through the N3 or O7 atom (Figure 6.8c) would obviously lead to such spectral changes. This assumption was further supported by the calculated negative charge densities on the nitrogen and oxygen atoms in the N3 deprotonated form, as well as the ν(Ag–O) and ν(Ag–N) stretching modes observed at about 214 cm^{-1} in the SERS spectra at pH 7, 6 and 5.

In conclusion, the SERS surface selection rules, in combination with the DFT results, indicated the chemisoption of both enolate forms to the AgNP surface in a perpendicular orientation or significantly tilted from the surface normal at all pH values. The N3-deprotonated form was the dominant tautomer in the adsorbate state, most probably attached through the O7 atom (at all pH values) or N3 atom (at acidic pH) to the metal surface.

Pavel *et al.* showed in a recent study that 5-fluorouracil undergoes a similar soft adsorption (i.e. no significant structural changes) on the metal surface of different Ag island films prepared by metal vapour deposition at room temperature and under high vacuum conditions (35, 11, 8.7 and 6.5 nm thick) [46]. Microlitre droplets of an aqueous 5-fluorouracil solution (10^{-3} mol l^{-1}) were deposited on the Ag island films and allowed to air dry at room temperature. Subsequent washing in deionized water removed the excess or non-binding agents. The Ag island films were imaged before and after the analyte deposition with atomic force microscopy. The 5-fluorouracil spots showed no changes after washing, indicating that the analyte solution mediated a stabilization of the film during drying. The 35 nm thick film provided the best SERS signal under non-resonant conditions using 514.5 nm laser excitation as for colloidal AgNPs (Figure 6.10a). The addition of 5-fluorouracil solution to the Ag colloid or Ag films left unchanged the absorption maxima of the two SERS substrates (i.e. at 408 and 664 nm). These absorption maxima correspond to the excitation of the transverse collective electron resonances of the metal particles. Both deprotonated forms, which were previously detected on colloidal AgNPs at neutral pH, were also present in this case. However, the interaction mechanism of 5-fluorouracil species was found to be substrate dependent. As can be noticed from Figure 6.10b, the SERS spectrum obtained on the Ag film differs significantly from the one recorded on the Ag colloid (major peak broadening and wavenumber shifts). A different SERS spectral pattern may be the result of changes in the molecular identity and/or a different bonding geometry of the 5-fluorouracil species on the Ag film. A detailed analysis of the SERS spectra excluded the first option [46]. The analyte molecules likely lie flat on the much larger Ag clusters or are significantly tilted form the surface normal. The strong chemisorption took place through the carbonyl groups and the π-electrons (Figure 6.10b inset). The soft adsorption of the anticarcinogenic drug to the metallic surface and the resistance of the Ag-complexed species to washing demonstrated the advantage of films over colloids and makes them very attractive for future pharmaceutical applications where washing is a necessity.

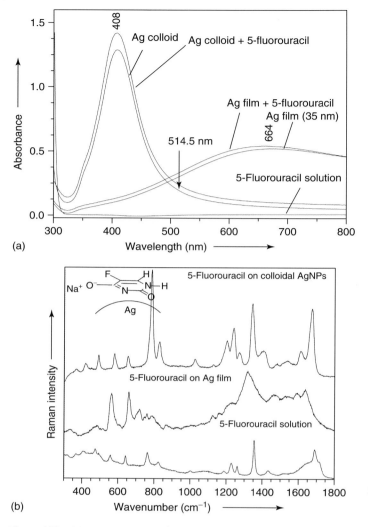

Figure 6.10 (a) UV-VIS spectra of 5-fluorouracil aqueous solution, Ag colloid, Ag film (35 nm thick), and 5-fluorouracil solution on the mentioned SERS substrates; (b) Ordinary Raman of 5-fluorouracil solution at neutral pH (10^{-1} mol/L) and its SERS spectra (10^{-3} mol/L) on two different substrates: Ag film and Ag colloid. Excitation line: 514.5 nm. The inset shows the proposed orientation of the N3-deprotonated form 5-fluorouracil to the Ag film at neutral pH.

6.4.2
β-Carotene

Similar SERS substrates were used by Sokolov *et al.* to record the SERS spectra of several types of biomolecules, including β-carotene, at concentrations of

10^{-7} mol l^{-1} without modifying the native conformation of the analyte molecules [47]. Ag island films of various thicknesses (between 30 and 200 Å) were prepared by metal deposition on glass slides and had extinction maxima in the 450–660 nm region of the electromagnetic spectrum. It should be noted that the authors could not obtain a pronounced Raman signal from the β-carotene solution deposited on glass slides under the experimental conditions used for SERS measurements. The Ag island films with extinction maxima near 520 nm provided the largest SERS enhancement factor (i.e. the films for which the distance between the silver nanoaggregates did not exceed their size). Significant differences such as upshifts in band position and changes in relative intensity were observed in the SERS spectra of β-carotene (10^{-7} mol l^{-1}) adsorbed directly on the Ag island films as compared to the RR spectra of its solution (2×10^{-6} mol l^{-1}). The absorption maximum of the lowest energy transition in β-carotene was near 480 nm. Therefore, the SERS spectrum of β-carotene was obtained in pre-resonance conditions with the 514.5 nm excitation line. These differences between the SERS and RR spectra vanished when β-carotene molecules were spaced away from the surface of the Ag film. Only a small decrease in the signal intensity of about 35% was observed when the β-carotene molecules were removed from the Ag surface at a distance of 5 nm using two monolayers of stearic acid molecules. This suggested that a short-range enhancement mechanism – though present to some extent when β-carotene is adsorbed directly on the surface – did not play a major role in the SERS enhancement mechanism for β-carotene. Rather, the enhancement mechanism is primarily long-range electromagnetic in origin. Therefore, this study has demonstrated that the appearance of chemical or electromagnetic enhancement mechanism depends not only on the nature of the adsorbed drug but also on its electronic absorption properties.

It is worth mentioning the two techniques used for the sample deposition onto Ag films in this study. In the first method, the SERS-active film was placed in a quartz cuvette containing the sample solution, and the SERS spectrum was collected directly by focussing the laser beam on the film. This technique was effective for samples with a comparatively low quantum yield of fluorescence in the visible region (e.g. β-carotene). For samples with a high quantum yield of fluorescence in the visible region (e.g. doxorubicin), the first 'wet method' was ineffective due to the strong interference from the fluorescence of the non-adsorbed molecules in the sample solution. In this case, the 'dry method' of sample deposition was used. A drop of the sample solution was deposited on the island film and allowed to adsorb for 2–3 min. The non-adsorbed species were then removed by using a filter paper. Under these conditions, the fluorescence from the adsorbed molecules was effectively quenched though the complexation with the metal surface. The Ag island films covered with monolayers of stearic acid molecules were prepared according to the following steps: (i) monolayers of stearic acid were produced on a subphase of pure water and (ii) these were then transferred to Ag-coated glass slides at a surface pressure of a 3×10^{-2} N m^{-2} by standard Langmuir–Blodgett transfer techniques.

6.5
Concluding Remarks

The examples provided in this chapter demonstrate SERS as a very promising analytical technique in the pharmaceutical industry. The fast, label-free and non-invasive nature of the SERS-based technique together with its high molecular specificity and sensitivity makes SERS highly suitable for quality control (QC) processes of pharmaceuticals such as QC identification, final goods QC, trace impurity detection and in-line process control.

References

1. Cîntă-Pînzaru, S., Pavel, I., Leopold, N. and Kiefer, W. (2004) Identification and characterization of pharmaceuticals using Raman and surface-enhanced Raman scattering. *J. Raman Spectrosc.*, **35**, 338–346.
2. Williams, A.C. (2001) Handbook of Raman spectroscopy: from the research laboratory to the process line, in *Practical Spectroscopy Series* (eds I.R. Lewis, G. Howell and M. Edwards), Marcel Dekker, New York, pp. 575–592.
3. Asher, S.A. and Johnson, C.R. (1984) Raman spectroscopy of a coal liquid shows that fluorescence interference is minimized with ultraviolet excitation. *Science*, **225**, 311–313.
4. Chalmers, J.M. and Griffiths, P.R. (2002) *Handbook of Vibrational Spectroscopy*, John Wiley & Sons, Ltd, Chichester Baffins Lane, UK.
5. Nie, S.M. and Emery, S.R. (1997) Probing single molecules and single nanoparticles by surface-enhanced Raman scattering. *Science*, **275**, 1102–1106.
6. Wooley, P.S., Keely, B.J. and Hester, R.E. (1996) Surface-enhanced resonance Raman spectra of water-insoluble tetraphenylporphyrin and chlorophyll a on silver hydrosols with a dioxane molecular spacer. *Chem. Phys. Lett.*, **258**, 501–506.
7. Pavel, I., Cîntă, S., Venter, M., Deak, A., Haiduc, I., Rösch, P., Cozar, O., Iliescu, T. and Kiefer, W. (2000) Vibrational behavior of transition metal cupferronato complexes. Raman and SERS studies on nickel(II) cupferronato complexes. *Vib. Spectrosc.*, **23**, 71–76.
8. Huang, L., Tang, F., Shen, J., Hu, B., Meng, Q. and Yu, T. (2001) A simple method for measuring the SERS spectra of water-insoluble organic compounds. *Vib. Spectrosc.*, **26**, 15–22.
9. Gardenas-Trivino, G., Vera, V. and Muňoz, C. (1998) Silver colloids from nonaqueous solvent. *Mater. Res. Bull.*, **33**, 645–653.
10. Han, M.Y., Quek, C.H., Huang, W., Chew, C.H. and Gan, L.M. (1999) A simple and effective chemical route for the preparation of uniform nonaqueous gold colloids. *Chem. Mater.*, **11**, 1144–1147.
11. Gift, A., Shende, C., Inscore, F.E., Maksymiuk, P. and Farquharson, S. (2004) Five-minute analysis of chemotherapy drugs and metabolites in saliva: evaluating dosage. *Proc. SPIE Int. Soc. Opt. Eng.*, **5261**, 135–141.
12. Farquharson, S., Inscore, F.E., Gift, A.D. and Shende, C.S. (2006) Method and apparatus for rapid extraction and analysis, by surface-enhanced Raman spectroscopy (SERS), of drugs in saliva. US Patent Appl. Publ., U.S. Ser. No. 967,486, p. 14.
13. Li, Y.S., Vo-Dinh, T., Stokes, D.L. and Wang, Y. (1992) Surface-enhances Raman analysis of *p*-nitroaniline on vacuum evaporation and chemically deposited silver-coated alumina substrates. *Appl. Spectrosc.*, **46**, 1354–1357.
14. Taguenang, J.M., Kassu, A., Sharma, A. and Diggs, D. (2007) Surface enhanced

Raman spectroscopy on the tip of a plastic optical fiber. *Proc. SPIE Int. Soc. Opt. Eng.*, **6641**, 66411X/1–66411X/6.

15. Smith, M., Stambaugh, K., Smith, L., Son, H.J., Gardner, A., Cordova, S., Posey, K., Perry, D. and Biris, A.S. (2009) Surface-enhanced vibrational investigation of adsorbed analgesics. *Vib. Spectrosc.*, **49**, 288–297.

16. Wang, Y., Li, Y.-S., Zhang, Z. and An, D. (2003) Surface-enhanced Raman scattering of some water insoluble drugs in silver hydrosols. *Spectrochim. Acta Part A*, **59**, 589–594.

17. Andronie, L., Cîntă-Pînzaru, S. and Cozar, O. (2009) The paracetamol adsorption behavior monitored by Raman and surface-enhanced Raman spectroscopy. *AIP Conference Proceedings, Mathematics Physics and Applications*, vol. 1131, pp. 191–197.

18. Huba, K. and Istvan, A. (2006) Drug excipients. *Curr. Med. Chem.*, **13**, 2535–2563.

19. Lee, Y.-H., Farquharson, S., Kwon, H., Shahriari, M. and Rainey, P. (1999) Sol–gel chemical sensors for surface-enhanced Raman spectroscopy. *Proc. SPIE Int. Soc. Opt. Eng.*, **3537**, 252–260.

20. Farquharson, S. and Maksymiuk, P. (2003) Simultaneous chemical separation and surface-enhanced Raman spectral detection using silver-doped sol–gels. *Appl. Spectrosc.*, **57**, 357–482.

21. Farquharson, S., Gift, A.D., Maksymiuk, P. and Inscore, F.E. (2004) Rapid dipicolinic acid extraction from *Bacillus* spores detected by surface-enhanced Raman spectroscopy. *Appl. Spectrosc.*, **58**, 351–354.

22. Lee, P.C. and Meisel, D. (1982) Adsorption and surface-enhanced Raman of dyes on silver and gold sols. *J. Phys. Chem.*, **86**, 3391–3395.

23. Thorley, F.C., Baldwin, K.J., Lee, D.C. and Batchelder, D.N. (2006) Dependence of the Raman spectra of drug substances upon laser excitation wavelength. *J. Raman Spectrosc.*, **37**, 335–341.

24. Binev, I.G., Vassileva-Boyadjieva, P. and Binev, Y.I. (1998) Experimental and ab initio MO studies on the IR spectra and structure of 4-hydroxyacetanilide (paracetamol), its oxyanion and dianion. *J. Mol. Struct.*, **447**, 235–246.

25. Kumazawa, T., Seno, H., Lee, X.-P., Ishii, A., Watanabe-Suzuki, K., Sato, K. and Suzuki, O. (1999) Extraction of methylxanthines from human body fluids by solid-phase microextraction. *Anal. Chim. Acta*, **387**, 53–60.

26. Peica, N., Andronie, L.M., Cîntă Pînzaru, S. and Kiefer, W. (2004) Buffered *versus* unbuffered aspirin species monitored by Raman and surface-enhanced Raman spectroscopy. *Studia Univ. Babes-Bolyai – Phys.*, 3 XLIX, 57–62.

27. Snow, R.W., Guerra, C.A., Noor, A.M., Myint, H.Y. and Hay, S.I. (2005) The global distribution of clinical episodes of *Plasmodium falciparum* malaria. *Nature*, **434**, 214–217.

28. Frosch, T., Küstner, B., Schlücker, S., Szeghalmi, A., Schmitt, M., Kiefer, W. and Popp, J. (2004) In vitro polarization-resolved resonance Raman studies of the interaction of hematin with the antimalarial drug chloroquine. *J. Raman Spesctrosc.*, **35**, 819–821.

29. Wood, R.B., Langford, S.J., Cooke, B.M., Lim, J., Glenister, F.K., Duriska, M., Unthank, J.K. and McNaughton, D. (2004) Resonance Raman spectroscopy reveals new insight into the electronic Structure of β-Hematin and malaria pigment. *J. Am. Chem. Soc.*, **126**, 9233–9239.

30. Cîntă-Pînzaru, S., Peica, N., Küstner, B., Schlücker, S., Schmitt, M., Frosch, T., Faber, J.H., Bringmann, G. and Popp, J. (2006) FT-Raman and NIR-SERS characterization of the antimalarial drugs chloroquine and mefloquine and their interaction with hematin. *J. Raman Spectrosc.*, **37**, 326–334.

31. Ridley, R.G. (2002) Medical need, scientific opportunity and the drive for antimalarial drugs. *Nature*, **415**, 686–693.

32. Buller, R., Peterson, M.L., Almarsson, R.N. and Leiserowitz, L. (2002) Quinoline biding site on malaria pigment crystal: a rational pathway for antimalarial drug design. *Cryst. Growth Des.*, **2**, 553–562.

33. Leed, A., DuBay, K., Ursos, L.M.B., Sears, D. and de Dios, A.C. (2002) Solution structures of antimalarial drug-heme complexes. *Biochemistry*, **41**, 10245–10255.
34. Pagola, S., Stephens, P.W., Bohle, D.S., Kosar, A.D. and Madsen, S.K. (2000) The structure of malaria pigment beta-haematin. *Nature*, **404**, 307–310.
35. Kneipp, K., Kneipp, H., Itzkan, I., Dasari, R.R. and Feld, M.S. (1999) Surface-enhanced Raman scattering: a new tool for biomedical spectroscopy. *Curr. Sci.*, **77**, 915–924.
36. Knasmüller, S., Steinkellner, H., Majer, B.J., Nobis, E.C., Scharf, G. and Kassie, F. (2002) Search for dietary antimutagens and anticarcinogens: methodological aspects and extrapolation problems. *Food Chem. Toxicol.*, **40**, 1051–1062.
37. Graham, M.A., Lockwood, G.W., Greenslade, D., Brienza, S., Bayssas, M. and Gamelin, E. (2000) Clinical pharmacokinetics of oxaliplatin: A critical review. *Clin. Cancer Res.*, **6**, 1205–1218.
38. Goodman, L.A. and Gilman, A. (1996) *The Pharmacological Basis of Therapeutics*, 9th edn, MacGraw-Hill, New York.
39. Dorman, D.C., Coddington, K.A. and Richardson, R.C. (1990) 5-Fluorouracil toxicosis in the doc. *J. Vet. Intern. Med.*, **4**, 254–257.
40. Kim, H.O., Ahn, S.K., Alves, A.I., Beach, I.W., Jeong, L.S., Choi, B.G., Roey, P.V., Schinazi, R.F. and Chu, C.K. (1992) Asymmetric synthesis of 1,3-dioxolane-pyrimidine nucleosides and their anti-HIV activity. *J. Med. Chem.*, **35**, 1987–1995.
41. Seela, F. and Muth, H.-P. (1988) Synthese von 7-Desaza-2′,3′-didesoxyguanosin durch Desoxygenierung seines 2′-Desoxy-β-D-ribofuranosids. *Liebigs Ann. Chem.*, **3** 215–219.
42. Coll, M., Saal, D., Frederick, C.A., Aymami, J., Rich, A. and Wang, A.H.-J. (1989) Effects of 5-fluorouracil/guanine wobble base pairs in Z-DNA: molecular and crystal structure of d(CGCGFG). *Nucleic Acids Res.*, **17**, 911–923.
43. Pavel, I., Cota, S., Cîntă-Pînzaru, S. and Kiefer, W. (2005) Raman, surface enhanced Raman spectroscopy, and DFT calculations: a powerful approach for the identification and characterization of 5-fluorouracil anticarcinogenic drug species. *J. Phys. Chem. A*, **109**, 9945–9952.
44. Creighton, J.A. (1988) in *Spectroscopy of Surface* (eds R.J.H. Clark and R.E. Hester), John Wiley & Sons, Inc., New York.
45. Moskovits, M. (1982) Surface selection rules. *J. Chem. Phys.*, **77**, 4408–4416.
46. Pavel, I., Cota, S., Cîntă-Pînzaru, S. and Kiefer, W. (2006) SERS substrate-dependent interaction of the anticarcinogenic drug 5-fluorouracil with silver. *Part. Sci. Technol.*, **24**, 301–309.
47. Sokolov, K., Khodorchenko, P., Petukhov, A., Nabiev, I., Chumanov, G. and Cotton, T.M. (1993) Contributions of short-range and classical electromagnetic mechanisms to surface-enhanced Raman scattering from several types of biomolecules adsorbed on cold-deposited island films. *Appl. Spectrosc.*, **47**, 515–522.

7
SERS and Separation Science
Alison J. Hobro and Bernhard Lendl

7.1
Introduction

Analysis of samples from such diverse fields as medicine and biology, environmental and pollution science or food analysis can all require a separation step to isolate the compounds of interest from the usually complex matrix they are present in. Electrophoresis- and chromatography-based separation techniques include gas and liquid chromatography (LC) as well as capillary zone electrophoresis and electrokinetic chromatography, all of which are routinely used to separate different analytes in a sample, the particular choice of techniques being dependent on the nature of the given analytical problem. Often, these techniques are coupled to detection systems such as UV absorption or fluorescence emission spectroscopy, which, while being highly sensitive, provide little molecule-specific information. Such information can be provided through hyphenation to mass spectrometry or vibrational spectroscopic techniques such as infrared absorption and Raman scattering. However, this can be problematic, as infrared detection can be severely affected by strong solvent absorption and conventional Raman spectroscopy does not always provide sufficiently strong signals especially considering online detection in these hyphenated systems.

The enhancement factors generally associated with SERS, enhancements 10^2-10^6 and reports of single-molecule detection [1, 2], make SERS an appealing technique to couple with conventional separation techniques. SERS provides many of the advantages of conventional Raman spectroscopy, especially the information-rich spectra for identification and potential quantification of samples, while compensating for the weak signals generally associated with Raman scattering. Here we discuss a number of conventional electrophoresis- and chromatography-based separation techniques, namely, capillary electrophoresis (CE), LC, gas chromatography (GC) and thin layer chromatography, in the context of their hyphenation with SERS. We end with three more unusual applications of separation science coupled to SERS.

There are several issues that must be taken into account when coupling SERS to the various separation techniques. In many cases, the separation step requires

Surface Enhanced Raman Spectroscopy: Analytical, Biophysical and Life Science Applications. Edited by Sebastian Schlücker
Copyright © 2011 WILEY-VCH Verlag GmbH & Co. KGaA, Weinheim
ISBN: 978-3-527-32567-2

an interaction of the analyte mixture with a stationary phase, or controlled use of electro-osmotic flow. In order for the separation and subsequent SERS detection to be effective, surfaces involved in the separation step (capillaries, stationary phase, etc.) must be clean and this usually means that separation should be independent of the detection step to prevent any contamination of the separation step apparatus with the SERS-active media. The isolation of the separation and detection steps is usually achieved by combining the analyte and SERS-active media after the separation step. Separation processes also produce transient peaks and it is important that retention of separate analytes on the same part of the SERS surface is avoided, something that can be overcome by spotting of the analyte on fresh areas of the SERS surfaces through the use of protective coatings or constant addition of SERS-active colloids to the separated samples.

Many separation methods involve flowing streams (e.g. mobile phases in capillaries; see also Chapter 8) and the effects this has on SERS as a detection method was explored by Pothier and Force [3, 4] in the context of potential coupling to such separation systems. Their system, consisting of three electrodes (a 4 mm outer diameter silver working electrode, a platinum foil auxiliary electrode and a saturated calomel electrode), is shown in Figure 7.1. Their initial experiments, using pyridine as a test substrate, showed that, through the manipulation of the various electrode potentials, they could measure SERS spectra of their analyte

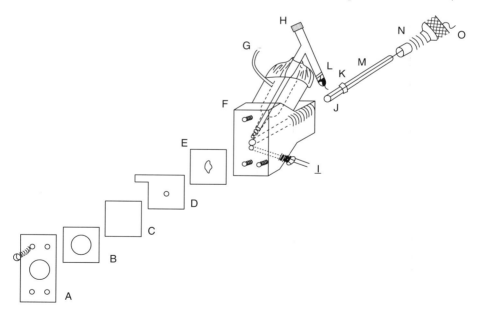

Figure 7.1 SERS flow cell of 30 µl: (A) metal window frame; (B) rubber gasket; (C) 2 cm × 2 cm × 1 mm glass window; (D) platinum foil auxiliary electrode; (E) 30 µl gasket; (F) Plexiglas cell; (G) exit tubing; (H) calomel reference electrode; (I) inlet tubing; (J) silver working electrode; (K) rubber O-ring; (L) silver wire lead; (M) 4 mm glass tubing; (N) knurled nut and (O) copper lead [3].

and then remove it from the electrode in less than 10 s, ready for the next SERS spectrum collection. Moving on to DNA bases, they found that a more cathodic potential was required to effectively remove all the analytes from the working electrode in between SERS measurements [3], and this was linked to the SERS intensity of the individual analytes [4]. Under stopped-flow conditions, their detection limits were 175, 233 and 211 pmol for adenine, thymine and cytosine, respectively [3].

7.2
SERS and Capillary Electrophoresis (CE)

Most capillary electrophoresis SERS (CE–SERS) systems are off-line methods where the eluted substances are deposited onto a SERS-active surface for analysis at a later point in time. However, one of the earliest reports of CE–SERS [5] was for online analysis and used running buffers containing silver colloids as the SERS substrate. A measurement window was introduced into the capillary through removal of the polyimide coating, approximately 15 cm from the inlet end, and SERS spectra were collected through this window using a confocal Raman microspectrometer. Careful optimization of the experimental parameters allowed detection of 5×10^{-6} M riboflavin and 1×10^{-8} M rhodamine 6G solutions, as shown in Figure 7.2. It was noted that the quality of the SERS signals obtained could be affected by the laser power used, suggesting that high laser powers may lead to degradation of the analyte or the SERS substrate, and that the capillary required cleaning between experimental runs to avoid the build-up of colloid on the capillary wall, which also contributed to signal degradation [5].

The first off-line CE–SERS measurements were reported in the same year by He et al. [6]. Here, an SERS substrate is rastered underneath a capillary tip in order to collect the elutants. Once dried, the path can be retraced by a Raman spectrometer in order to collect SERS spectra along this line. Such a setup, shown in Figure 7.3, separates the separation and detection steps and provides several advantages, allowing the full length of the capillary to be used for separation and the rastering of the substrate under the capillary tip effectively converting the temporal separation of the capillary into spatial resolution along the SERS surface. He et al. employed four different SERS-active substrates (Au–Li Ag substrates, Ag colloid monolayers, sputtered Ag substrates and Au–Li Ag–Ag film substrates) and showed effective separation and SERS detection for a mixture of *trans*-1,2-bis(4-pyridyl)ethylene and N,N-dimethyl-4-nitrosoaniline. The application to biological materials was exemplified by the separation of tyrosine and tryptophan, while the same method was also applied to the separation and identification of chlorophenols. As highlighted by the authors, because of the difference in surface affinities and cross sections for different analytes, calibration steps will be necessary for any quantification, but CE–SERS is an effective separation and identification method for multiple analytes even when CE separation may be incomplete.

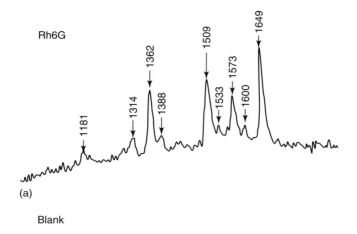
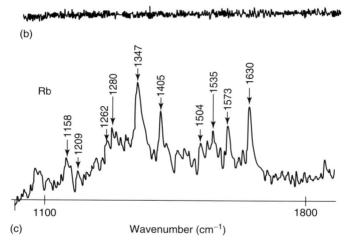

Figure 7.2 CE–SERS spectra: (a) SERS spectrum of 1×10^{-8} M rhodamine 6G injected with an elution time of 9 min; (b) blank spectrum between the two analytes of interest at a time of 11 min and (c) SERS spectrum of 5×10^{-6} M riboflavin injected with an elution time of 13 min. CE conditions: electrokinetic injection 1 kV for 10 s, running voltage 2.5 kV, distance to measurement window 15 cm. SERS conditions: 514.5 nm, 200 mW, 1 s acquisition time [5].

DeVault and Sepaniak [7] used a modified version of electrospray to spatially focus the effluent from a CE capillary onto an SERS-active substrate, in this case silver colloid solutions sprayed onto roughened glass slides, for off-line CE–SERS detection. Their experiments highlighted the importance of careful buffer selection for the CE separation as, while there are a large range of buffers available for separating charged analytes, not all of them will be compatible with an SERS surface. The example noted by DeVault and Sepaniak is that of erythrosin B in the presence of a borate buffer. Under such conditions, the fluorescence levels,

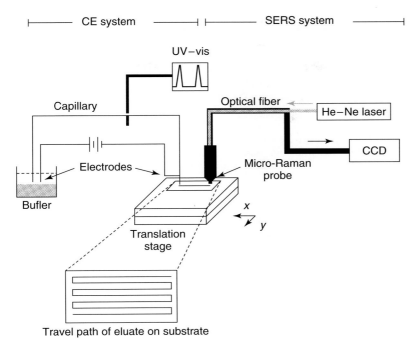

Figure 7.3 CE–SERS apparatus used by He et al. An SERS-active substrate is rastered underneath the capillary tip to collect the CE eluants. The deposited trail can then be followed by a Raman spectrometer to collect SERS spectra of the deposited analytes [6].

which are normally quenched when a fluorescent compound comes into contact with silver nanoparticles, is still present, indicating that erythrostatin B is repelled by the borate molecules on the silver surface. Finally, they also highlight the advantages of performing such experiments off-line, in that division of separation and detection means that the SERS-active surface can be optimized after CE elutant deposition and that exposure times can be increased, increasing the SERS signals and, consequently, the limits of detection.

More recently, Connatser et al. [8] used a micro-fluidic device (shown in Figure 7.4) with an integrated SERS-active surface to separate and detect a mixture of riboflavin and resorufin. Physical vapour deposition of silver onto polydimethylsiloxane (PDMS) was used to create partially embedded 3D silver clusters which provided the SERS-active surface. This material was then used to cover the open channels created by positive molding of the micro-fluidic device such that the micro-fluidic channel had one wall made from the sensing material. These surfaces were susceptible to degradation on contact with water but the degradation process was generally on the order of hours and so did not significantly affect the performance of the separation and detection experiments performed. The authors did note the possibility that some buffers may hasten the degradation of the silver

Figure 7.4 Depiction of integrated micro-fluidic device used by Connatser et al. under the 10× objective of a Raman microspectrometer. The inset shows an SEM image of the silver–PDMA nanocomposite at approximately 90K magnification [8].

surface and, as such, buffer selection should also take into account their reactivities with silver. Determination of the sample was carried out by averaging 33 spectra from the centre of an electrophoretic band and the authors calculated that they injected less than 10 fmol of each sample into the micro-fluidic device in a typical experiment.

Dijkstra et al. [9] tested a number of substrates: etched silver foil, vapour-deposited silver film, a silver oxylate pre-coated silica thin layer chromatography (TLC) plate and a silica TLC plate to which silver colloids and poly(L-lysine) were added after analyte deposition, for at-line coupling of CE with SERS and SERRS. They suggested that the coupling of CE with SERRS was more straightforward than the coupling to SERS, due to the higher intensities obtained with SERRS. They showed that the SERRS spectra of two closely related dyes, crystal violet and basic fuchsin, were independent of the nature of the SERRS-active substrate used. The most intense SERRS spectra were obtained using the post-deposition added colloids and, when using this substrate, the deposited effluent was immediately sorbed by the SERRS-active substrate, allowing the speed the table (which is moved during the analyte deposition) to be reduced. The SERRS intensities obtained using an etched silver foil and a vapour-deposited silver film as substrates were considerably lower but gave sufficient sensitivities and, given the potential for automated measurements, unlike the manual post-deposition addition of colloid, the authors felt these substrates also warranted further consideration.

Most recently, a novel method for online CE–SERS has been developed by Leopold and Lendl [10], where silver SERS substrates are formed directly in the capillary by laser-induced silver substrate (LISS) growth in the presence of silver nitrate and citrate. The authors used two systems, one in which the detection window was formed by inserting a fused silica capillary between two sections of the capillary, and the other in which the detection window was formed directly in the capillary by burning the polyimide capillary coating. In both cases, the LISS was formed on the capillary wall of the detection window and the analytes adsorbed and desorbed onto the freshly formed silver surface at the focal point of the Raman laser. Their system provided both retention times and structural information on the separation of rhodamine 6G and 4-(2-pyridylazo)resorcinol. The formation of the SERS surface inside the capillary provided a number of advantages in that new substrates could be formed by moving the Raman laser to another position along the capillary, preventing memory effects and that the formation of the LISS does not affect the separation process as they are formed after this step.

7.3 SERS and Liquid Chromatography (LC)

Coupling of LC and SERS was first reported by Freeman *et al.* [11]. HPLC effluent was monitored by a UV–visible detector before flowing into tubing, where it was mixed with NaCl-activated citrate-reduced silver colloids before reaching a flow cell and the spectrometer. The authors found that the use of frits or Teflon tubing in parts of the setup in contact with the silver sol caused blockages and memory effects as a result of colloid aggregation and deposition. SERS measurements were taken by trapping each eluting analyte in the flow cell under stopped-flow conditions and taking a normal spectral scan. By doing so, SERS measurements could be carried out in two ways, either by taking a full spectral scan or, if prior information was available, by monitoring the intensity of an SERS band characteristic for the analyte of interest. The inclusion of the UV–vis spectrometer provided information regarding the linear flow velocity of each elutant, which was necessary to trap each analyte in the flow cell. The authors noted that, with improvements in Raman detection systems to allow real-time Raman data collection, the UV–visible absorption spectrometer could be removed from the system. Overall, the system was reported to have a dynamic range over 3 orders of magnitude, a linear response between 100 ppb and 50 ppm and a reproducibility of 1%, overcoming the poor sensitivity of coupling LC with conventional Raman spectroscopy.

Another HPLC–SERS study, focussing on the separation and detection of purine bases in real time, investigated the potential of such measurements in flowing liquids [12]. Silver colloids were chosen as the SERS-active substrate because of the ease of adding such a substrate to a continuously flowing system. One of the problems with obtaining a sufficiently good SERS spectrum for identification in flowing systems is due to the short irradiation times usually experienced by the analyte in such situations. Sheng *et al.* established that the required irradiation time for a good SERS spectrum was affected by the concentration and nature of the analyte species, the pH and temperature of the analysed solution and the condition of the silver colloid used as the SERS-active substrate. They exploited the advantages of raising the temperature, namely, an increase in colloidal aggregation, with the experiments performed at 65 °C to maximize the SERS signals obtainable in a flowing system. The connection between the HPLC and the SERS elements of the experiment was also deemed to be important, with a small length of tubing required to minimize the dispersion of the eluted purine analytes while still allowing for complete mixing between the eluent and the silver colloid. While the system allowed for the separation and real-time spectral identification of the purine bases analysed, the continuous build-up of the Ag colloid on the inner wall of the Raman cell used led to tailing in the SERS-detected chromatograms. This phenomenon, the memory effect, was most noticeable for adenine, which adsorbed most strongly on the colloid surface, producing the greatest levels of aggregation and deposition. Therefore, it was necessary to rinse the Raman capillary used with 6 M HNO_3 after each run.

The memory effect, caused by the build-up or deposition of colloidal particles on the inside of flow cells (see also Chapter 8), tubing, and so on, is a significant problem for measurements performed under flowing conditions. Cabalin *et al.* [13, 14] overcame this problem through the use of a windowless flow cell. This flow cell consisted of two chromatographic stainless steel tubes mounted on an aluminium frame. After the chromatographic eluent and colloidal solutions were mixed, they passed into this flow cell for spectral measurement with the draining liquid supported by the surface tension formed between the two stainless steel tubes [13, 14]. This flow cell has been used to study drugs [13] and human urine [14] samples, with no evidence of peak tailing or memory effects visible in the SERS-detected chromatograms.

Kennedy *et al.* [15] also addressed the additional problems of previous requirements of electrochemical surface replenishment required by electrodes [3, 4] or the requirement for post-column silver colloid deposition when analysing samples in flow. They developed a cascade flow cell for SERS detection in conjunction with both HPLC and flow injection systems. The SERS-active substrate was a modified silver foil that was polished with optical fibre polishing paper before etching with 30% nitric acid. The substrates were then rinsed in ethanol and coated in a 1 nM 1-propanethiol/ethanol solution. These substrates could be slotted into the flow cell, allowing replacement of the SERS-active substrate as necessary. They found that, for HPLC–SERS, the nature of the LC mobile phase and the hydrophobicity of the analyte compound both influenced the intensity of the SERS signals obtained and in some cases this meant that the SERS detection would only be marginally useful for HPLC hyphenation [15].

The group of Schneider used HPLC–SERS for the separation and detection of illicit substances, both in isolation and in bodily fluids [16, 17]. In both studies, they used a gelatin-based silver halide dispersion which forms a silver surface upon interaction with the probe laser light. This photolytic formation of the SERS-active surface is thought to reduce the potential for surface contamination prior to analyte adsorption. Their first experiments [16] concentrated on illicit drugs, namely, cocaine, heroin, ecstasy and amphetamine, and they were able to identify both the drugs and derivatives, such as diacetylmorphine, but also a number of additional substances such as paracetamol and caffeine. In a second paper [17], they analysed urine samples from two dead people who had ingested drugs prior to death and were able to use HPLC–SERS to identify a cocktail of illicit and legal substances, including caffeine, dioxepine, dihydrocodeine and methadone. Additionally, they were able to separate and identify substances such as codeine, carbamazepine and fats from the blood of one of the individuals. In both these studies, the use of a methanol-based buffer gave similar performance to the more conventional acetonitrile-based buffer solution, while giving a reduced background signal. The addition of the SERS detection method provided structural information, rather than solely relying on the elution time, for the identification of each compound and afforded a further benefit through the ability to differentiate between two co-eluting substances [16, 17].

SERRS has also been coupled to HPLC as a secondary, off-line detection method for the separation and identification of nitrophenol compounds. 2-Nitrophenol, 4-nitrophenol, 2,4-dinitrophenol and 4,6-dinitrocresol were separated using reverse-phase HPLC, and the fractions were collected and made basic through the addition of KOH. Finally, a roughened silver electrode was dipped into each fraction for 10 min prior to SERRS spectral acquisition. Compared to the resonance Raman spectra, the SERRS spectra gave intense analyte bands with little contribution from the solvent. The SERRS spectra obtained were unique for each analyte and, additionally, the spectral differences between the individual analytes were more marked in the SERRS spectra compared with the resonance Raman spectra [18].

Soper et al. [19] have reported the use of SERRS as an off-line detection method for LC coupled to TLC (LC–TLC–SERRS). Here, LC effluent was deposited onto a TLC plate. Citrate-reduced silver colloids could be added to the LC effluent prior to deposition, or could be added to the TLC plate, post analyte deposition. The TLC plate was mounted on an X–Y stage to allow deposition of the LC effluent at different positions on the TLC plate, which could then be probed with a Raman spectrometer to collect SERRS spectra. Such a system provides a number of advantages. The majority are associated with all off-line techniques, namely, the ability to analyse the deposited analytes by more than one technique, the potential to measure characteristic spectra associated with adsorbed analytes and the lack of complications associated with measurements in flowing liquid. However, an additional advantage of coupling LC with TLC is the potential to subject the analyte mixture to two distinct separation stages. However, such an advantage was not explored here, as the research was restricted to a proof of concept. The authors indicated that post-deposition addition of colloid was more successful, as this removed the potential problem of colloidal aggregation within the LC system and associated tubing as well as allowing for the removal of the LC solvent, and potential interferents, from the deposited spots prior to SERRS analysis.

Most recently, Carrillo-Carrión et al. used micro-LC to separate pesticides [20] and nucleic acid bases [21] using post-deposition of silver-quantum dots (Ag-QDs) nanoparticles as the SERS-active surface. In their first study of pesticides, the LC-separated analytes were deposited onto a CaF_2 plate via a flow-through microdispenser in separate spots. The Ag-QDs were then added to each drop and allowed to dry before SERS measurements. Here, the limits of detection were 0.1 and 0.2 ng for the pesticides studied but, as the authors point out, not all of the deposited analyte spot is measured during the SERS analysis and so the limits of detection can be significantly lower [20]. In their second study, the microdispenser deposited each eluant into wells on a plate. This was especially useful for the

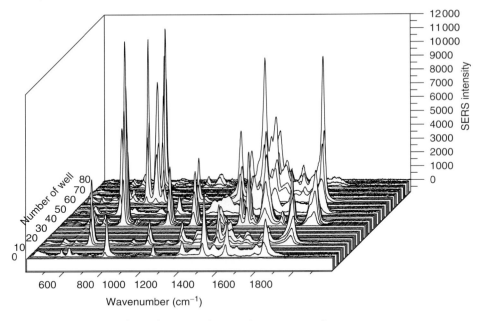

Figure 7.5 3D electropherogram showing the separation of adenine, cytosine, hypoxanthine, thymine and xanthine with micro-LC, deposited via a microdispenser into wells on a plate to which silver-quantum dots were added [21].

detection of analytes whose optimum SERS detection occurs at different pH values, as the pH of each well could be adjusted independently. By doing so, the authors were able to separate and detect a complex mixture of purine and pyrimidine bases and their derivatives [21], as shown in Figure 7.5.

7.4
SERS and Gas Chromatography (GC)

Roth and Kiefer [22] were the first to investigate the coupling of GC and SERS in 1994. They used two different approaches. In the first method, the gaseous eluates were introduced into a liquid silver sol and pumped through a flow cell to facilitate online SERS detection. Such a method, once the gaseous analyte has been introduced into the flowing silver sol, is analogous to LC–SERS. In the second method, the eluates were condensed onto a moving solid thin layer chromatographic plate that had been previously coated with a silver colloid solution. The analyte spot was then moved, stepwise, under the focus of the Raman laser for SERS detection. Such a method can be performed online and also permits the analyte spots to be moved in and out of the laser focus as desired, allowing longer examination of the analyte spots as necessary.

Carron and Kennedy [23] also used modified SERS substrates as a detection method for GC and highlighted their potential for hyphenation with other separation methods. They used an SERS surface prepared from 0.1 mm silver foil etched in 30% nitric acid onto which they then added a 1-propanethiol coating. This coating served two purposes: protecting the silver SERS substrate from oxidation, and creating a specific chemical environment to facilitate the adsorption of their test molecules: benzene, toluene, ethylbenzene and o,m,p-xylenes. These chemicals can be difficult to separate using GC alone, and adding SERS as a detection method allowed the poorly resolved ethylbenzene, p-xylene, m-xylene and o-xylene to be spectroscopically differentiated.

In 2005, Heaps and Griffiths [24] measured off-line GC–SERS spectra. GC elutants were deposited onto a vapour-deposited silver surface on a ZnSe plate. This plate was liquid-nitrogen-cooled and held in a vacuum chamber to prevent the condensation of water vapour from the atmosphere. After deposition, the slide was warmed to room temperature and then SERS spectra of the elutants collected. Owing to the nature of the experimental setup, the study was limited to analytes that were sufficiently volatile to elute from the GC column but were not volatile enough to evaporate from the SERS surface during the warming process. Despite this limitation, they were able to easily detect 3 ng of caffeine. From their comparisons of GC–Raman and GC–SERS, the authors also predicted the detection limits for such GC–SERS systems and suggested that, potentially with an optimized SERS substrate, detection limits of 1 pg may be possible [24].

7.5
SERS and Thin Layer Chromatography (TLC)

Although TLC in isolation has a low limit of detection, and the hyphenation with Raman spectroscopy did not yield promising results [25], coupling with SERS alleviates this problem [26]. The first report of TLC–SERS came from Sequaris and Koglin in 1987 [25]. They separated a range of nucleic purine derivatives using silica gel HPTLC plates, which were then sprayed until wet with a borohydride-reduced silver colloid. The purine spots were identified by the coloured spots arising on the TLC plate surface as a result of colloidal aggregation induced by the analyte molecules. The spectra obtained showed differences from the conventional Raman spectra, with different relative band intensities due to the SERS effect but some band broadening and shifting in the HPTLC–SERS spectra.

A study of a number of TLC and HPTLC plates for hyphenation with Raman spectroscopy also considered the potential of SERS analysis. Using citrate-reduced silver colloids, which had been pre-aggregated through the addition of sodium perchlorate, the authors were able to record clear micro-FT SERS spectra of the carotenoid, crocetin, at 10^{-5} M concentration and calculated that this corresponded to an analysed mass of 0.02 fg [27]. Other experimental factors, including the choice of metal used for the SERS surface, development method and laser excitation wavelength, that affect the success of TLC–SERS measurements have been discussed in more detail by Kocsis *et al.* [28] and Horvarth *et al.* [29].

As would be expected, the nature of the SERS surface has a great deal of influence on the success of a TLC–SERS hyphenated detection system. A recent study [26] compared silver film over nanosphere (AgFON) and silica gel-coupled citrate-reduced Ag colloids as SERS substrates for hyphenation with TLC. They found that, although AgFON substrates should provide better spectra in theory, in practice these substrates are easily compromised by carbon contamination, as the affinity of the carbon for the silver SERS surface was greater than that of their organic dye analytes. In comparison, their silica gel–sliver colloid substrates provided better enhancement due not only to the fact that citrate molecules on the surface of these colloids is more easily displaced than the carbon contamination but also to the fact that the range of colloid particle sizes will give rise to a range of different local surface plasmon resonances (some of which will be tuned to the analyte absorption) and that the nanoparticle aggregates will also give rise to 'hot spots'.

TLC–SERS has also been used to study the composition of artists' dyes using silica gel-coupled citrate-reduced Ag colloids as the SERS-active substrate [26]. The experiments of Brosseau *et al.* showed that, while two different dyes, alizarin and purpurin, had very similar structures and polarities and therefore migrated to the same spot during TLC, SERS spectra taken from this region were able to distinguish alizarin on the leading edge and purpurin at the trailing edge of the spot.

7.6
Other Separation Methods

Han *et al.* [30] have recently developed a protein detection method based on a combination of western blotting and SERS detection. Western blotting is a technique in which proteins are separated by gel electrophoresis or isoelectric focussing electrophoresis and then electroblotted onto a nitrocellulose sheet. This nitrocellulose sheet is then usually analysed via immunochemical detection methods. However, Han *et al.* have stained the nitrocellulose with colloidal silver to facilitate SERS analysis of each of the protein spots. Their initial experiments were able to detect as little as 2 ng per protein band for myoglobin and bovine serum albumin, highlighting the potential of this technique for low-concentration protein analysis. A schematic of their protocol is shown in Figure 7.6.

An integrated separation and SERS detection micro-fluidic device was reported by Connatser *et al.* [31], as shown in Figure 7.7. Designed patterns for substrates were written into an e-beam-sensitive layer using e-beam lithography and then coated with 25 nm of silver or gold via vapour deposition. These SERS-active substrates where then sealed within a PDMS micro-channel, the

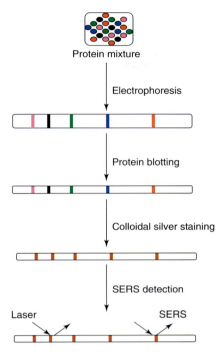

Figure 7.6 Procedure for protein detection based on a combination of western blotting and SERS detection [30].

walls of which were treated with tetraethyl orthosilicate and ethylamine to reduce the adsorption of the analytes onto non-substrate walls. Separation performed in the channel was achieved by applying a potential to the reservoirs containing buffer and sample solutions. After separation, determined as the point when all the analytes had moved past the laser-induced fluorescence (LIF) detection window, the voltages were terminated and the micro-fluidic device moved to a Raman spectrometer for SERS analysis. The authors found that the limits of detection, reported as 7×10^{-8} M for *para*-aminobenzoic acid (*p*-ABA), were close to but not as good as for LIF, something they believe will improve with further development of the SERS-active surfaces included in the micro-fluidic device.

Becker *et al.* [32] have used a variation of free-flow electrophoresis, namely, isotachophoretic (ITP) focussing, in conjunction with SERS detection to highlight the potential of such a system for separation studies. ITP involves placing the analyte under investigation between two (leading and terminating) electrolyte solutions. The nature of the electrolyte solutions is chosen so that they form an electrophoretic mobility gradient through the separation chamber. Applying an electric field generates a gradient in electric field strength, causing the analyte ions to stack in the order of their electrophoretic mobilities. Becker *et al.* showed that, through careful control of citrate-reduced silver colloids, such SERS-active surfaces could be added to solutions within an ITP microchip for the successful

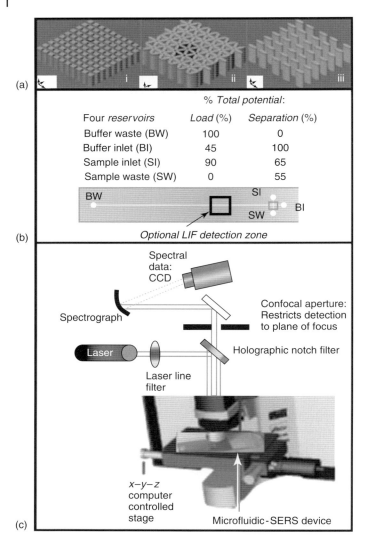

Figure 7.7 (a) Cartoon depictions of electron beam lithographically created patterns integrated for proof-of-concept next-generation SERS substrates (250 nm high pillars capped with silver) in micro-fluidic devices. (b) The four buffer zones to which relative potentials are applied to actuate sample loading and separation. (c) The functional Raman and translational stage scheme with the microfluidic-SERS device on the stage [31].

SERS enhancement of the Raman signal of the analyte, myoglobin. Although no separation of different analyte molecules was performed, the authors demonstrated that, through the application of an electric field, detection of myoglobin was restricted to 2 out of 64 outlet channels, highlighting the potential electrophoretic mobility-based separation.

7.7
Conclusions

Coupling of separation methods with SERS detection has been constantly developing since the early experiments in the late 1980s. Many of the problems of memory effects and SERS surface replenishment requirements have been overcome, and the recent application of such systems to real separation requirements rather than test samples means a move from the research lab to real-life applications. Miniaturization of the separation and detection components used in such separation-SERS coupled systems will lead to reduced sample volume requirements, especially as the area of the SERS surface probed by the laser is small. In future, such systems will routinely be used to separate and identify real-life multi-component samples with detection limits on the order of a few picograms or less.

References

1. Nie, S. and Emory, S.R. (1997) Probing single molecules and single nanoparticles by surface-enhanced Raman scattering. *Science*, **275**, 1102–1106.
2. Kneipp, K., Kneipp, H., Kartha, V.B., Manoharan, R., Deinum, G., Itzkan, R., Dasari, R.R. and Feld, M.S. (1998) Detection and identification of a single DNA base molecule using surface-enhanced Raman scattering (SERS). *Phys. Rev. E.*, **57**, R6281–R6284.
3. Pothier, N.J. and Force, R.K. (1990) Surface-enhanced Raman spectroscopy at a silver electrode as a detection system in flowing streams. *Anal. Chem.*, **62**, 678–680.
4. Pothier, N.J. and Force, R.K. (1992) Surface-enhanced Raman spectroscopy at a silver electrode as a real-time detector in flowing streams. *Appl. Spectrosc.*, **46**, 147–151.
5. Nirode, W.F., Devault, G.L., Sepaniak, M.J. and Cole, R.O. (2000) On-column surface-enhanced Raman spectroscopy detection in capillary electrophoresis using running buffers containing silver colloidal solutions. *Anal. Chem.*, **72**, 1866–1871.
6. He, L., Natan, M.J. and Keating, C.D. (2000) Surface-enhanced Raman scattering: a structure-specific detection method for capillary electrophoresis. *Anal. Chem.*, **72**, 5348–5355.
7. DeVault, G.L. and Sepaniak, M.J. (2001) Spatially focused deposition of capillary electrophoresis effluent onto surface-enhanced Raman-active substrates for off-column spectroscopy. *Electrophoresis*, **22**, 2303–2311.
8. Connatser, R.M., Riddle, L.A. and Sepaniak, M.J. (2004) Metal-polymer nanocomposites for integrated microfluidic separations and surface enhanced Raman spectroscopic detection. *J. Sep. Sci.*, **27**, 1545–1550.
9. Dijkstra, R.J., Gerssen, A., Efremov, E.V., Ariese, F., Brinkman, U.A.T. and Gooijer, C. (2004) Substrates for the at-line coupling of capillary electrophoresis and surface-enhanced Raman spectroscopy. *Anal. Chim. Acta*, **508**, 127–134.
10. Leopold, N. and Lendl, B. (2010) On-column silver substrate synthesis and SERS detection in capillary electrophoresis (CE). *Anal. Bioanal. Chem.* (accepted manuscript).
11. Freeman, R.D., Hammaker, R.M., Meloan, C.E. and Fateley, W.G. (1988) A detector for liquid chromatography and flow injection analysis using surface-enhanced Raman spectroscopy. *Appl. Spectroc.*, **42**, 456–460.
12. Sheng, R., Ni, F. and Cotton, T.M. (1991) Determination of purine bases by reversed-phase high-performance liquid chromatography using real-time

surface-enhanced Raman spectroscopy. *Anal. Chem.*, **63**, 437–442.

13. Cabalin, L.M., Ruperez, A. and Laserna, J.J. (1993) Surface-enhanced Raman spectrometry for detection in liquid chromatography using a windowless flow cell. *Talanta*, **40**, 1741–1747.

14. Cabalin, L.M., Ruperez, A. and Laserna, J.J. (1996) Flow-injection analysis and liquid chromatography: surface-enhanced Raman spectrometry detection by using a windowless flow cell. *Anal. Chim. Acta*, **318**, 203–210.

15. Kennedy, B.J., Milofsky, R. and Carron, K.T. (1997) Development of a cascade flow cell for dynamic aqueous phase detection using modified SERS substrates. *Anal. Chem.*, **69**, 4708–4715.

16. Sägmüller, B., Schwarze, B., Brehm, G., Trachta, G. and Schneider, S. (2003) Identification of illicit drugs by a combination of liquid chromatography and surface-enhanced Raman scattering spectroscopy. *J. Mol. Struct.*, **661–662**, 279–290.

17. Trachta, G., Schwarze, B., Sägmüller, B., Brehm, G. and Schneider, S. (2004) Combination of high-performance liquid chromatography and SERS detection applied to the analysis of drugs in human blood and urine. *J. Mol. Struct.*, **693**, 175–185.

18. Ni, F., Thomas, L. and Cotton, T.M. (1989) Surface-enhanced resonance Raman spectroscopy as an ancillary high-performance liquid chromatography detector for nitrophenol compounds. *Anal. Chem.*, **61**, 888–894.

19. Soper, S.A., Ratzlaff, K.L. and Kuwana, T. (1990) Surface-enhanced resonance Raman spectroscopy of liquid chromatographic analytes on thin-layer chromatographic plates. *Anal. Chem.*, **62**, 1438–1444.

20. Carrillo-Carrión, C., Lendl, B., Simonet, B.M. and Valcárcel, M. (2010) Coupling of µLC-UV and SERS detection with Ag-QDs NPs via a flow-through microdispenser (manuscript submitted).

21. Carrillo-Carrión, C., Lendl, B., Simonet, B.M. and Valcárcel, M. (2010) Off-line micro HPLC SERS detection for the quantification of nucleic acid bases in the low ng range (manuscript submitted).

22. Roth, E. and Kiefer, W. (1994) Surface-enhanced Raman spectroscopy as a detection method in gas chromatography. *Appl. Spectrosc.*, **48**, 1193–1195.

23. Carron, K.T. and Kennedy, B.J. (1995) Molecular-specific chromatographic detector using modified SERS substrates. *Anal. Chem.*, **67**, 3353–3356.

24. Heaps, D.A. and Griffiths, P.R. (2005) Off-line direct deposition gas chromatography/surface-enhanced Raman scattering and the ramifications for on-line measurements. *Appl. Spectrosc.*, **59**, 1305–1309.

25. Sequaris, J.-M.L. and Koglin, E. (1987) Direct analysis of high-performance thin-layer chromatography spots of nucleic purine derivatives by surface-enhanced Raman scattering spectroscopy. *Anal. Chem.*, **59**, 525–527.

26. Brosseau, C.L., Gambardella, A., Casadio, F., Crzwacz, C., Wouters, J. and Van Duyne, R.P. (2009) Ad-hoc surface-enhanced Raman spectroscopy methodologies for the detection of artist dyestuffs: thin layer chromatography-surface enhanced Raman spectroscopy and in situ on the fiber analysis. *Anal. Chem.*, **81**, 3056–3062.

27. Caudin, J.P., Beljebbar, A., Sockalingum, G.D., Angiboust, J.F. and Manfait, M. (1995) Coupling FT Raman and FT SERS microscopy with TLC plates for in situ identification of chemical compounds. *Spectrochim. Acta A*, **51**, 1977–1983.

28. Kocsis, L., Horvath, E., Kristof, J., Frost, R.C., Redey, A. and Mink, J. (2001) Effect of the preparation conditions on the surface-enhanced Raman-spectrometric identification of thin-layer-chromatographic spots. *J. Chromatogra. A*, **845**, 2303–2311.

29. Horvath, E., Katay, G., Tyihak, E., Kristof, J. and Redey, A. (2000) Critical evaluation of experimental conditions

influencing the surface-enhanced Raman spectroscopic (SERS) detection of substances separated by layer liquid chromatographic techniques. *Chromatographia*, **51**, S297–S301.

30. Han, X.X., Jia, H.Y., Wang, Y.F., Lu, Z.C., Wang, C.X., Xu, W.Q., Zhao, B. and Ozaki, Y. (2008) Analytical technique for label-free multi-protein detection based on Western blot and surface-enhanced Raman scattering. *Anal. Chem.*, **80**, 2799–2804.

31. Connatser, R.M., Cochran, M., Harrison, R.J. and Sepaniak, M.J. (2008) Analytical optimisation of nanocomposite surface-enhanced Raman spectroscopy/scattering detection in microfluidic separation devices. *Electrophoresis*, **29**, 1441–1450.

32. Becker, M., Budich, C., Deckert, V. and Janasek, D. (2009) Isotachophoretic free-flow electrophoretic focussing and SERS detection of myoglobin inside a minaturised device. *Analyst*, **134**, 38–40.

8
SERS and Microfluidics
Thomas Henkel, Anne März, and Jürgen Popp

8.1
Introduction

Conventional Raman spectroscopy has been established as a valuable tool for spectroscopy of organic and inorganic molecules. Because of the low efficiency of the Raman scattering process, this technique is limited to samples with a high concentration of analyte molecules. The utilization of the effect of surface enhancement of the Raman scattering increases the limits of detection up to 10 orders in magnitude. Surface-enhanced Raman scattering (SERS) benefits from this effect and enables investigation of small populations of molecules down to the single-molecule level. SERS has become a valuable method for the non-destructive investigation of small population of molecules, which meets the requirements for micro and trace analytics, metabolomics, biomedical applications and monitoring of contaminants in the food industry.

This way, SERS is well suited for accessing multidimensional information from the tiniest objects and analytical samples. For automation, these samples must be placed, metered and processed close to the microscopic detection area. Moreover, the required steps for the preparation of such small samples and objects must be realized close to the detection region. Microfluidic approaches offer a rich and highly customizable set of procedures and operation units for sample processing at the microscale. These procedures benefit from the small characteristic dimensions and the well-controlled operations at the microscale. For implementation of a particular protocol, these operation units can be combined in lab-on-a-chip devices that implement a complete analytical or microchemical process.

This chapter introduces the lab-on-a-chip technology and its application for SERS-based analytics. Recent application examples are given and discussed for the different microfluidic platforms.

Lab-on-a-chip devices for SERS typically implement a user-defined process protocol for sample preparation and delivery of the sample to a microscale optical detection window for the readout of information by Raman spectroscopy (Figure 8.1). This way, the combination of lab-on-a-chip technology with SERS provides a valuable tool for automation and improvements of the reproducibility of

Surface Enhanced Raman Spectroscopy: Analytical, Biophysical and Life Science Applications. Edited by Sebastian Schlücker
Copyright © 2011 WILEY-VCH Verlag GmbH & Co. KGaA, Weinheim
ISBN: 978-3-527-32567-2

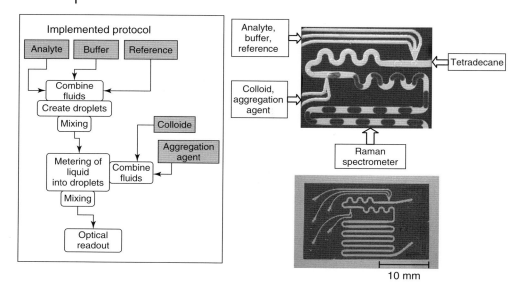

Figure 8.1 Example of a microfluidic device for droplet-based flow-through SERS analytics. All the required steps of the analytical protocol are implemented in the lab-on-a-chip device. In the first step, droplets are formed from the analyte solution, buffer and reference substance solution. After mixing, SERS-active colloids are metered into the droplet and the SERS spectra are measured on chip after binding the analyte molecules to the SERS-active colloidal substrate.

the measurements even for small analytical sample volumes and biological objects. Furthermore, microfluidic approaches for integrated sample enrichment contribute to a further increase in sensitivity. So, electrokinetic sample enrichment on SERS-active substrates can increase sensitivity by an additional eight orders in magnitude [1]. This way, a detection limit of 10 fmol l^{-1} could be reached for adenine.

8.2
Lab-on-a-chip Technology

The lab-on-a-chip concept extends the earlier concept of micro total analysis systems (μTASs), originally developed by Manz [2] in the early 1990s for general laboratory tasks. A lab-on-a-chip is a miniaturized device or system that implements single or multiple laboratory tasks. As a consequence, lab-on-a-chip devices are application specific and developed for a particular process protocol. For efficient development, all the selected operation units must be interoperable. Their characteristics must be predictable in order to allow a model-based system design. For cost-efficient fabrication, they should be manufactured by a common technology, preferably as disposables. The development process (Figure 8.2) starts with the analysis and abstraction of a user-defined process protocol. In the first step, this protocol is translated into a sequence of laboratory unit operations. Most common laboratory

Figure 8.2 Workflow for development and fabrication of lab-on-a-chip devices.

unit operations are metering of liquids, mixing, splitting, incubation, separation, extraction and heating and cooling. During the model-based design process, the operation units are selected, scaled and arranged in a microfluidic network. Finally, the engineering design documents are prepared. First prototypes are prepared for initiation, application development and design optimization. After successful implementation, cost-efficient technologies for mass production are selected.

Initiation and operation of lab-on-a-chip devices require a generic systems platform that provides facilities for flow management, Raman microscopy, optical inspection and process monitoring as well as for mounting the microfluidic chip device with respect to the optical axis of the observation system (Figure 8.3). For the development and initiation process itself, a highly customizable system is required, which can be adapted to the particular measurement tasks. After initiation and protocol development, the requirements of a specialized operating platform are experimentally confirmed. So, the development of an optimized platform for the particular lab-on-a-chip-based procedure can be started in parallel with the protocol optimization and the development of technologies for mass fabrication of the device.

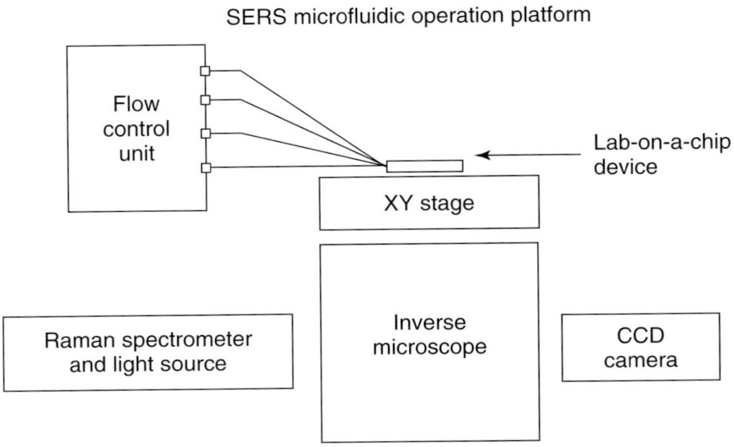

Figure 8.3 Generic setup for SERS experiments in lab-on-a-chip devices. The most common setups are based on an inverse microscope that is equipped with a Raman spectrometer and a CCD camera for process monitoring. Flow control is done using programmable syringe pumps, pressure-based flow control units or electrokinetically driven flow.

8.3
Microfluidic Platforms and Application for SERS

Microfluidics deals with microscale flow dynamics and the development and investigation of technical systems for flow control and sample manipulation at the microscale with characteristic dimensions ranging from the sub-millimetre down to the sub-micrometre scale. Numerous microfluidic devices and systems for micro flow control and fluid processing, including valves, pumps, micromixers, heat exchangers and sensors for hydrodynamic parameters, have been developed and reported during the last two decades. The complex and often device-specific fabrication procedures are difficult for the free combination of different types of such systems into single, monolithic systems as preferred for the lab-on-a-chip technology. From this point of view, the microfluidic platforms approach was introduced by Haeberle and Zengerle in 2007 [3], which characterizes these devices as operation units and assigns them to microfluidic platforms. They give the following definition: "A microfluidic platform provides a set of fluidic unit operations, which are designed for easy combination within a well defined (and low cost) fabrication technology. The platform allows the implementation of different application specific systems (assays) in an easy and flexible way, based on the same fabrication technology."

In their article, they give an overview on the different microfluidic platforms and their operation units. Each of these microfluidic platforms provides a set of interoperable and combinable operation units that can be used to implement a user-defined protocol as a sequence of laboratory unit operations. Microfluidic platforms with high impact are characterized by a large number of available operation units for the implementation of more or less complex process protocols. The next section introduces the reader to these platforms and characterizes them for their applicability for SERS spectroscopy.

8.3.1
Capillary-Driven Test Stripes

Capillary-driven test stripes – also known as lateral flow assays were introduced in the 1960s. Recently, a large collection of such test stripes is available as quick tests or for point-of-care applications and for the detection of pollutants in water and environmental samples. A comprehensive review on the application of lateral flow assays has been made by Posthuma-Trumpie et al. [4]. The sample fluid is applied on the loading frame of the stripe and transported to the detection window by capillary forces (Figure 8.4). During transport, the fluid may be filtered. Buffer components, reagents, analyte-specific antibody/enzyme conjugates or probe molecules that are deposited at distinct positions of the stripe are delivered into the fluid according the requirements of the implemented protocol. At the detection position, the analyte conjugates are collected and used for a staining procedure. For optical readout, a staining procedure is performed inside the detection window, which displays the result of the assay as a change in colour or transparency. Colloidal nanoparticle

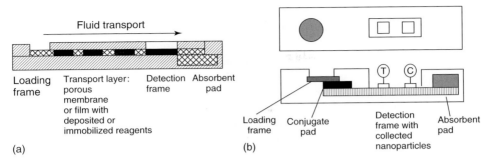

Figure 8.4 Capillary-driven test stripes: (a) Basic design and (b) implementation of a nanoparticle labelled lateral flow immunoassay [4].

probes have been successfully used as labels for the implementation of staining procedures [5] (see also Chapters 11 and 12). Because of the SERS activity, such particle labels can be directly used for readout by Raman spectroscopy utilizing the SERS effect.

A specialized Raman spectrometer for the readout of the detection frame of the stripe is the only required equipment. No additional facilities for fluid management and sample application are necessary. From the end users' point of view, the system must be compact and portable to meet the requirements of point-of-care diagnostics and field analytics.

SERS nanotags [6] for application in lateral flow assays have been developed that exhibit a specific Raman signature. They are made by linking SERS-active label molecules with gold or silver nanoparticles (Figure 8.5). These tags are

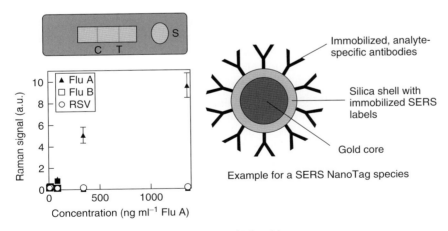

Figure 8.5 Application of SERS nanotags in multiplexed lateral flow immunoassays for parallel detection of Flu A, Flu B and RSV viruses. For multiplex detection, the analyte-specific Raman signatures of the nanotags were detected and used for quantitation [6].

further linked to antibodies as recognition molecules. During incubation, the analyte molecule binds to the nanotag. So the analyte is labelled with the nanotag signature, which can be read out by a Raman spectrometer. In lateral flow assays, these conjugates are transported to the detection window and collected for readout by surface-immobilized capture antibodies.

Freeman and colleagues [6] have reported on the development of encapsulated gold and silver nanotags and their application in lateral flow immunoassays for the multiplexed detection, discrimination and quantitation of different viruses by SERS readout of the assay. Gold nanotags with different Raman signatures were synthesized and cross-linked with antibodies for the detection of inactivated influenza viruses Flu A and Flu B and for the respiratory syncytial virus (RSV) (Figure 8.5). Test stripes are prepared with a mixture of the nanotag conjugates and loaded to the conjugate pad. Analyte-bound species are collected in a single stripe of the detection window with an immobilized mixture of second antibodies against all analytes, followed by multiplexed readout and quantitation of the analytes based on the SERS signatures of the nanotag labels.

8.3.2
Microfluidic Large-Scale Integration and PDMS Microchannels

The microfluidic large-scale integration (LSI) platform utilizes a system of stacked elastomer layers, each of them with integrated microchannels. The stack can be prepared on top of a solid support, for example, glass, silicon or a polymer. The technique for the preparation of 3D topologies and microchannels by replication of a master geometry by moulding using UV-curable polydimethyl siloxane (PDMS) was introduced by the group of Whitesides in 1996 [7]. Microchannels can act either as fluid channels for transport of sample fluids or as control channels that are used for pneumatic or fluidic actuation of integrated valves and membranes [8] (Figure 8.6). While PDMS has been approved and established as the material of choice for the preparation of these devices, the platform is not limited to this elastomer. The platform itself provides a large collection of operation units for valving, pumping, metering of liquids, sorting and switching of flows and mixing of fluids that can be externally controlled by a pneumatic or fluidic control system. Although it offers a powerful platform for the implementation of complex process protocols, no example has been reported for SERS applications that comprehensively makes use of the potential of this platform. In contrast, single-layer microchannel systems with integrated functional structures for fluid metering, mixing and optical readout, prepared by PDMS replication technology, form the most often used microfluidic devices for SERS application. All these approaches can be extended by the addition of control layers to implement more comprehensive analytical protocols in a single-chip device. Therefore, all SERS approaches that use PDMS microchannels are summarized under this microfluidic platform.

A standard Raman setup as given in Figure 8.3 can be directly used for SERS measurements. For operation of the control channels, the system must be extended

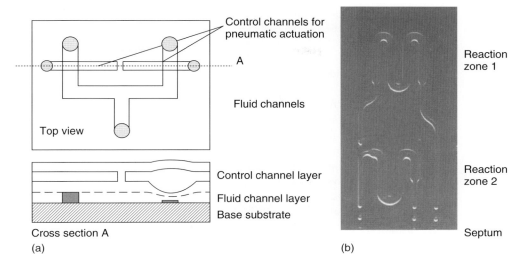

Figure 8.6 PDMS microchannel systems and LSI platform: (a) Sketch of a multilayer system with fluid channels and pneumatically actuated control channels for valving and throttling. Application of pressure to the control channels seals the fluid channel section below. (b) Single-layer PDMS chip device with two reaction zones and needle ports with integrated septums.

with a pressure or flow control unit for actuation of the integrated control channels and control facilities.

Surface-immobilized SERS substrates have been prepared on the elastomer surface [9] by deposition of thin silver films into nanowells for the detection of rhodamine 6G and adenosine. SERS substrates formed by the deposition of silver composites and fabrication of three-dimensional nanostructures by electron beam lithography [10] have been applied for the detection of rhodamine 6G, resorufin, *para*-aminobenzoic acid, *ortho*-phenanthroline, 6-amino-2-naphthoic acid and terephthalic acid. Other approaches are the deposition of nanoparticles (see also Chapter 2) or nanostructured noble-metal films. All kinds of planar SERS substrates can be integrated by clamping the PDMS microchannel layer on top of a base support, equipped with an SERS-active substrate. Because of the good chemical resistance of PDMS, regeneration and cleaning procedures for the SERS substrate may be implemented even with aggressive reagents or polar organic solvents.

Colloidal SERS substrates and aggregation reagents can be added to the analytical sample on demand and efficiently mixed using integrated static micromixers or by actuation. Care must be taken to prevent the adsorption of colloids and nanoparticle/analyte aggregates at the microchannel walls and optical detection windows. This causes baseline shifts and limits the time for detection. A cleaning step between measurements is employed for long-term or serial measurements. This particular procedure can be integrated into LSI devices.

Numerous applications of PDMS microdevices have been reported for sensitive quantitation. A selection is given below: [11]; nicotine (0.1 ppm) [12]; methyl parathion pesticides (0.1 ppm) [13]; crystal violet (5 nmol l^{-1}), mitoxantrone (0.1 pmol l^{-1}) and detection of the pesticides [14].

8.3.3
Centrifugal Microfluidics

The centrifugal microfluidics platform uses rotational forces (radial force, Coreolis force) and acceleration forces in combination with capillary and inertial forces for fluid management in microfluidic networks [15, 16]. Microfluidic networks are arranged radial symmetrically on a disc (Figure 8.7). The sample fluid is loaded in the central region of the disc and transported by centrifugal forces through the microchannel system to the outer region. Backflow to the central region can be realized by capillary forces. Fluidic anchors, defined by geometry or local wetting properties, are used to stop fluid motion at low rotational frequencies while releasing flow at higher frequencies. Siphon structures allow the collection of fluids and the release of the collected fluids after the structure is filled. Typically realized in CD-ROM geometry, standard equipment and components from CD/DVD/Blue-ray

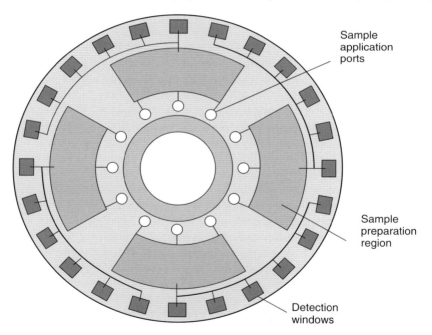

Figure 8.7 Schematic representation of a microfluidic disc for centrifugal microfluidics. Samples and reagents are loaded in the central region. Reagents may be pre-packaged in cavities of the sample preparation unit. Samples are processed in the sample preparation region and aliquoted into the detection window cavities that are optionally preloaded with analyte-specific detection reagents. This way, multiple parameters can be measured for each sample in a single run. Multiple units can be arranged on a single disc for parallel processing of different samples. (Four units are shown on the sketch.)

technology for rotational force control and optical readout of information can be applied for the implementation of portable lab-on-a-chip-based solutions for microanalytics and diagnostics. Moreover, the highly automated and efficient CD manufacturing process can be used for the cost-efficient mass fabrication of disposable discs by the injection/compression moulding process. For fabrication, only the front plate of the injection/compression mould, which normally contains a nickel master with the digital data, has to be substituted by a front plate containing the microfluidic structures. Because of the availability of polymer materials with optimized optical parameters even for short wavelengths down to 405 nm for the Blue-ray technology, the centrifugal microfluidics platform provides superior prerequisites for optical readout of information. Development of packaging technology for preparation of reagent-preloaded discs is one of the recent challenges for this platform.

The centrifugal microfluidics platform provides a comprehensive set of scalable operation units for valving, aliquoting and metering of liquids into samples; for sedimentation and filtering of particles; for washing of resins; for routing of fluids either into the waste or the sample channel; for thermal management, for example, for polymerase chain reaction (PCR); for optical readout by fluorescence, absorption and chemiluminescence measurements; and for spectroscopic readout [17].

For SERS measurements, the system must be equipped with a rotational stage with programmable rotation velocity and acceleration instead of the $x-y$ stage, described in the initial setup.

Lee and co-workers [18] report on the development of an optofluidic CD platform for parallel processing of up to 12 tests with a total of 84 detection positions in a single analysis run. The disc has been realized as a compound system from a glass wafer and a PDMS microchannel system, prepared by replication of a silicon master. Surface-immobilized SERS-active substrates were prepared on top of the glass substrate inside the detection frame by aggregation and precipitation of SERS-active gold nanoparticle suspension. Additional enhancement of the SERS signal was realized by repeated application of the analyte solution on the SERS-active sites, followed by a drying process in order to saturate all of the binding sites with the analyte. This way, a detection limit of 1 nmol l^{-1} was reached for the model analyte rhodamine 6G after 30 binding cycles and an acquisition time of 10 s per spot.

8.3.4
Electrokinetic Platform

The electrokinetic platform is characterized by the application of electric fields for flow control and actuation of particles and charged molecules in fluids at the microscale. Because of the small characteristic dimensions in microfluidics, short interaction distances allow the implementation of strong electric fields or field gradients, even at low voltages below the electrochemical potential of the electrodes. This way, electrochemical redox processes at integrated electrodes can be suppressed. Three basic electrokinetic effects (Figure 8.8), namely, electro-osmotic flow (EF), electrophoresis (EP) and dielectrophoresis (DEP) can be applied and

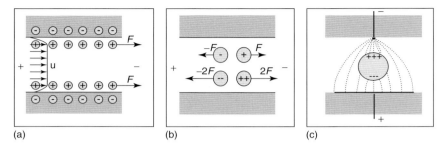

Figure 8.8 Electrokinetic transport effects: (a) electro-osmotic flow (EOF); (b) electrophoresis (EP) and (c) dielectrophoresis (DEP).

combined. EF is driven by the transport of ions in the electric double layer on the surface of the microchannel wall in an electric field. Electrophoretic transport relative to the surrounding fluid is the result of Coulomb forces between charged species in the applied electric field. DEP is the result of charge separation in an intrinsically uncharged particle in an inhomogeneous alternating current (AC) electric field. Direction and magnitude of dielectrophoretic forces strongly depend on the electrical properties of the particle, the surrounding fluid and the frequency of the AC electric field. The effect is widely applied in microfluidics for the separation and manipulation of particles and cells [19].

For application of electrokinetic effects in SERS measurements, the platform needs to be complemented with components for control of electrode potentials and electrical interfacing of the chip device.

Functionality of surface-immobilized SERS substrates can be doubled by their application as integrated electrodes. Therefore, the SERS-active nanostructures can be prepared either directly as electrodes or by immobilization of SERS-active nanoparticles on the top of the electrodes. Charged analyte molecules can be transported to and enriched in the electrode space. This way, a strong local enrichment of the analyte concentration at the SERS-active substrates can be realized, resulting in a dramatic increase in sensitivity for the detection of charged analyte molecules.

Colloidal substrates typically consist of charged nanoparticles. These nanoparticles can be actuated by EP and DEP. This way, they can be transported in fluids independent of the fluid flow. Because of the laminar flow in microchannels, a parallel co-flow of fluids with multiple composition can be realized. Electrokinetic forces now can be applied to transport the nanoparticles between these fluids in order to implement a chemical or analytical protocol. Moreover, they can be collected and attracted at distinct sites of the microchannel for enrichment and readout.

Erickson et al. have reported on the electrokinetic sample actuation and enrichment at nanowells for improved efficiency of DNA hybridization assay to nanoparticles and readout of the hybridization assays by SERS [20]. The key operating unit consists of a microchannel section, in which the bottom face is equipped with nanowells with integrated electrode structures and the top face contains the opposite electrode structure. Gold nanoparticles can be collected into the nanowells by electrokinetic actuation with an applied voltage of 1 V for 5 s or delivered into the volume by the application of a rejection potential for 5 s. While being attracted, the sample fluid between the electrodes can be exchanged and washing procedures applied in order to remove non-specifically bound DNA molecules from the nanoparticles. This way, active transport of the nanoparticles in the electrode space can be utilized to increase the binding of complementary analyte DNA from the sample fluid to the gold nanoparticles by hybridization. For the reported assay, the analyte DNA was labelled with TAMRA (carboxy-tetramethyl-rhodamine) and the TAMRA SERS signature was read out from the nanoparticles. Up to 80 bind/release cycles were applied before the final wash to remove non-specifically bound DNA from the attracted nanoparticles. Analyte DNA could be detected down to concentrations of 30 pmol l^{-1}.

An approach for analyte sample enrichment on SERS-active substrates, which impressively demonstrates the potential of this approach, was reported by the group of Lee [1, 21]. The SERS-active substrate was prepared at the bottom surface of a microchannel. A wired electrode was placed over the SERS substrate at the top of the microchannel. Adenine, which exhibits a positive charge at pH 7.4, was chosen as model analyte. By application of an electrical field of -0.6 V cm^{-1}, electrokinetic transport of the positively charged adenine molecules to the negatively charged SERS substrate collects the analyte molecules from the volume between the SERS substrate and the top electrode on the surface of the SERS substrate. This way, a pre-concentration from a 10 fmol l^{-1} solution of adenine was reached to a level that corresponds to the signal of a 1 µmol solution of adenine without electrokinetic pre-concentration. In summary, an additional increase in detection limits by eight orders in magnitude has been experimentally confirmed for adenine by electrokinetic pre-concentration.

8.3.5
Droplet-Based Microfluidics

The application of segmented flow in conventional flow-through analysis systems was introduced in the 1960s. Because of the progress in microsystems technology and microfluidics of multiphase flows, microchannel systems with low capillary number could be prepared, in which the transport phenomena are mainly dominated by interfacial forces. Since 2001 [22], droplet-based microfluidics has become a major discipline in microfluidics and lab-on-a-chip technology. Application of the droplet-based microfluidics platform for Raman spectroscopy was introduced by Cristobal in 2006 [23]. The first SERS measurements were published by Popp and co-workers [24].

In general, sample volumes are processed in microchannel systems as individual droplets, where each of the droplets may represent an individual sample (Figure 8.9). Microfluidic operation units are available for droplet generation, metering of liquid

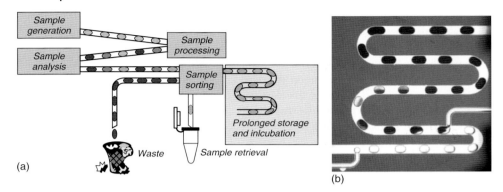

Figure 8.9 Droplet-based microfluidics platform. (a) Typical workflow for assay implementation by droplet-based microfluidics. (b) Lab-on-a-chip device for droplet generation and dosing of reagent into droplets. A deep coloured dye (ferric rhodanide) is formed by gradient metering of ferric ions into droplets, containing ammonium rhodanide solution. Limited mixing due to the axisymmetric droplet internal flow is observed for droplets, where only a small amount of ferric ions has been added. This can be overcome by mixing in winding channel sections as shown in Figure 8.1.

into droplets, controlled droplet fusion, stacking of droplets from different sources, merging and aliquoting of droplets and collection of the dispersed phase of multiple droplets by phase separation [25]. Interfacially generated forces can be used to design smart operation units that automatically perform their operation without any external control [26]. Moreover, such multistate devices can be used to implement droplet-based logical operations and combine them into droplet-based logic processors [27].

Different basic approaches are pursued for implementation of droplet-based systems and for retaining of the individuality of droplets during the process.

8.3.5.1 Straight Plug-Flow Concept

Transport of droplet trains as plugs that seal the given microchannel geometry completely is the basic assumption behind this concept [28]. Droplets are prevented from bypass flows and the volume of the separation fluid between droplets remains constant for the whole process. Contact between adjacent droplets, which may cause coalescence, is avoided. Reliability can be improved by the formation of a permanent lubrication layer of the separation fluid on the microchannel surface. This requires an appropriate selection of the fluids and microchannel surface modifications.

8.3.5.2 Surfactant-Stabilized Sample Droplets

Surfactants may be added to the system to generate repellent droplet interfaces. This suppresses droplet coalescence even if droplets come into contact with each other. In this case, droplets must not seal the microchannel completely. For controlled pair-wise coalescence of droplets, either external forces [29, 30] or integrated strictures [31] can be applied.

8.3.5.3 Processing as Foams in Microchannel Systems

The third approach is based on processing droplet trains as foams in microchannel networks, where the separation fluid forms thin lamellas between the sample droplets. The self-organization of the foams can be controlled by the shape of the microchannel. Multiple patterns may be interconverted [32]. A special set of microfluidic operation units exists for this particular process model for droplet generation [33] and electrocoalescence of droplets [29], extending the application spectrum for this approach.

8.3.5.4 Conclusion

Droplet-based microfluidics provides a variety of methods and operation units for the implementation of analytical protocols [34]. While not mandatory, interface properties can be optionally controlled by application of surfactants. This way, droplet size becomes independent of the microchannel cross section and the set of available operation units for droplet processing is extended.

For readout of information by SERS/Raman spectroscopy, the microfluidic platform setup depicted in Figure 8.3 is applied. For synchronization of data acquisition with droplet flow, additional facilities for recognition of droplet interfaces are required. This can be realized either as a fibre probe, which catches light reflexes from the excitation laser beam at the droplet interface in the transmission mode, or by implementation of an image-based sensing approach.

One of the main challenges in long-term SERS experiments using colloidal SERS substrates is to avoid baseline shifts caused by the deposition of colloid particle/analyte conjugates on the walls of the optical detection windows. Therefore, a fluid composition can be selected that forms a permanent, self-regenerating lubrication layer of the continuous phase on the surface of the microchannel walls, preventing the walls from coming into direct contact with the sample droplets containing the colloid. The concept was introduced by Popp and co-workers [24] for long-term measurements of concentration fluctuations. No baseline shifts were observed for continuous measurement over an 8 h period.

Further developments towards the application of droplet-based SERS as a quantitative method for ultrasensitive measurement of drug and metabolite concentrations have led to the application of internal standards [35]. Therefore, isotope-edited analyte molecules have been applied as an internal standard. Because of the changes in the atomic mass, the vibrational wavenumber is shifted and the Raman bands of the isotope-labelled analyte molecules can be distinguished from the unlabelled ones (Figure 8.10). This approach overcomes the changes in SERS signal amplification due to ageing of colloids and inter-batch variability. The first application examples were demonstrated for the heteroaromatics nicotine and pyridine.

Both direct and competitive assays have been implemented by droplet-based SERS using colloidal SERS substrates. In direct assays, the analyte directly binds to the SERS-active substrate. The band intensities in the analyte spectra are directly utilized as a measure of the analyte concentration or its ratio to an internal standard molecule. For competitive approaches, the SERS-active colloid is preloaded with an SERS-active probe molecule that is released upon binding of the target analyte.

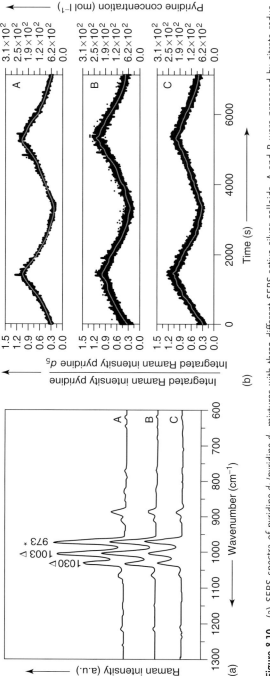

Figure 8.10 (a) SERS spectra of pyridine-d_0/pyridine-d_5 mixtures with three different SERS-active silver colloids. A and B were prepared by citrate reduction and C was obtained by reduction with sodium borohydride. All colloids achieve different SERS enhancement. Bands marked with an asterisk can be assigned to the isotope-labelled standard. Measurements of programmed variations of the analyte concentration at a fixed concentration of the internal standard of 1.9×10^{-2} M were performed for all these three colloids. The pyridine-d_0 band intensity was normalized to the band intensity of the internal standard pyridine-d_5. As shown in (b), identical concentration profiles have been obtained for all these independent experiments.

It provides a powerful approach for the quantitation of SERS-inactive analytes. This approach was applied by deMello and co-workers for the ultrasensitive quantitation of mercury ions using rhodamine preloaded colloidal gold nanoparticles as SERS substrates [36] in a droplet-based microfluidic approach.

8.4
Summary

Actually, lab-on-a-chip platforms and microfluidic approaches are under evaluation for the integration of microscale sample preparation and SERS spectroscopy for qualitative and quantitative analytics. A total of 56 publications is found in the Website of Science database for the topical query for the combination 'microfluid* AND SERS' (Figure 8.11). Fifty-four of these articles have been published since 2005. Therefore, we can conclude that scientific research on this field has started with the development and experimental evaluation of concepts for miniaturized SERS analytics. During this short period, sample implementations for the various microfluidic platforms have been investigated and characterized with respect to sensitivity and reliability using model analytes. Meanwhile, SERS has become available as a quantitative analytical method. These methods benefit from the signal enhancement by the SERS effect by 8–10 orders in magnitude compared to standard Raman scattering. Electrokinetic pre-concentration and collection of analyte molecules on the SERS substrates in microfluidic devices offers a further amplification of eight orders in magnitude. So, detection limits down to

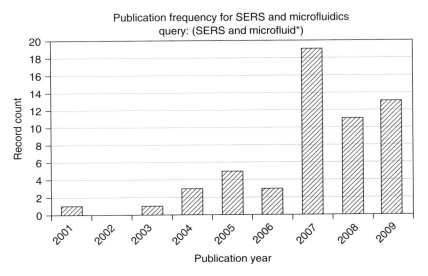

Figure 8.11 Annual publication frequency of the application of SERS in microfluidic devices. (Source: ISI Web of Science, Thomson Reuters, New York.)

the femtomolar range have been experimentally confirmed by combination of microfluidics with SERS technology.

As a result, the demand for development of instrumentation has grown. Target applications in the fields of biomedical diagnostics, process control, clinical chemistry, biotechnology and trace analytics have been identified. Because of the high sensitivity and the detailed information about the vibrational spectra of the analyte provided by SERS, the method has become available for ultrasensitive label-free detection of organic compounds at physiologically active concentrations. It can be expected that SERS in combination with lab-on-a-chip technology will allow the design and implementation of flexible analytical systems that will outperform established approaches with respect to sensitivity, reliability, sample throughput and mobility.

References

1. Cho, H., Lee, B., Liu, G.L., Agarwal, A. and Lee, L.P. (2009) Label-free and highly sensitive biomolecular detection using SERS and electrokinetic preconcentration. *Lab. Chip.*, **9**, 3360–3363.
2. Manz, A., Graber, N. and Widmer, H.M. (1990) Miniaturized total chemical-analysis systems – a novel concept for chemical sensing. *Sens. Actuators. B Chem.*, **1**, 244–248.
3. Haeberle, S. and Zengerle, R. (2007) Microfluidic platforms for lab-on-a-chip applications. *Lab. Chip.*, **7**, 1094–1110.
4. Posthuma-Trumpie, G.A., Korf, J. and van Amerongen, A. (2009) Lateral flow (immuno) assay: its strengths, weaknesses, opportunities and threats. A literature survey. *Anal. Bioanal. Chem.*, **393**, 569–582.
5. Snowden, K. and Hommel, M. (1991) Antigen-detection immunoassay using dipsticks and colloidal dyes. *J. Immunol. Methods*, **140**, 57–65.
6. Doering, W.E., Piotti, M.E., Natan, M.J. and Freeman, R.G. (2007) SERS as a foundation for nanoscale, optically detected biological labels. *Adv. Mater.*, **19**, 3100–3108.
7. Xia, Y.N., Kim, E., Zhao, X.M., Rogers, J.A., Prentiss, M. and Whitesides, G.M. (1996) Complex optical surfaces formed by replica molding against elastomeric masters. *Science*, **273**, 347–349.
8. Quake, S.R. and Scherer, A. (2000) From micro- to nanofabrication with soft materials. *Science*, **290**, 1536–1540.
9. Liu, G.L. and Lee, L.P. (2005) Nanowell surface enhanced Raman scattering arrays fabricated by soft-lithography for label-free biomolecular detections in integrated microfluidics. *Appl. Phys. Lett.*, **87**, 074101-1–074101-3.
10. Connatser, R.M., Cochran, M., Harrison, R.J. and Sepaniak, M.J. (2008) Analytical optimization of nanocomposite surface-enhanced Raman spectroscopy/scattering detection in microfluidic separation devices. *Electrophoresis*, **29**, 1441–1450.
11. Quang, L.X., Lim, C., Seong, G.H., Choo, J., Do, K.J. and Yoo, S.K. (2008) A portable surface-enhanced Raman scattering sensor integrated with a lab-on-a-chip for field analysis. *Lab. Chip.*, **8**, 2214–2219.
12. Jung, J.H., Choo, J., Kim, D.J. and Lee, S. (2006) Quantitative determination of nicotine in a PDMS microfluidic channel using surface enhanced Raman spectroscopy. *B. Korean Chem. Soc.*, **27**, 277–280.
13. Lee, D., Lee, S., Seong, G.H., Choo, J., Lee, E.K., Gweon, D.G. and Lee, S. (2006) Quantitative analysis of methyl parathion pesticides in a polydimethylsiloxane microfluidic channel using confocal surface-enhanced Raman spectroscopy. *Appl. Spectrosc.*, **60**, 373–377.

14. Abu-Hatab, N.A., John, J.F., Oran, J.M. and Sepaniak, M.J. (2007) Multiplexed microfluidic surface-enhanced Raman spectroscopy. *Appl. Spectrosc.*, **61**, 1116–1122.
15. Madou, M., Zoval, J., Jia, G.J., Kido, H., Kim, J. and Kim, N. (2006) Lab on a CD. *Annu. Rev. Biomed. Eng.*, **8**, 601–628.
16. Nolte, D.D. (2009) Invited review article: review of centrifugal microfluidic and bio-optical discs. *Rev. Sci. Instrum.*, **80**, 101101.
17. Ducree, J., Haeberle, S., Lutz, S., Pausch, S., von Stetten, F. and Zengerle, R. (2007) The centrifugal microfluidic bio-disc platform. *J. Micromech. Microeng.*, **17**, 103–115.
18. Choi, D., Kang, T., Cho, H., Choi, Y. and Lee, L.P. (2009) Additional amplifications of SERS via an optofluidic CD-based platform. *Lab. Chip.*, **9**, 239–243.
19. Kaler, K. and Pohl, H.A. (1978) Continuous biological cell separation using dielectrophoresis. *B. Am. Phys. Soc.*, **23**, 285–285.
20. Huh, Y.S., Chung, A.J., Cordovez, B. and Erickson, D. (2009) Enhanced on-chip SERS based biomolecular detection using electrokinetically active microwells. *Lab. Chip.*, **9**, 433–439.
21. Cho, H., Long, Y.T. and Lee, L.P. (2007) Study on biomolecules by electrokinetic concentration-based SERS amplification. *Biophys. J.*, **90**, 337A–337A.
22. Nisisako, T., Torii, T. and Higuchi, T. (2002) Droplet formation in a microchannel network. *Lab. Chip.*, **2**, 24–26.
23. Cristobal, G., Arbouet, L., Sarrazin, F., Talaga, D., Bruneel, J.-L., Joanicot, M. and Servant, L. (2006) On-line laser Raman spectroscopic probing of droplets engineered in microfluidic devices. *Lab. Chip.*, **6**, 1140–1146.
24. Ackermann, K.R., Henkel, T. and Popp, J. (2007) Quantitative online detection of low-concentrated drugs via a SERS microfluidic system. *ChemPhysChem*, **8**, 2665–2670.
25. Teh, S.Y., Lin, R., Hung, L.H. and Lee, A.P. (2008) Droplet microfluidics. *Lab. Chip.*, **8**, 198–220.
26. Kielpinski, M., Malsch, Dl., Gleichmann, N., Mayer, G. and Henkel, T. (2008) Application of self-control in droplet-based microfluidics. Proceedings of the 6th ICNMM, June 23-25, 2008, ASME 2008, Darmstadt, pp. 1565–1570.
27. Prakash, M. and Gershenfeld, N. (2007) Microfluidic bubble logic. *Science*, **315**, 832–835.
28. Henkel, T., Bermig, T., Kielpinski, M., Grodrian, A., Metze, J. and Köhler, J.M. (2004) Chip modules for generation and manipulation of fluid segments for micro serial flow processes. *Chem. Eng. J.*, **101**, 439–445.
29. Priest, C., Herminghaus, S. and Seemann, R. (2006) Controlled electrocoalescence in microfluidics: targeting a single lamella. *Appl. Phys. Lett.*, **89**, 134101-1–134101-3.
30. Baroud, C.N., de Saint Vincent, M.R. and Delville, J.P. (2007) An optical toolbox for total control of droplet microfluidics. *Lab. Chip.*, **7**, 1029–1033.
31. Tan, Y.-C., Ho, Y. and Lee, A. (2007) Droplet coalescence by geometrically mediated flow in microfluidic channels. *Microfluid. Nanofluid.*, **3**, 495–499.
32. Surenjav, E., Priest, C., Herminghaus, S. and Seemann, R. (2009) Manipulation of gel emulsions by variable microchannel geometry. *Lab. Chip.*, **9**, 325–330.
33. Chokkalingam, V., Herminghaus, S. and Seemann, R. (2008) Self-synchronizing pairwise production of monodisperse droplets by microfluidic step emulsification. *Appl. Phys. Lett.*, **93** 254101-1–254101-3.
34. Köhler, J.M., Henkel, T.H., Grodrian, A., Kirner, T., Roth, M., Martin, K. and Metze, J. (2004) Digital reaction technology by micro segmented flow - components, concepts and applications. *Chem. Eng. J.*, **101**, 201–216.
35. März, A., Ackermann, K.R., Malsch, D., Bocklitz, T., Henkel, T. and Popp, J. (2009) Towards a quantitative SERS approach - online monitoring of analytes in a microfluidic system with isotope-edited internal standards. *J. Biophotonics.*, **2**, 232–242.

36. Wang, M., Benford, M., Jing, N., Cote, G. and Kameoka, J. (2009) Optofluidic device for ultra-sensitive detection of proteins using surface-enhanced Raman spectroscopy. *Microfluid. Nanofluid.*, **6**, 411–417.

9
Electrochemical SERS and its Application in Analytical, Biophysical and Life Science

Bin Ren, Yan Cui, De-Yin Wu, and Zhong-Qun Tian

Raman spectroscopy, as a vibrational spectroscopy, can record fingerprint spectra from electrodes and provide much insight into a variety of surface and interfacial processes at the molecular level: for example, qualitatively determining surface bonding, conformation and orientation. Raman spectroscopy invariably uses lasers from the ultraviolet (UV) to the near infrared (NIR). More importantly, the technique can be applied *in situ* to investigate solid–liquid, solid–gas and solid–solid interfaces of both fundamental and practical importance. The technique can be used flexibly to study porous electrode materials of high surface area, to which many surface techniques are not applicable. Therefore, Raman spectroscopy is among the most promising methods for use in electrochemistry. The major disadvantage of Raman spectroscopy is its very low detection sensitivity. However, surface-enhanced Raman scattering (SERS) can improve the sensitivity significantly by several orders of magnitude for roughened surfaces of many metals including noble and transition metals.

This chapter will first introduce some fundamental background and features of electrochemical surface-enhanced Raman scattering (EC-SERS), followed by a detailed description of the experimental setup for electrochemical Raman spectroscopy and preparation of SERS substrates. The emphasis will be on how to obtain reliable information of very sensitive bio-related systems. Some examples, varying from a model molecule of benzene to real biomolecules, such as dopamine, NADH, DNA and cytochrome *c* (cyt *c*), will be shown to demonstrate how to apply EC-SERS for bio-related application. Finally, prospects and further developments of EC-SERS application in bio-related systems will be discussed.

9.1
Electrochemical Surface-Enhanced Raman Spectroscopy

The first SERS spectra were obtained from an electrochemical cell when Fleischmann *et al.* tried to roughen the Ag electrode in order to increase the surface area and hence the number of adsorbed pyridine molecules on the laser spot [1]. This phenomenon was soon demonstrated to be due to the surface enhancement effect

Surface Enhanced Raman Spectroscopy: Analytical, Biophysical and Life Science Applications. Edited by Sebastian Schlücker
Copyright © 2011 WILEY-VCH Verlag GmbH & Co. KGaA, Weinheim
ISBN: 978-3-527-32567-2

related to the rough surface, which was later named SERS [2, 3]. This landmark discovery has opened up a great opportunity to design highly sensitive surface diagnostic techniques applicable to not only electrochemical interfaces but also biological and other ambient ones.

In the mid-1990s, an important breakthrough in the EC-SERS was made by which substantial surface Raman enhancements could be imparted to the VIIIB transition metals of importance for electrochemistry and catalysis. Tian's group developed several surface roughening procedures and demonstrated that SERS could be directly generated on pure Pt, Ru, Rh, Pd, Fe, Co and Ni electrodes, and their surface enhancements ranged in general from one to three orders of magnitude [4]. Since the early 2000s, the replacement randomly roughened surfaces with well-controlled nanostructures of both coinage (e.g. Au, Ag and Cu) and transition metals has introduced the latter as a very promising class of highly SERS-active substrate [5]. To date, several molecular-level investigations by Raman spectroscopy on diverse adsorbates on various material electrodes have been realized.

9.2
Features of Electrochemical Surface-Enhanced Raman Spectroscopy

The EC-SERS system generally consists of nanostructured electrodes and electrolytes. Like other branches of SERS, the basic research on EC-SERS can be broadly divided into two activities: characterization and identification. For identification, the emphasis is on demonstrating whether SERS is able to detect target species in terms of sensitivity and selectivity. For characterization, one should have a more comprehensive understanding of SERS regarding the enhancement mechanism and the surface selection rule. Thereby, one will be able to evaluate the relative contribution of the enhancement mechanisms to the total enhancement and use it to understand how the surface species are interacting with the SERS substrates. Beyond the analytical aspect, EC-SERS has paid more attention to the characterization of the physical aspect and also made great efforts to provide really meaningful information for revealing the adsorption configuration and the reaction mechanism for electrochemistry [6].

The SERS mechanisms have been described in detail by Etchegoin [7] (see also Chapter 1). We just want to mention that both electromagnetic and chemical enhancement can be influenced to some extent by changing the applied electrode potential, that is, the Fermi level of metal and dielectric constant of interfacial electrolyte, in an EC-SERS system. The former, in particular, can be strongly tuned by the potential, leading to a drastic change of the interfacial structure and property. This makes EC-SERS one of the most complicated systems in SERS but also endows the latter with more flexibility. In the following part, we will briefly introduce the features of EC-SERS, so that one may be able to fully harness the advantage of EC-SERS for bio-related application.

9.2.1
Electrochemical Double Layer of EC-SERS Systems

As shown in Figure 9.1, when a light beam illuminates a nanostructured electrode surface to excite the SERS process, a strong optical electric field is established in the electrochemical double layer region. As a consequence, there are two kinds of electric fields with distinctively different properties, the alternating electromagnetic (EM) field and the static electrochemical (EC) field, coexisting in the electrochemical system. The EC field could be quite strong, as the potential drop occurs mainly over the compact layer with about 1 nm thickness or less and the diffuse layer [8]. It influences effectively the interaction (or bonding) between the metal and the adsorbate, the surface orientation of the adsorbate and the structure of the double layer, which may in turn cause the redistribution of the surface-localized optical electric field.

By adjusting the electrode potential, the density and polarity of the surface charge can be changed. When the electrode potential is more positive than the potential of zero charge (PZC), the surface will be positively charged and the water molecules interact with the surface via the negatively charged O end (Figure 9.1a). Meanwhile, other anionic species in the electrolyte may also approach the surface and the molecules of interest may interact with the electrode surface. Taking the pyridine molecule as an example, it can repel the surface water and interact with the surface via both the π-orbital and lone-pair orbital of N atom in a tilted configuration. When the electrode potential is moved to the negative side of the PZC, the interaction of the electrode with the O end of water becomes weaker and that with the H end becomes stronger (Figure 9.1b). Meanwhile, other cations are adjacent to the surface. Under this condition, pyridine may interact with the surface via the lone-pair orbital of N atom, resulting in a vertical configuration. With further negative movement of the electrode potential, the interaction gets weaker

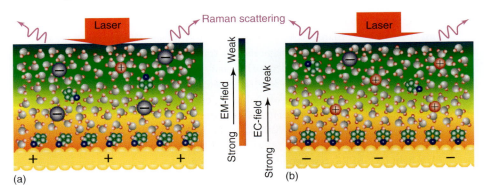

Figure 9.1 Schematic diagrams of the electrochemical interfaces exhibiting SERS and with the coexistence of electromagnetic field and the electric field at the electrode potentials (a) positive or (b) negative to the potential of zero charge [9].

and changes from chemisorbed to physisorbed type, or the pyridine molecule will even be desorbed from the surface.

9.2.2
Potential-Dependent SERS Spectral Characters

Another important and common feature of EC-SERS is that the SERS intensity strongly depends on the electrode potential. The change of the electrode potential may result in a change in the coverage and/or the adsorption orientation of the molecule, both of which will lead to a change in the SERS intensity. It is also possible that, even when the coverage or orientation does not change, the SERS intensity may still change with the potential. The following two factors may mainly account for this phenomenon: the change of the bonding interaction of the molecule with the surface and/or the photon-driven charge transfer (CT) mechanism. A change in bonding strength may affect the geometric and electronic structure of the molecule, which eventually leads to a change in the spectral features of the molecule. In the case of the photon-driven CT effect, one may expect two CT directions, metal to molecule and molecule to metal, which depend on the electronic structure of the adsorbed molecules and the electrode materials as well as the excitation wavelength [10].

9.2.3
Electrode Materials and Excitation Energy Dependence

It is well known that the optical property of a material is closely associated with its electronic structure. The coinage metals as the most SERS-active substrates, which have the prominent optical property due to free electrons, support the effective surface plasmon resonance (SPR). But the SPR effect is severely quenched for Cu and Au due to the d–sp electronic interband transitions when the excitation energies are higher than 2.2 eV, whereas the interband transition of silver occurs at an energy higher than 3.8 eV, corresponding to a wavelength shorter than 326 nm.

In comparison with the coinage metals, transition metals (VIIIB element group) of practical importance in electrochemistry have very different electronic band structures, where the Fermi level is located at the d band and the interband excitation occurs in the whole of visible region. The interband electronic transition depresses the SPR quality of transition metals considerably [11]. Therefore, it reduces the effectiveness of SPR to show intense SERS, as observed in many experiments with visible excitation. It is of interest that the SERS excited by the UV light (UV-SERS) has been observed unambiguously only from some transition metals so far, for example Rh, Pd, Co, Ni and Al, but not from the typical SERS-active substrates of coinage metals [12, 13].

Owing to the strong chemisorption interaction, the Raman scattering intensities of the vibrational modes of the adsorbate can be enhanced under certain conditions. It should be noted that the binding interaction of the same kind of molecules is generally stronger on transition metal surfaces than that on coinage metals. When a strong chemical bond is formed, it will not only change the electronic structure of the adsorbate itself, but also influence to some extent the surface electronic structure. This may cause a shift of the SPR wavelength and lead to a change of the local optical electric field at the metal surface [14]. Moreover, the change in the electronic distribution of the adsorbed molecule even at its electronic ground state may cause different enhancements for different vibrational modes [15].

9.2.4
Electrolyte Solutions and Solvent Dependence

When the electrolyte is changed, it will not only change the double layer structure of the electrochemical interface, but also influence the electrochemical reaction rates and even the reaction window. For example, the double layer will be compressed or expanded with the increase or decrease of the concentration of the electrolyte, respectively. Depending on the electrolyte, non-specific or specific adsorption may occur on the electrode surface. The specific adsorbed ions, due to the strong interaction with the metal surface, will possibly induce a shift of the SPR bands. The electrolyte ion may also be coadsorbed with the adsorbates in a competitive or induced way.

A change of the solvent will change the optical property of the electrochemical system. When organic solvents or ionic liquids are used, the electrochemical window will be drastically expanded and a marked change of the surface physical and chemical properties may be expected. They not only change the oxidation potential of the metal electrode, but also eliminate the hydrogen evolution of water. When the solvent is changed from water to organic or ionic liquid solvents, the SPR wavelength of the metal nanostructures will be red shifted due to the increase in the refractive index of the solvent.

9.2.5
The Electrochemically Influenced SERS Enhancement

As mentioned earlier, the SERS effect in the electrochemical systems is still contributed by the EM and CT enhancements, but it may be modulated by the factors driven by the applied potential. The surface charge density of metal electrodes can be tuned by changing the applied potential with respect to the PZC, resulting in a shift in the SPR wavelength [16, 17]. Applying a positive potential to the electrode surface will lead to a damping and a red shift of the plasmon resonance band. On the contrary, a negative potential will result in an increase and blue shift of the plasmon resonance band.

The potential- and metal-dependent SERS intensity observed in many EC-SERS systems gives strong evidence that the chemical enhancement is cooperative with the EM enhancement, as illustrated in Figure 9.2 [9], where HOMO and LUMO denote the highest occupied molecular orbital and the lowest unoccupied molecular orbital of the adsorbed molecule, respectively. $\psi_g(V_i)$ and $\psi_{CT}(V_i)$ denote the molecular adsorption ground state and the photon-driven CT excited state formed from a filled level of the metal to the LUMO of the adsorbed molecule at an applied

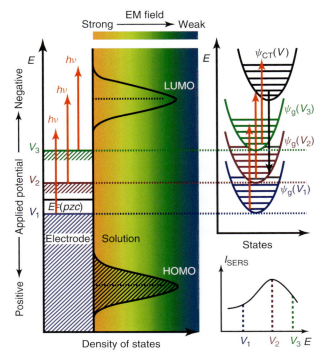

Figure 9.2 Schematic diagrams of the photon-driven charge transfer from a metal electrode to an adsorbed molecule in the EC-SERS system. (Left) The conceptual model of the energy levels changed with the electrode potential in the CT process. (Right) The relevant energy states involved in the electronic levels and the vibrational levels in the CT process. (Right bottom) The corresponding SERS intensity–potential profile. V_i denotes the applied potential [9].

potential V_i. The energy position of the CT state is assumed to be independent of the potential. The SERS signal first increases and then decreases with the negative movement of the electrode potential. This change of the SERS intensity could be explained through the photon-driven CT from the metal to the molecule using the concept of energy levels or energy states. As can be seen on the left part of the figure (energy level concept), at an applied potential of V_1, it is insufficient to produce the photon-driven CT states on the surface with an excitation energy of $h\nu$. As a result, the change of the relative SERS intensity could be determined mainly by the bonding effect. When the potential is changed to V_2, because of the increase of the Fermi level of the metal electrode, the excitation energy matches the required CT energy, leading to a resonance-like Raman scattering and a significant enhancement in the SERS intensity of the relevant vibrational modes. However, if the potential is further negatively shifted to V_3, the excitation energy does not match an ideal resonance condition. Therefore, the contribution of CT states to the SERS signal will decrease.

The right part of Figure 9.2 presents a conceptual picture relevant to not only the energy level but also the energy states with the vibrational coordinates. It displays clearly the change of the surface vibrational energy of the adsorbate with the electrode potential. It should be noted that each energy level represents the total energy of the combined system of the adsorbed molecule and the interacting metal electrode (molecule/metal) in the excited or ground electronic state that can be influenced by the potential change.

From above, one may conclude that EC-SERS is among the most complicated SERS systems. On the other hand, it gives more flexibility to investigate the interfacial structure and mechanism of complex systems by changing the experimental conditions, such as the electrode material, electrolyte, solvent, electrode potential and temperature, in fields of both SERS and electrochemistry. Besides the characterization aspect, if one chooses a potential at which the biomolecule shows the highest signal intensity, one will be able to significantly improve the detection sensitivity.

9.3 Experimental Techniques of EC-SERS

In a typical EC-SERS study, the electrochemical system is investigated while the electrode potential is changed and the spectral response (including the intensity and wavenumber change or even the appearance of new bands) is recorded. In most cases, the experimental data will be interpreted by analysing the intensity or wavenumber change of some characteristic bands (vibrational modes), which may directly reflect a change in the surface coverage, orientation, structure, composition and morphology and sometimes may indicate the involvement of a certain kind of surface enhancement mechanism. Therefore, it is very essential to calibrate both the collection efficiency and wavenumber accuracy using one of the following standards: $520.6\,\text{cm}^{-1}$ of Si wafer, $1333.2\,\text{cm}^{-1}$ band of diamond, 2229.4 and $3072.3\,\text{cm}^{-1}$ of benzonitrile [18].

Before attempting a new EC-SERS study, it is important to first obtain the normal Raman spectrum of the species in its original form, such as the pure liquid, solid or even some standard samples of the expected products, and the Raman spectrum of the solution to be used in EC-SERS. When the concentration is too low, one may consider increasing the concentration to improve the quality of the spectrum. Then, these good-quality spectra will serve as the references for interpreting the EC-SERS result. If the spectrum is too complex to make assignments, an isotopic substitution of a specific atom or theoretical calculations of normal modes may be helpful. Before an EC-SERS study, it will be very helpful to measure an electrochemical cyclic voltammogram to obtain the characteristic potential for the *in situ* EC-SERS study.

9.3.1 Experimental Setup

Figure 9.3 displays the experimental setup for *in situ* EC-SERS. It includes a laser to excite the SERS of samples, a Raman spectrometer to disperse and detect the

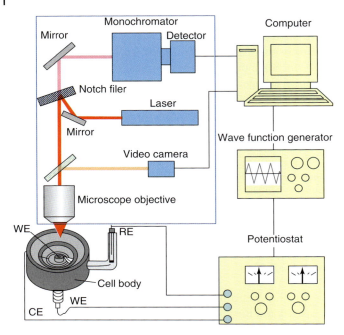

Figure 9.3 Diagram showing the experimental setup for EC-SERS, which includes a Raman spectrometer (in the block), potentiostat, computer, wave function generator and an EC-SERS cell. WE, working electrode; CE, counter electrode; RE, reference electrode [9].

Raman signal, a computer to control the Raman instrument for data acquisition and manipulation, a potentiostat or galvanostat to control the potential or current of the working electrode and an EC-SERS cell to carry out the reaction. In the case of time-resolved studies, it may be necessary to have a wave function generator to generate various kinds of potential/current controls over the electrode and to trigger the detector accordingly to acquire the time-resolved SERS signal.

9.3.2
EC-SERS Cell Design

The sample cell is the core component of the EC-SERS experimental setup. It is relatively simple with a flexible design for investigating aqueous electrolytes. A typical cell configuration used in our lab is shown in the left-bottom corner of Figure 9.3. It consists of a conductive SERS-active working electrode, an inert counter electrode (usually a Pt wire ring) to form a closed circuit and a reference electrode (usually the saturated calomel electrode abbreviated as SCE or a Ag/AgCl electrode) to indicate the potential of the working electrode. The three electrodes should be assembled in a suitable relative geometric position to allow both efficient Raman and accurate electrochemical measurements. For example, the reference electrode should be placed in a compartment with a Luggin capillary tip placed very close to the working electrode to ensure an accurate control of the potential. An optically transparent quartz or glass window may be used to protect the solution or electrode from being contaminated and interfered by the ambient atmosphere and to allow high throughput.

9.3.3
Improving the Detection Sensitivity

Detection sensitivity has always been a key concern to optical methods for probing electrode surfaces because normally only monolayer or even sub-monolayer species (about 10^{15} per cm^2) are presented, which is a challenge especially for weak Raman scatterers. Thus, how to increase the detection sensitivity of EC-SERS becomes a very important issue in the application of Raman spectroscopy in electrochemistry. There are several possible ways to improve the detection sensitivity: (i) use a high numerical aperture (NA) microscope objective with a long working distance to increase the collection efficiency; (ii) increase the laser power as far as it is below the damage threshold because the Raman signal is proportional to the laser power; (iii) use the potential difference method to extract the very weak interfacial signal from the bulk signal; and (iv) select a proper wavelength to utilize the resonance Raman effect. By properly treating the electrode surface, one may be able to obtain very strong SERS from Ag, Au and Cu surfaces and mild SERS signal from Pt, Rh, Pd, Fe, Co and Ni surfaces.

9.3.4
Preparation of SERS-Active Electrode Surfaces

It is well known from previous SERS studies of Au, Ag and Cu that a necessary, but not sufficient, requirement for a large surface enhancement is some form of surface roughness. With the development of nanoscience and nanotechnology, the substrates that can be used for SERS have also been expanded from the traditional massive metal electrodes and metal colloidal sols to template-fabricated substrates and nanoparticle-assembled electrodes for the larger enhancement.

Generally, the electrodes for the EC-SERS study are made by sealing a metal rod into an inert Teflon sheath. To improve the reproducibility of the experiment, the electrode surface should be first mechanically polished with alumina powders down to 0.3 μm, rinsed with ultrapure water and sonicated to remove any adhering alumina. One may also use severe hydrogen evolution or electrochemical cleaning to remove surface impurities. Then the electrode may be used for electrochemical oxidation and reduction or to be dispersed with a layer of nanoparticles and used as an EC-SERS substrate. For other types of substrates, the readers may refer to our previous review [9].

9.3.4.1 Electrochemical Oxidation and Reduction Cycles (ORCs)
Since Ag and Au are dominantly used for bio-related analysis, here we only describe the method for Ag and Au used in the visible to NIR region. The experimental

variables in the oxidation and reduction cycles (ORCs) include the oxidation and reduction potentials, the type of potential–time function and the amount of charge passed during the oxidation step, which may be different using different electrolytes. All potentials mentioned below are referred to the SCE reference electrode.

Ag Electrode A mechanically polished mirror-finished silver electrode is immersed in 0.1 M KCl solution with its potential kept at −0.25 V for a while to reduce the existing surface oxide. Then the potential is pulsed to +0.18 V for 8 s for oxidation. After that, the potential is moved negatively to keep the reduction current density at 3 mA cm^{-2} until a complete reduction of the dissolved silver and the formed silver chlorides. Finally, the potential is moved back to −0.25 V for complete reduction, and the electrode will now appear light grey to light yellow. To obtain a Ag substrate free of chloride, the electrode should be held at −0.60 V in a 1 M NaClO$_4$ solution. Afterwards, the potential is stepped from −0.60 to +0.55 V for oxidation for 3–5 s. Then, the potential is switched back to −0.10 V for complete reduction. The colour of the electrode will be dark green to dark brown.

Au Electrode To obtain an SERS-active Au substrate, one may follow the ORC procedure given by Weaver and co-workers [19]. The Au electrode is first electrochemically cleaned in 0.1 M H$_2$SO$_4$ solution in the potential range −0.25 to 1.5 V. After rinsing with ultrapure water, the Au electrode is kept at −0.3 V in 0.1 M KCl until the current stabilizes. Then, the potential is scanned to 1.2 V at 1 V s^{-1} and kept for 1.2 s for oxidation, then scanned back to −0.3 V at 0.5 V s^{-1} and kept for 30 s for reduction. The cycle is repeated for about 15 min, and the final potential should be −0.3 V to ensure a reduction state of the electrode. This roughening process results in a Au surface with a brown appearance.

9.3.4.2 Preparation of SERS Substrates Using Metal Nanoparticles

Electrochemically roughened SERS substrates have a rather broad distribution of roughness, which is unfavourable for understanding the interfacial structure and maximizing the SERS activity. Alternatively, with nanotechnology, it is now possible to synthesize or fabricate metal nanostructures of various shapes and sizes with a narrow size distribution. The nanoparticles assembled on an electrically conductive substrate can significantly improve the surface uniformity of EC-SERS substrates. Therefore, use of nanoparticle sol or assembled nanoparticles as SERS substrates has been booming in recent years.

The optical properties of metal nanoparticles critically depend on the shape, size and composition of the particles (see also Chapters 1 and 2). Gold nanoparticles with sizes smaller than 20 nm can be directly prepared using Frens' method and used as seeds for synthesizing larger nanoparticles [20]. Nanoparticles larger than 20 nm but smaller than 100 nm can be prepared according to a two-step seed-mediated method by adding 25 mM HONH$_3$Cl into the 12 nm Au sol while stirring followed by the addition of a certain amount of 2.5 mM HAuCl$_4$ to the above solution under stirring for several minutes [21]. Nanoparticles with sizes larger than 100 nm can be synthesized using a three-step seed-mediated method, starting from

12 nm seeds to obtain 45 nm Au seeds to avoid the aggregation of nanoparticles. Thereby, nanospheres with a diameter ranging from about 12 to 200 nm with very uniform distribution can be routinely obtained, with the 130 nm nanospheres showing the strongest SERS with 632.8 nm excitation [22]. Normally, transition metal nanoparticles show relatively weak SERS activity in comparison with Ag and Au nanoparticles. In order to improve the SERS activity, one may utilize the SERS strategy of coating a thin layer of transition metal over Au nanoparticles [23, 24].

Figure 9.4a–f shows the SEM images of monodisperse Au nanoparticles with increasing sizes of 16 ± 2, 50 ± 3, 80 ± 5, 120 ± 8, 135 ± 10 to 160 ± 15 nm, respectively. The shape and size distribution of the nanoparticles become broader with increasing particle size. For electrochemical purpose, the nanoparticles should be supported on a conductive substrate for EC-SERS measurement. After synthesis, the monodisperse, spherical Au nanoparticles should be first centrifuged several times to remove excess reactants and surfactants. Then, the cleaned sol is dropped on a smooth, conductive substrate, such as the glassy carbon (GC), Au or Pt electrode surface. Afterwards, the electrode is dried in a desiccator for about 20 min. The thickness of the nanoparticles on the substrate is controlled by the amount of the particles dropped on the GC surface. Usually, the nanoparticles are drop-coated several times in order to get a uniform multilayer. The typical SEM image of such a Au nanoparticle-coated GC electrode is shown in Figure 9.4g. It can be seen that the film is ordered and fairly uniform over an area as large as 1 mm^2. These well-prepared and reproducible Au nanoparticles on the GC electrode are ready for the systematic SERS and electrochemical SERS studies.

9.3.5
SERS Substrate Cleaning

No matter what kind of electrodes are used, special attention has to be paid to remove the impurities from the surface to eliminate any possible artifacts that

Figure 9.4 SEM images of Au nanoparticles with different sizes: (a) 16 nm, (b) 50 nm, (c) 80 nm, (d) 115 nm, (e) 130 nm, (f) 160 nm. (g) A glassy carbon surface covered with Au nanoparticles with a diameter about 100 nm [22].

may interfere with the measurement and to release the surface sites occupied by impurities, especially in trace analysis. In practice, there are several ways to realize this goal. Nanoparticles synthesized using wet chemistry can be cleaned by several cycles of centrifugation. The electrochemical substrates can be cleaned using plasma or ozone cleaning and electrochemical cleaning by potential cycling, hydrogen evolution or other strong adsorbates (e.g. CN^-, Cl^-, alkanethiols) followed by desorption or electrochemical oxidation.

Iodide, which is an ion with a strong adsorption on Au and Ag, can be used to remove the surface contaminants. The adsorbed iodide is then electrochemically oxidized to form the weakly adsorbed IO_3^-, which can be easily cleaned by rinsing to obtain a clean substrate (see Figure 9.5a [25]). As can be seen from Figure 9.5b,c, the background existing in the uncleaned surface has obviously disappeared after the cleaning. As the oxidation potential of Au is very close to that of I^-, the oxidation of I^- will lead to the partial dissolution of Au, which will result in a decrease of the signal by 50%. To clean the Ag surface, a negative potential can be applied to remove the I^- adsorbed on the surface by desorption. Substrate cleaning is a prerequisite to the SERS detection of weakly adsorbed molecules and unknown samples. How to develop a simple and effective cleaning method while retaining the SERS activity is vitally important for the wide application of SERS.

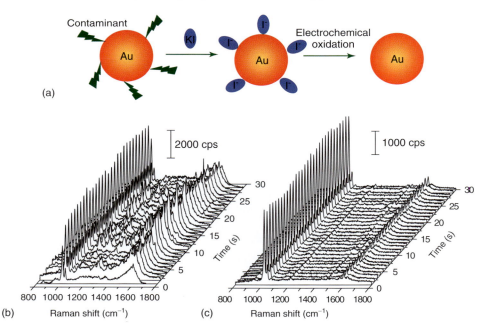

Figure 9.5 (a) The scheme of the cleaning process; (b) time-dependent SERS signals of pyridine obtained at the same point of the substrate before any treatment and (c) after treated with iodide followed by the electrochemical oxidation [25].

9.3.6
An Approach to Reliable SERS Measurement on Bio-related Systems by the Defocussing Method

A common experience of SERS experimentalists using a micro-Raman system is that the SERS signal will keep changing with time with extended illumination under a laser power in the order of milliwatts, due to the tightly focussed laser spot on the surface. The use of a movable sample can solve this problem but will add much trouble to the experiment. A common strategy to reduce this effect is to decrease the laser power, which, of course, will lead to a proportional decrease of the Raman intensity (Figure 9.6a, lower line). However, in a normal micro-Raman system, if we slightly tune the sample off the focal plane, the laser power density will be reduced significantly but the collection efficiency will not deteriorate as fast (Figure 9.5a). From Figure 9.6, we can easily find that at the same laser power density of 1% of the original power density, the signal intensity retains over 80% of the original signal by the defocussing method (by about 5 μm away from the focal plane), but the signal intensity decreases linearly to 1% by lowering the laser power. By focusing the laser at about 25 μm away from the focal plane of the microscope, the laser power density on the sample is reduced to about 0.043% of the ideal focus case. However, the Raman signal obtained still remains about 13% of the maximum signal. This result is extremely meaningful for those systems with very weak Raman signal or that are sensitive to light [26]. Figure 9.6b demonstrates the striking effect of the defocussing method by using oligonucleotide-modified gold surface as an example, and the spectra were obtained with the same acquisition time. The spectra became complicated and deteriorated when the sample was brought into focus. It is quite obvious that the oligonucleotides had already decomposed when the laser was in focus. We can easily conclude that a smaller focus did not always help to improve the signal intensity as well as the signal-to-noise ratio due to the decomposition

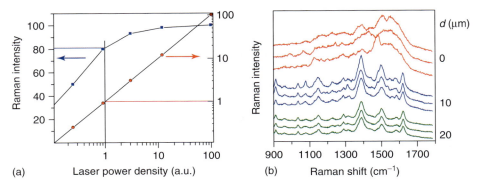

Figure 9.6 (a) The dependence of normalized Raman intensity on the laser power density obtained by the defocussing method (upper) and by lowering the laser power (lower). (b) Effect of defocussing on the SERS signal of an oligonucleotide-modified SERS-active gold surface. At each defocussing distance, three spectra were acquired. All the spectra were obtained with 1 s acquisition time [26].

of the surface species, which produces the Raman and fluorescence signals giving rise to the noisy background. This example convincingly proves that, by using only the defocussing method, one can effectively lower the laser power density at the sample while collecting much stronger Raman signals than by simply lowering the laser power at constant focus size. Apparently, a high NA objective will have a much obvious effect, and a large confocal hole size will improve the throughput and the Raman signal. Recently, there has been an increase in the use of the line focussing method, which expands the laser focus from a spot to a line and can also significantly lower the laser power density without lowering the overall laser power but retaining a reasonably good signal.

9.4
Applications of EC-SERS

Since the first SERS (also EC-SERS) report of pyridine adsorption on the roughened Ag surface, thousands of papers have been published on various aspects of SERS, including over 500 papers dealing with EC-SERS. Because of the limitation of space, we are not able to cover all these in this chapter. But these studies can be easily classified into the following categories: (i) detailed investigation of the interaction of a molecule on the surface under an electrochemical control – we will use benzene and DNA adsorption on the surface as examples; (ii) improved sensitivity of the molecules on the surface by controlling the potential and (iii) indirect detection of the target species by using the strong SERS signal of the Raman tag of dye molecules with potential control.

9.4.1
Model System – Benzene Adsorption and Reaction on Transition Metal Surfaces

As the simplest aromatic molecule and with a clear assignment of all the vibrational modes, benzene has received much attention concerning its reaction and adsorption behaviour on metal surfaces. However, the seemingly simple system is by no means simple [27].

For example, when the potential is more negative than -0.5 V, the Raman spectrum in Figure 9.7a(C) shows the characteristics of cyclohexane, indicating the hydrogenation of benzene in the presence of surface hydrogen on the Pt surface in this potential region. At 1.2 V, the Raman spectrum (in Figure 9.7a(E)) shows the characteristic of liquid chlorobenzene of different substitution numbers, in addition to the 992 cm^{-1} band from benzene, indicating the chlorination of benzene at positive potentials on the roughened Pt surface in the chloride solution.

In the potential region between -0.5 and 0.6 V, the SERS signal from the adsorbed benzene can be observed. Figure 9.7b shows a set of SERS spectra from the roughened Pt electrode. At -0.5 V, several Raman peaks at around 991, 1012, 1043, 1271, 1539 and 1595 cm^{-1} can be clearly seen, related to benzene vibrations. Meanwhile, two peaks at around 310 and 341 cm^{-1} and a peak at around

9.4 Applications of EC-SERS

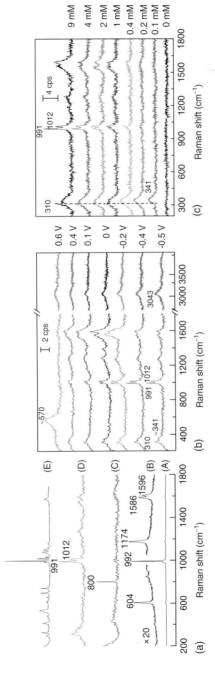

Figure 9.7 (a) Normal Raman spectra of liquid benzene (A and B), and surface Raman spectra in 0.1 M KCl + 9 mM benzene solution at (C) −1.0 V, (D) −0.5 V and (E) 1.4 V. (b) Potential- dependent surface-enhanced Raman spectra on a roughened Pt surface in 0.1 M NaF solutions + 9 mM benzene. (c) Concentration dependence of surface-enhanced Raman spectra in 0.1 M NaF solution with benzene concentration indicated in the figure [27].

3043 cm^{-1} are observed. Their intensities decrease quickly with the positive shift of the electrode potential and no obvious Raman shift can be discerned. When the electrode potential is positively shifted to -0.2 V, the analysis of the surface Raman spectra becomes difficult and unreliable due to the laser-induced carbonization of surface adsorbates. At 0.6 V, the signal of the adsorbed benzene vanishes as a result of the oxidation of the platinum surface. The bands at 310 and 1012 cm^{-1} have the same potential dependence. Meanwhile, the band at 341 cm^{-1} does not change much with potential until the oxidation of the surface.

The results from a concentration-dependent Raman study depicted in Figure 9.7c will help to understand the process. The Raman spectrum is featureless over the wavenumber range 200–1800 cm^{-1} when the solution is free of benzene. At a benzene concentration of about 0.1 mM, a broad band at around 341 cm^{-1} can be clearly seen, and the peak at around 1012 cm^{-1} is very weak. At a concentration of 0.4 mM, the 1012 cm^{-1} peak becomes obvious and a weak shoulder at 310 cm^{-1} appears. Meanwhile, the intensity of the peak at 991 cm^{-1} increases steadily with increasing benzene concentration up to 9 mM. We can see from Figure 9.7c that the intensity of the 991 cm^{-1} peak increases almost linearly with increasing benzene concentration, but, at the same time, the wavenumber is almost the same as that in solution (992 cm^{-1}), indicating the physisorbed nature of benzene and its vibrational bands should be sensitive to the applied potential. The 309 and 1012 cm^{-1} peaks follow essentially the same trend with concentration and are due to the Pt–C band and benzene ring breathing mode with one of hydrogens of benzene lost to form C_6H_5Pt. The broad peak at about 341 cm^{-1} cannot find its companion in the high wavenumber range. This species has a low coverage on the surface, as the charge of hydrogen adsorption/desorption is almost identical to the surface free of this species as revealed by cyclic voltammetry. Studies with a Au core Pt shell (Au@Pt) nanoparticles-assembled electrode, which can provide a higher SERS enhancement, reveal a peak at around 872 cm^{-1} assigned to the benzene ring breathing vibration for benzene adsorbed parallel to the Pt surface, which accompanies the 341 cm^{-1} band.

9.4.2
SERS for Studying Biological Molecules

SERS has been used for studying DNA and constituents, protein and constituents, lipids and related materials as well as some small molecules, such as citric acids, aromatic carboxylic acids and nicotinamide adenine dinucleotide (NAD). Information concerning the structure of the molecule itself, orientation of the molecule on the surface and the potential-dependent behaviour has been provided. If the molecules, such as heme proteins, porphyrins, and flavin derivatives, have an electronic absorption band in the region of the laser excitation wavelength, surface-enhanced resonance Raman scattering (SERRS) can be used to achieve extremely high detection sensitivity up to the single-molecule level (see also Chapters 4, 10 and 11). It should be emphasized that, when SERS or any other surface technique is applied, care should be taken to ensure that the biological molecule keeps its activity and does not get denatured when adsorbing on or bonding to the surface.

9.4.2.1 SERS Study of the Adsorption Behaviour of NADH

The adsorption of NAD^+ on a gold electrode serves as an example for the application of Raman spectroscopy to biological molecules [28]. The SERS spectra, varying with the electrode potential, as well as the normal solution Raman spectra are given in Figure 9.8A. The 745 and 1335 cm^{-1} bands are from adenine, and 1035 cm^{-1} is the characteristic band from the pyridine ring of nicotinamide. With the electrode potential changing negatively, several significant changes can be observed. At potentials negative of -0.25 V, the signal at about 1335 cm^{-1} becomes weaker, while that at 1035 cm^{-1} becomes stronger. The adenine band shifts suddenly from 735 to 745 cm^{-1} when the potential moves across -0.25 V. Thus, it is reasonable to assume, as depicted in Figure 9.8B, that at potentials positive of -0.25 V, the adenine moiety of NAD^+ is adsorbed through the N1 atom and the amino group, whereas the pyrimidine ring of the adenine moiety is in direct contact with the electrode, reflected by the significant increase of the adenine band wavenumber from 735 to 745 cm^{-1}. This configuration results in a closer distance of the nicotinamide moiety to the electrode surface in a vertical conformation, giving a clear band at 1035 cm^{-1}. At potentials negative of -0.25 V, the increase in the intensity of the 1335 cm^{-1} band corresponding to the vibration of N7–C5 atoms indicates that the orientation changes from the N1–N7 to the adsorption of N7 atom, with the N7–C5 bond normal to the electrode surface. This change, in turn,

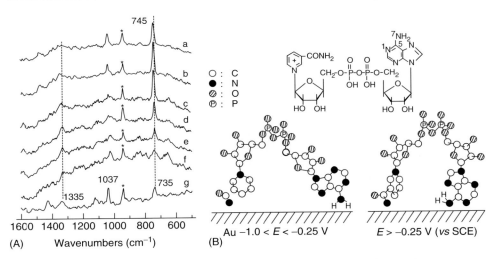

Figure 9.8 (A) SERS spectra of a gold electrode in the presence of 2 mM NAD^+ in a phosphate buffer solution with 0.1 M $NaClO_4$ (pH 7) at (a) 0.25; (b) 0; (c) -0.25; (d) -0.5; (e) -0.75 and (f) -1.0 V vs SCE, together with the Raman spectrum of 100 mM NAD^+ in the solution. The signal labelled with an asterisk is from 0.1 M ClO_4^- in the solution. (B) The molecular structure and the proposed model of NAD^+ adsorption at an Au surface at various potentials [28].

increases the distance of the nicotinamide to the surface, resulting in a decrease of the intensity of the 1335 cm^{-1} band. However, at potentials negative of -1.0 V, the SERS signal disappears as a result of the desorption of the NAD$^+$ from the Au surface.

9.4.2.2 SERS Study of Single-Stranded and Double-Stranded DNA on Gold Surfaces

The interaction of DNA molecules with the electrode surface is very interesting from both fundamental and application points of view. It will help a better design of genechip, which inevitably involves the immobilization and hybridization of DNA on the chip. The hybridization efficiency depends on the coverage and orientation of oligonucleotide probes on the substrate. Therefore, it is essential to understand the interaction of DNA with the substrate.

The potential-dependent SERS spectra of single-stranded DNA (ssDNA) and double-stranded DNA (dsDNA) immobilized on Au substrates have been carried out in a wide potential range from -0.4 to 0.6 V (vs SCE) and are shown in Figure 9.9a,b, respectively [29]. According to the potential dependence of the spectra in the region of the two bands at about 723 and 1447 cm^{-1} related mainly to the adenine ring vibration, one could divide the potential into three ranges: (i) -0.4 to 0 V (for ssDNA), 0.2 V (for dsDNA), (ii) 0.2 V (for ssDNA), 0.4 V (for dsDNA) and (iii) 0.6 V. At the far negative potentials of region (i), the band intensities and band positions keep constant. In the potential region (ii), almost all band intensities decrease substantially and band positions exhibit a blue shift. The band intensities reach a minimum at about 0.6 V. For ssDNA, the band at 723 cm^{-1} moves to 729 cm^{-1} and its intensity decreases by 24% and the intensity of the band at 1447 cm^{-1} decreases by 40%. Both bands have contributions from the out-of-plane modes. For the dsDNA, the two bands at 725 and 1448 cm^{-1} are very sensitive to potential change and their intensities decrease by 50 and 60% at about 0.4 V, respectively. Since the desorption does not occur in the potential region investigated, the substantial decrease of band intensities at more positive potentials may indicate the orientation change of helices. According to the surface selection rules of SERS, the out-of-plane ring modes can be used to determine the orientation of molecules on the surface: when the orientation changes from a more flat to more vertical orientation, the intensity of the out-of-plane mode should decrease. Therefore, the above result suggests a change of a flattened ring orientation to a tilted/vertical ring orientation of adenine when the potential is made more positive. The potential at which conversion occurs is different for ssDNA and dsDNA, the former at 0.2 V and the latter at 0.4 V, which is probably due to the difference in their inflexibility and charges. Moreover, the orientation changes of ssDNA and dsDNA are reversible upon varying the potential.

9.4.2.3 EC-SERS Study of Cytochrome c on a DNA-Modified Gold Surface

Cyt c has special biological function (to transfer electrons between membrane-bound enzyme complexes of cyt c reductase and cyt c oxidase), and its structure, function and redox properties have attracted much interest (see also Chapter 10). The interactions of negatively charged DNA coated on an electrode with positively

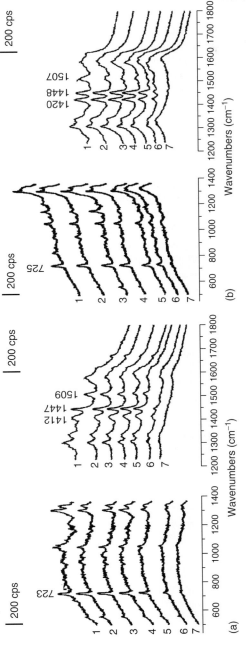

Figure 9.9 The electrochemical *in situ* SERS spectra of ssDNA (a) and dsDNA (b) on a gold substrate in 10 mM Tris + 10 mM NaCl (pH 7.2) buffer at (1) −0.4, (2) −0.3, (3) −0.2, (4) 0, (5) 0.2, (6) 0.4, (7) 0.6 V in the direction of 1 → 7, respectively [29].

charged cyt c can mimic charge-interaction domains of the membrane-bound and membrane-associated proteins under the physiological condition. The information obtained can also be used for the development of a biosensor. Since time-resolved SERRS study of cyt c will be a special chapter in this book (Chapter 10), we introduce only the SERRS study of the interaction of cyt c with DNA here [30].

Figure 9.10 shows the *in situ* SERR spectra of cyt c adsorbed on a DNA-modified Ag electrode in the wavenumber range of 1300–1800 cm^{-1}, providing rich information on the oxidation, spin and ligation state. At 0.12 V, at which the adsorbed cyt c is fully oxidized, the ν_3 region displays a peak at 1504 cm^{-1} (clearly seen in the inset), which is typical for a six-coordinated low-spin configuration of the oxidized (6cLSOx) haem in the native state, while a shoulder at about 1490 cm^{-1} is characteristic of a five-coordinated high-spin configuration of the oxidized (5cHSOx) haem in the non-native state. It indicates the coexistence of two different haem structures in the fully oxidized cyt c adsorbed on the DNA-modified electrode. At −0.08 V, the adsorbed cyt c is fully reduced. The ν_3 region displays only one sharp peak at 1493 cm^{-1} with good symmetry, which is typical for a six-coordinated low-spin configuration of the reduced (6cLSRed) haem in the native state. These results suggest a redox-dependent coordination equilibrium for cyt c adsorbed on the DNA-modified electrode. Upon decreasing the potential, the portion of the native 6cLSRed steadily increases at the expense of the non-native 5cHSOx. Moreover, this potential-dependent conformational transition is reversible (data not shown), indicating that the secondary structure of the adsorbed cyt c keeps nearly unchanged although the haem pocket is disturbed. On considering that a

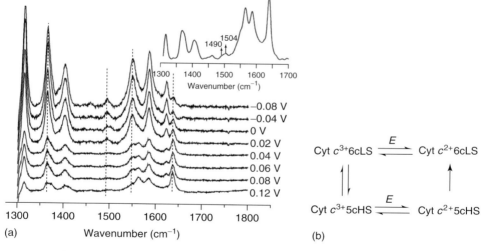

Figure 9.10 (a) *In situ* SERR spectra of cyt c on a DNA-modified Ag electrode. Inset: expanded view of the spectrum at 0.12 V. (b) Reaction scheme of the redox process of the adsorbed cyt c on a DNA-modified metal electrode [30].

concomitant change of the coordination state will significantly reduce the probability of the electron transfer due to a much higher reorganization energy, the redox process of cyt c adsorbed on DNA-modified metal electrode can be seen in Figure 9.10b. Previous *in situ* SERR and RR spectroscopic studies have shown that, when cyt c is electrostatically adsorbed on the electrode, the native state of the adsorbed cyt c, B1 (in which axial ligands of the haem iron, His18 and Met80 stabilize a 6cLS state), is in equilibrium with a new conformational state, B2 (non-native cyt c, such as 5cHS, which lacks the Met80 ligand). The formation of the conformational state B2 is induced by both the local field that results from the Coulombic interactions between the positively charged lysine groups on the surface of cyt c and the negatively charged DNA molecules, and the external field caused by the potential drop across the electrode/electrolyte interface. Both electric fields can affect the electron transfer rate of the adsorbed cyt c.

9.4.3
EC-SERS as a Method to Improve the Detection Sensitivity of Dopamine

As stated earlier, with the potential control, the surface state of the SERS substrate can be tuned to improve the affinity of the SERS substrate to the species to be detected so that the SERS signal can be improved. Here we use EC-SERS detection of dopamine as an example [31]. The SERS spectrum in the aromatic ring stretching region of dopamine in a pH 7.2 buffer is shown in the inset of Figure 9.11a. The spectrum shows intense bands at 1269, 1331, 1424 and 1479 cm^{-1}. The intense 1479 cm^{-1} dopamine band is assigned as v_{19b}. The carbon–oxygen stretching, at about 1270 cm^{-1}, is almost as intense. These two facts indicate that dopamine may be adsorbed on the silver electrode through metal–oxygen bonds, that is, bidentate. Since the 1479 cm^{-1} band gives the highest signal, it is used as the signal for quantitative dopamine measurements. The dependence of the dopamine 1479 cm^{-1} band on the electrode potential is shown in Figure 9.11a. The intensities of the other bands are proportional to that of the 1479 cm^{-1} band intensity at all potentials. The steady-state cathodic current is also shown in the figure. Spectral intensity reaches a maximum near the PZC for silver at −0.9 V and declines at extremely negative potentials, which will cause irreversible carbon reduction at the electrode as well as hydrogen evolution. Therefore, with potential of maximum intensity obtained in the potential-dependent Raman spectrum, the concentration dependence of dopamine SERS intensity can be studied at −0.9 V in order to maximize the signal intensity. The peak heights of SERS spectra obtained at different dopamine concentrations are shown in Figure 9.11b as the working curve. The curve has a log–log plot slope of 0.74, at concentrations below 10^{-5} M. At high concentrations, the working curve approaches a limiting value, presumably corresponding to surface saturation. This kind of working curve is very typical in concentration-dependent SERS measurement and can be used for determining the concentration of dopamine in the solution.

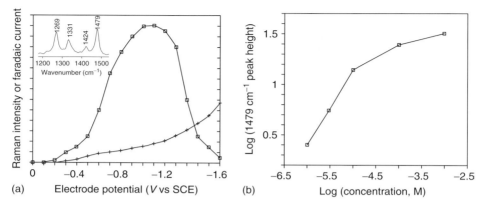

Figure 9.11 (a) Electrode potential dependence of the dopamine 1479 cm^{-1} band in SERS (inset: SERS spectrum of 1 mM dopamine in solution of pH 7.2 obtained at −0.9 V vs SCE). The electrode current is shown in the bottom line. (b) Concentration dependence of the dopamine 1479 cm^{-1} SERS intensity [31].

9.4.4
Discrimination of Mutations in DNA Sequences by Electrochemical Melting Using SERS as Probing Signal

As stated earlier, the charge of an electrode surface will change with a change of the electrode potential, which will affect the configuration of the molecule on the surface. Using this property, Bartlett's group at the University of Southampton proposed an electrochemical melting technique using SERS as the probing signal, named *SERS-E*melting [32].

The ssDNA probe sequence with about 25 bases was used, whose length makes the DNA duplexes formed after hybridization to locate well within the double layer range to feel the effect of the electrode potential. The orientation of the attached dsDNA can be reversibly switched from near the surface at potentials close to the PZC to perpendicular to the surface at potentials negative of the PZC. A negative surface charge will decrease the stability of the dsDNA at the electrode surface and enhance melting. To detect with SERS, the target sequences were labelled with dye molecules with a very strong resonance Raman signal. Therefore, after hybridization the substrates will show the SERRS signal of the dye, which will change with the variation of the electrode potential. This strategy is right a type of indirect detection with SERS label but controlled by the electrode potential.

The SERS-*E*melting experiments were carried out in a custom-designed micro-Raman cell at room temperature with an electrode potential control. The substrate used for the SERS-*E*melting is a gold sphere segment void (SSV) surface made by electrodeposition of a gold layer with a thickness of 480 nm through a close-packed template of 600 nm diameter polystyrene spheres assembled on an evaporated gold surface, followed by removal of the template spheres (Figure 9.12e). The SSV surface was then modified with ssDNA probe sequences and protected with mercaptohexanol to reduce non-specific binding of target DNA at the gold surface (Figure 9.12a,b). Then the target sequences labelled with dye molecules showing resonance Raman signal were hybridized, which places the SERS label to the surface (Figure 9.12c). The cell was flushed several times with 10 mM TRIS buffer, pH 7, and the surface was equilibrated at the open circuit potential for several minutes. Thereafter, the SSV electrode was potentiostatted and the potential decreased in steps from −0.2 to −1.5 V vs Ag/AgCl to measure the SERS signal at each potential step after a fixed time interval of 250 s (Figure 9.12d). As the potential is taken cathodic, the band at 1500 cm^{-1} characteristic of the Texas Red label first increases and then decreases in intensity, eventually disappearing at the most cathodic potential, showing the typical SERS-Emelting profiles. The first derivatives of the sigmoidal curve fits to the melting curves (inset) were used to define the melting potentials (Em) (see inset of Figure 9.12f). The profiles for the perfect match and the two mutations are clearly distinguishable with shifts in the melting potentials (ΔEm) of 110 and 60 mV, respectively. As expected, the mutation with the triple deletion is the least stable and shows the most positive melting potential. The authors also demonstrate the excellent reusability of these surfaces for DNA detection and mutation discrimination. A typical SERS-Emelting analysis in the present unoptimized form took less than an hour. Although this work appears to be more demonstrative, clearly further work is required to understand the full details of the Emelting process. But the good reproducibility, high sensitivity and accuracy make SERS-Emelting a simple method to discriminate single-nucleotide polymorphisms.

9.5
Perspectives

As has been shown, EC-SERS is one of the most complex systems and operates with the synergetic effect of SERS and electrochemistry. All EC-SERS-active systems must possess nanostructures and the SERS activity is critically dependent on the configuration and composition of nanostructures as well as the applied electrode potential, by which some new insights on the SERS phenomena can be gained. On the other hand, SERS is a powerful tool to characterize surface molecules and can provide the fingerprint information of their molecular bonds and molecule–surface bond for the electrochemical interfaces. However, so far the EC-SERS process still lacks a complete microscopic understanding, such as the SERS behaviour of water in different potential regions. It is also necessary to further expand the substrate, surface and molecule generalities of EC-SERS. We briefly describe some future developments in terms of the EC-SERS application in bio-related systems.

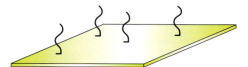
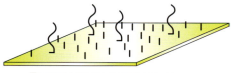

(a) Immobilization on surface

(b) Treated with mercaptohexanol to prevent non-specific binding

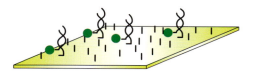
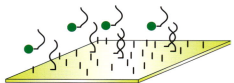

(c) Labelled target sequence hybridized placing the label close to surface

(d) Potential ramp applied to determine melting potential

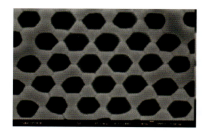
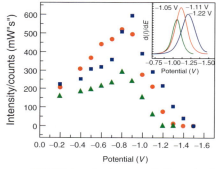

(e) Gold segment void surfaces made from polystyrene spheres

(f) SERS intensity change with the potential and the E-melting curve (inset)

Figure 9.12 (a–d) Experimental procedures for SERS-Emelting experiments. (e) SEM image of a gold sphere segment void (SSV) surfaces made by electrodeposition of 480 nm of gold layer through a close-packed template of 600 nm diameter polystyrene spheres assembled on an evaporated gold surface. (f) The change of the SERS intensity of the 1500 cm^{-1} peak of Texas Red with potential for three different DNA probes, wild type (no mutation, solid square), 1653C/T (single point mutation, solid circle) and ΔF 508 mutation (triple deletion, solid triangle). The inset shows the first-order derivatives of the sigmoidal fits of the intensity curves to manifest the melting potential [32].

9.5.1
Ordered Nanostructured Electrode Surfaces

SERS substrates prepared by various methods to date are far from uniform and therefore the reproducibility of the SERS experiments is still a major concern. The main trend is to further develop some simple and effective methods to fabricate well-ordered nanostructured electrode surfaces by electrochemical and/or chemical methods, such as template methods [33, 34]. It will provide a great opportunity for not only improving the reproducibility and optimizing the SERS activity of existing materials but also searching for a new generation of SERS materials, such as bimetallic and alloy materials. Theoretical simulation will guide effective experimental designs of nanostructures tailored with the special size, shape and inter-particle spacing. It is advantageous to utilize the template method and nanoimprinting technique to reproducibly fabricate nanostructured conductive substrates. Core–shell (or sandwiched) nanostructures, for example consisting of a highly SERS-active metal and a non-SERS-active semiconductor or even a conducting polymer, may extend the EC-SERS to semiconductors and even the whole range of materials in electrochemistry.

9.5.2
EC-SERS Study of Cell under Culturing Condition

Up to now, there have been some demonstrations of using SERS to detect living cells. These studies can be classified into three types: (i) to prepare a solid SERS substrate, and then drop the living cells on the substrate to study the chemical composition of the cell membrane; (ii) to prepare Ag or Au sols first and incubate them together with the cells to allow the ingestion of nanoparticles into the cell and the study of the composition of subcellular organelles (see also Chapter 13); (iii) to label nanoparticles with some strong Raman tags and then protect the nanoparticles by an inert layer of silica or polymers and further with some specific binding group to interact with the cell (see also Chapter 12). Apparently, among these methods, the latter two strategies are impossible to be combined with the electrochemical control of the substrate, but can be used to detect the change of the species under the electrochemical perturbation. However, by using the first method, it will be possible to change the charge of the surface by changing the electrode potential so as to increase the affinity of molecules of different charge to the surface with the opposite charge and therefore to give a signal.

It is of minor biological importance to investigate living cells by exposing them to air. Special measures have to be taken in order to extend the healthy life of the cell. For this purpose, most of previous SERS studies were done by dropping the cells over the SERS substrate and protecting them with solution and sealing [25]. Unfortunately, the cells are not in their living environment and natural state under such a condition and lack nutrition supply. As a result, the lifetime of a healthy cell is generally only about 1–2 h. The ultimate goal of using SERS for cell detection will be to investigate the cell under the cell culture condition. This goal can be realized with a microscopic cell incubation stage that provides the necessary optical window for microscopic observation and keeps the critical incubation conditions, for example, nutrition supply and maintenance of temperature and CO_2 partial pressure. By keeping the cell under the incubation condition, the cells in different cell cycles can be investigated in detail with the help of electrochemical SERS.

9.5.3
Integration of EC-SERS with Microfluidic Devices

Microfluidic devices are capable of integrating the sample preparation, reaction, separation, detection and cell culturing, sorting and dissociation into devices of 1 cm or even smaller. Microfluidic devices offer the unique advantages with combinatorial and high throughput handling and investigation of the systems of interest: for example, minimal sample requirement, reduced reaction time, ease of use, improved product conversion and reduced waste generation (see also Chapter 8). Recently, there have been some demonstrative experiments on SERS detection on microfluidic chips. If the advantage of EC-SERS can be taken to further enhance the detection sensitivity of the system or modulate the interaction of the species with the surface, the advantages of microfluidic device and the EC-SERS can be fully integrated. The key challenges are on the reliable preparation of a clean and stable SERS-active substrate inside the microfluidic channels.

9.5.4
Applications of EC-SERS in Biosciences and Biosensors

Although we have demonstrated some biological application of EC-SERS, the advantages of EC-SERS have not been fully exploited. If we can properly control the electrode potential, we can tune the surface enhancement to the maximum and the surface to a state suitable for biomolecule interaction, which will not only enhance the sensitivity of SERS but also the selectivity for biodetection. Furthermore, simultaneous detection by both SERS and EC methods may provide additional information for complex biosystems. Applying EC-SERS to biodetection by making full use of its advantage will be a very important direction and opportunity for EC-SERS with the growing demand for lowering the limit of detection and higher selectivity for diagnostic techniques in biological and biomedical applications.

It should also be finally pointed out that, in the application of SERS for biological systems, one should pay additional attention to the issue of selectivity, because all these samples have a very complicated composition. Other species in the sample, especially some proteins, will be competitively adsorbed on and deteriorate the SERS substrate, leading to unsuccessful measurement. Therefore, a proper design of the strategy with a high selectivity is highly desirable for SERS application in bio-related systems.

Acknowledgements

This work was supported by National Basic Research Program of China (973 Program Nos. 2009CB930703, 2007CB935603 and 2007DFC40440), Natural Science Foundation of China (20620130427, 20825313 and 20827003). Whenever the work from the authors' group is mentioned, it is the great contribution of the self-motivated and hard-working students.

References

1. Fleischmann, M., Hendra, P.J., and McQuillan, A.J. (1974) Raman spectra of pyridine adsorbed at a silver electrode. *Chem. Phys. Lett.*, **26**, 163–166.
2. Jeanmaire, D.L. and Van Duyne, R.P. (1977) Surface-enhanced Raman spectroelectrochemistry, Part 1. heterocyclic, aromatic, and aliphatic amines adsorbed on the anodized silver electrode. *J. Electroanal. Chem.*, **84**, 1–20.
3. Albrecht, M.G. and Creighton, J.A. (1977) Anomalously intense Raman spectra of pyridine at a silver electrode. *J. Am. Chem. Soc.*, **99**, 5215–5217.
4. Tian, Z.Q., Ren, B., and Wu, D.Y. (2002) Surface-enhanced Raman scattering: from noble to transition metals and from rough surfaces to ordered nanostructures. *J. Phys. Chem. B*, **106**, 9463–9683.
5. Tian, Z.Q., Ren, B., Li, J.F., and Yang, Z.L. (2007) Expanding generality of surface-enhanced Raman spectroscopy with borrowing SERS activity strategy. *Chem. Commun.*, 3514–3534.
6. Tian, Z.Q. and Ren, B. (2003) in *Encyclopedia of Electrochemistry*, vol. **3** (eds P. Unwin, A.J. Bard and M. Stratmann), Wiley-VCH Verlag GmbH, Weinheim, pp. 572–659.
7. Le Ru, E.C. and Etchegoin, P.G. (2009) *Principles of Surface-enhanced Raman Spectroscopy*, Elsevier, Amsterdam.
8. Wieckowski, A. (1999) *Interfacial Electrochemistry: Theory, Experiment, and Applications*, Marcel Dekker, New York.
9. Wu, D.Y., Li, J.F., Ren, B., and Tian, Z.Q. (2008) Electrochemical surface-enhanced Raman spectroscopy of nanostructures. *Chem. Soc. Rev.*, **37**, 1025–1041.
10. Birke, R.L. and Lombardi, J.R. (1988) in *Spectroelectrochemistry – Theory and Practice* (ed. R.J. Gale), Plenum, New York, pp. 263–348.
11. Tian, Z.Q., Yang, Z.L., Ren, B., Li, J.F., Zhang, Y., Lin, X.F., Hu, J.W., and Wu, D.Y. (2006) Surface-enhanced Raman scattering from transition metals with special surface morphology and nanoparticle shape. *Faraday Discuss.*, **132**, 159–170.
12. Ren, B., Lin, X.F., Yang, Z.L., Liu, G.K., Aroca, R.F., Mao, B.W., and Tian, Z.Q. (2003) Surface- enhanced Raman scattering in the ultraviolet spectral region: UV-SERS on rhodium and ruthenium electrodes. *J. Am. Chem. Soc.*, **125**, 9598–9599.
13. Doerfer, T., Schmitt, M., and Popp, J. (2007) Deep-UV surface-enhanced Raman scattering. *J. Raman Spectrosc.*, **38**, 1379–1382.
14. Kreibig, U. and Vollmer, M. (1995) *Optical Properties of Metal Clusters*, Springer, Berlin.
15. Wu, D.Y., Hayashi, M., Lin, S.H., and Tian, Z.Q. (2004) Theoretical differential Raman scattering cross-sections of totally-symmetric vibrational modes of free pyridine and pyridine-metal cluster complexes. *Spectrochim. Acta Part A*, **60**, 137–146.
16. Novo, C., Funston, A.M., Gooding, A.K., and Mulvaney, P. (2009) Electrochemical charging of single gold nanorods. *J. Am. Chem. Soc.*, **131**, 14664–14666.
17. Ali, A.H. and Foss, C.A. (1999) Electrochemically induced shifts in the plasmon resonance bands of nanoscopic gold particles adsorbed on transparent electrodes. *J. Electrochem. Soc.*, **146**, 628–636.
18. McCreery, R.L. (ed.) (2000) *Raman Spectroscopy for Chemical Analysis*, John Wiley & Sons, Inc., New York.
19. Gao, P., Gosztola, D., Leung, L.W.H., and Weaver, M.J. (1987) Surface-enhanced Raman scattering at gold electrodes: dependence on electrochemical pretreatment conditions and comparisons with silver. *J. Electroanal. Chem.*, **233**, 211–222.
20. Frens, G. (1973) Controlled nucleation for the regulation of the particle size in monodisperse gold suspensions. *Nat. Phys. Sci.*, **241**, 20–22.
21. Brown, K.R. and Natan, M.J. (1998) Hydroxylamine seeding of colloidal Au nanoparticles in solution and on surfaces. *Langmuir*, **14**, 726–728.

22. Fang, P.P., Li, J.F., Yang, Z.L., Li, L.M., Ren, B., and Tian, Z.Q. (2008) Optimization of SERS activities of gold nanoparticles and gold-core– palladium-shell nanoparticles by controlling size and shell thickness. *J. Raman Spectrosc.*, **39**, 1679–1687.
23. Li, J.F., Yang, Z.L., Ren, B., Liu, G.K., Fang, P.P., Jiang, Y.X., Wu, D.Y., and Tian, Z.Q. (2006) Surface-enhanced Raman spectroscopy using gold-core platinum-shell nanoparticle film electrodes: toward a versatile vibrational strategy for electrochemical interfaces. *Langmuir*, **22**, 10372–10379.
24. Zou, S. and Weaver, M.J. (1998) Surface-enhanced Raman scattering on uniform transition-metal films: toward a versatile adsorbate vibrational strategy for solid-nonvacuum interfaces? *Anal. Chem.*, **70**, 2387–2395.
25. Li, M.D., Cui, Y., Gao, M.X., Luo, J., Ren, B., and Tian, Z.Q. (2008) Clean substrates prepared by chemical adsorption of iodide followed by electrochemical oxidation for surface-enhanced Raman spectroscopic study of cell membrane. *Anal. Chem.*, **80**, 5118–5125.
26. Lin, X.M., Cui, Y., Xu, Y.H., Ren, B., and Tian, Z.Q. (2009) Surface-enhanced Raman spectroscopy: substrate-related issues. *Anal. Bioanal. Chem.*, **394**, 1729–1745.
27. Ren, B., Liu, G.K., Lian, X.B., Yang, Z.L., and Tian, Z.Q. (2007) Raman spectroscopy on transition metals. *Anal. Bioanal. Chem.*, **388**, 29–45.
28. Cotton, T.M. (1988) in *Advances in Spectroscopy*, vol. 16 (eds R.J.H. Clark, R.E. Hester and H.J.H Clark), John Wiley & Sons, Ltd, Chichester, pp. 91–153.
29. Dong, L.Q., Zhou, J.Z., Wu, L.L., Dong, P., and Lin, Z.H. (2002) SERS studies of self-assembled DNA monolayer– characterization of adsorption orientation of oligonucleotide probes and their hybridized helices on gold substrate. *Chem. Phys. Lett.*, **354**, 458–465.
30. Jiang, X., Wang, Y., Qu, X., and Dong, S. (2006) Surface-enhanced resonance Raman spectroscopy and spectroscopy study of redox-induced conformational equilibrium of cytochrome c adsorbed on DNA-modified metal electrode. *Biosens. Bioelectron.*, **22**, 49–55.
31. Lee, N.S., Hsieh, Y.Z., Paisley, R.F., and Morris, M.D. (1988) Surface-enhanced Raman spectroscopy of the catecholamine neurotransmitters and related compounds. *Anal. Chem.*, **60**, 442–446.
32. Mahajan, S., Richardson, J., Brown, T., and Bartlett, P.N. (2008) SERS-melting: a new method for discriminating mutations in DNA sequences. *J. Am. Chem. Soc.*, **130**, 15589–15601.
33. Willets, K.A. and Van Duyne, R.P. (2007) Localized surface plasmon resonance spectroscopy and sensing. *Annu. Rev. Phys. Chem.*, **58**, 267–297.
34. Mahajan, S., Abdelsalam, M., Sugawara, Y., Cintra, S., Russell, A.E., Baumberg, J.J., and Bartlett, P.N. (2007) Tuning plasmon on nano-structured substrates for NIR-SERS. *Phys. Chem. Chem. Phys*, **9**, 104–109.

10
Electron Transfer of Proteins at Membrane Models

Peter Hildebrandt, Jiu-Ju Feng, Anja Kranich, Khoa H. Ly, Diego F. Martín, Marcelo Martí, Daniel H. Murgida, Damián A. Paggi, Nattawadee Wisitruangsakul, Murat Sezer, Inez M. Weidinger, and Ingo Zebger

10.1
Introduction

Electron transfer (ET) reactions represent key steps in biological energy transduction and conversion as well as in many enzymatic processes. Prominent examples are the oxidative phosphorylation in the respiratory chain and the light-driven water splitting in photosynthesis [1, 2]. These processes as well as many others take place at or in membranes involving integral or peripherally bound redox proteins. Membrane-based electron relays are based on a well-defined spatial arrangement of the redox partners with respect to each other, which is a prerequisite for efficient protein–protein ET processes. However, reaction conditions at membrane interfaces deviate substantially from those in the bulk since mobility of the reaction partners is restricted and dielectrics of the reaction medium are quite different. The most prominent difference, however, refers to the local electric fields. At membranes, electrical potentials arise from different ion concentrations at both sides of the membrane, the distribution of charged and uncharged lipid head groups and the alignment of molecular dipoles of lipids and water molecules in the membrane/solvent interfaces [3]. As a consequence, the potential follows a rather complex profile across the membrane. Particularly pronounced potential changes occur in the interfacial regions where local electric field strengths may approach 10^9 V·m^{-1}, which are likely to affect structures and processes of integral and membrane-attached proteins (Figure 10.1a).

These particular reaction conditions raise some concern whether conclusions drawn from studies of biological ET in solution fully hold for the membrane-based processes. In this respect, electrochemical systems have been suggested to represent appropriate model systems for biomembranes since, due to the similar potential distribution (Figure 10.1), proteins in the electrode/solution interface may experience comparable electrostatic forces as in the membrane/solution interface. Moreover, the electrochemical device allows employing surface-enhanced vibrational spectroscopies which, due to their high sensitivity, promise to provide

---

Surface Enhanced Raman Spectroscopy: Analytical, Biophysical and Life Science Applications. Edited by Sebastian Schlücker
Copyright © 2011 WILEY-VCH Verlag GmbH & Co. KGaA, Weinheim
ISBN: 978-3-527-32567-2

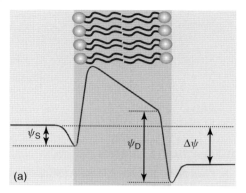
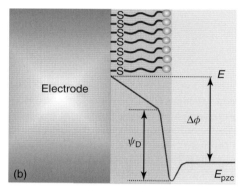

Figure 10.1 Schematic representations of (a) the potential distribution across a lipid bilayer and (b) an electrode-supported self-assembled monolayer of amphiphiles. ψ_S, ψ_D and $\Delta\psi$ denote the surface potential, dipole potential and transmembrane potential, respectively, whereas $\Delta\phi$ is the difference between electrode potential E and the potential of zero-charge E_{pzc}.

novel insight into the interfacial processes of redox proteins not accessible by other techniques. Since the pioneering work by Cotton *et al.* 30 years ago [4], methodological developments have reached a level that justifies these expectations, as will be outlined in this chapter.

This review focuses on the interfacial ET processes of the haem protein cytochrome *c* (Cyt *c*), which has turned out to be an ideal model protein to develop novel surface-enhanced vibrational spectroscopic techniques and to study fundamental processes in great detail. Most of the knowledge that has been accumulated on biological ET processes at interfaces is based on spectroscopic and electrochemical studies of Cyt *c*.

This chapter is organized as follows: First, we describe the analogies between biomembranes and coated electrodes, laying special emphasis on the interfacial electric field distribution and its consequences for protein binding. Second, we introduce the various surface-enhanced vibrational spectroscopic approaches that can be employed to analyse the structure and dynamics of immobilized proteins, in particular haem proteins. Subsequently, the individual puzzle pieces obtained by the various techniques are combined to provide a comprehensive description about the reaction mechanism and dynamics of Cyt *c* at interfaces under the influence of electric fields. Finally, the results are discussed in terms of their impact for understanding the biological functions of Cyt *c*.

10.2
Model Membranes and Membrane Models

The success of electrochemical systems as biomimetic devices has strongly profited from the recent development of strategies for immobilizing proteins on electrodes. Early experiments were restricted to bare metal electrodes, typically covered with chemisorbed anions, which frequently caused irreversibly denaturation of the adsorbed proteins. Nowadays, progress in surface coating by self-assembly of organic molecules allows biocompatible binding of proteins under preservation of the native structure. Particularly versatile strategies are based on ω-substituted mercaptanes which can form self-assembled monolayers (SAMs) on Au and Ag surfaces [5]. The quality of the SAM packing depends on the number of methylene groups x of the mercaptanes, and it is usually very good for $x \geq 10$ but decreases with smaller x. The head group of these amphiphiles, which may be regarded to be lipid analogues, can be chosen according to the properties of the target proteins to guarantee an efficient and, in many cases, uniform protein binding [6]. Thus, the electrode/SAM/solution interface represents a model for one-half of a lipid bilayer since it exhibits an analogous potential distribution due to the similar polarity profile (Figure 10.1b). As a consequence, local electric field strength at the interface of the charged head groups and the solution are likely to be comparable in strength.

Furthermore, the analogy of lipid bilayers and coated electrodes can be extended to the modulation of the local electric field. The crucial parameters are the concentration difference on both sides of the membrane ($\Delta \psi$) in the biological system and the difference between the electrode potential and potential of zero charge ($\Delta \phi$) in the electrochemical model. These analogies suggest that the effect of electric fields on the structure and processes of proteins is similar at membranes and at SAM-coated electrodes. The electrochemical system, however, offers the advantage that the electric field in the SAM/protein interface can be controlled in a defined manner and the interfacial potential distribution can be described by a simple electrostatic model [7, 8], based on an earlier proposal by Smith and White [9]. Accordingly, it is possible to provide an approximate description of the electric field strength at the SAM−protein interface, that is

$$E_{EF} = \frac{-\sigma_C(d_C)d_{RC} + \varepsilon_0\varepsilon_P(E - E_{pzc})}{\varepsilon_0(d_C\varepsilon_P + d_{RC}\varepsilon_C)} \quad (10.1)$$

where $\sigma_C(d_C)$ is the charge density on the SAM surface, which varies with SAM thickness d_C, and the quantities d_{RC}, ε_P and ε_C refer to the distance between the SAM and the redox centre, the dielectric constant in the protein and in the SAM, respectively. E and E_{pzc} are the electrode potential and the potential of zero charge (referred to the standard hydrogen electrode) and ε_0 is the permittivity. According to Equation 10.1, the electric field strength at the SAM surface increases with decreasing SAM thickness, increasing surface charge density and increasing difference between E and E_{pzc}.

To analyse the analogies between lipid bilayer/solution and electrode/SAM/solution interfaces, we have investigated the behaviour of Cyt c bound to both

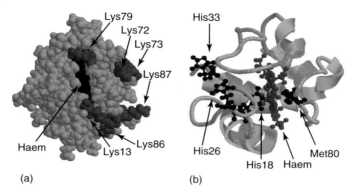

Figure 10.2 Representation of the structure of ferric horse heart Cyt c [11]. (a) View onto the front surface indicating the native axial ligands of the haem (Met80, His18) and the His ligands (His33, His26) replacing Met80 in the B2 state. (b) Space-filling model viewing onto the lysine-rich high-affinity binding domain.

types of devices. Cyt c exhibits a positively charged region around the exposed haem edge (Figure 10.2), which constitutes the docking site for interacting with the negatively charged binding domains of its natural membrane-bound redox partners. In this way, a transient protein–protein complex prior to inter-protein ET from cytochrome c reductase (CcR) to Cyt c and from Cyt c to cytochrome c oxidase (CcO) is formed [1]. The same protein region is likely to be involved in interactions with the mitochondrial membrane itself, which includes the negatively charged cardiolipin as one of the main lipid components [10].

Protein–membrane interactions are usually probed by using vesicles formed by phospholipids. Mono-anionic dioleoyl-phosphatidylglycerol (DOPG) allows preparation of large unilamellar vesicles with a diameter of about 150 nm which is about 50 times larger than the diameter of Cyt c [12]. Such vesicles, hence, can be considered to be a model for the extended negatively charged inner mitochondrial membrane. At low surface coverage, Cyt c binds via electrostatic interactions to form a stable protein–liposome complex which can be characterized by a variety of spectroscopic techniques. Among them, resonance Raman (RR) spectroscopy is particularly instructive since it provides detailed information about the structure of the haem cofactor on the basis of the so-called marker bands that are sensitive for the oxidation, spin and ligation states of the haem [13].

Comparing the RR spectra of Cyt c in solution and in the complex with DOPG liposomes thus allows identifying potential structural perturbations of the haem pocket. The RR spectrum of the ferric form of Cyt c in solution (Figure 10.3) reveals the characteristic marker band signature of an oxidized six-coordinated low-spin (6cLS) haem with the modes ν_3, ν_2 and ν_{10} at 1502, 1583 and 1634 cm^{-1}, respectively. Upon binding to DOPG vesicles, we note a broadening and an increased asymmetry of the band profiles, accompanied by small but clearly detectable wavenumber shifts and changes of the relative

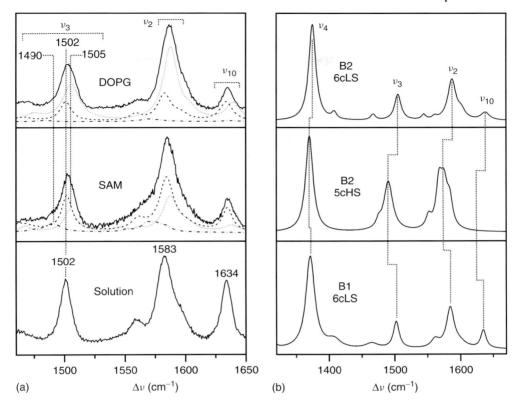

Figure 10.3 RR and SERR spectra of ferric Cyt c in solution, bound to phospholipid vesicles and immobilized on SAM-coated electrodes. (a): Bottom, RR spectrum of Cyt c in solution, pH 7.0; middle, SERR spectrum of Cyt c on an Ag electrode coated with C_1-COOH at $E = 0.37$ V; top, RR spectrum of Cyt c bound to DOPG liposome at a protein/lipid ratio of 0.02, pH = 7.0. The excitation line was 413 nm; further experimental conditions are given elsewhere [12, 16]. (b): Component spectra (including the oxidation marker band v_4) of the B1(6cLS), B2(5cHS) and B2(6cLS) states determined by a component analysis as described previously [14].

intensities. These findings indicate contributions of more than one species to the experimental RR spectrum. This is in fact confirmed by a detailed spectral analysis (component analysis) revealing the involvement of three states, the 'normal' ferric 6cLS form denoted as state B1 and two so-called B2 species, a five-coordinated high spin (5cHS – B2(5cHS)) and an additional 6cLS form (B2(6cLS)) that differs from the B1 state [14]. A comprehensive spectroscopic analysis of this species employing various optical and magnetic resonance spectroscopic techniques shows that the conformational changes that lead to the B2 states mainly refer to the tertiary protein structure and to the structure of the haem pocket. Furthermore, the B2 states were found to be identical to the intermediates formed in the first phase of protein unfolding where the Met80 ligand is removed from the haem iron and the coordination site either remains vacant (5cHS) or is occupied by His33 or His26 (6cLS) (Figure 10.2) [14, 15].

For Cyt c bound to DOPG vesicles, this structural transition is primarily controlled by the electric field. This conclusion can be drawn from the dependence of the B2/B1 conformational equilibrium constant $K_{B2/B1}$ determined as a function of the surface coverage by Cyt c molecules. It is found that $K_{B2/B1}$ decreases with increasing protein/lipid ratio, which in turn corresponds a compensation of the negative surface charges by the protein and thus to a shielding of the surface potential [12] (Figure 10.4). It should be noted that even at full monolayer coverage, corresponding to the protein/lipid ratio of about 0.1, there is still a substantial contribution of the B2 states.

Using a Ag electrode covered with carboxyl-terminated SAMs, the immobilized Cyt c can be studied by surface-enhanced resonance Raman (SERR) spectroscopy. We first focus on the SERR spectra obtained from Cyt c immobilized on short SAM built up by a mercaptane including only one methylene group. At an electrode potential of 0.37 V, where the ferric form of Cyt c prevails, the SERR

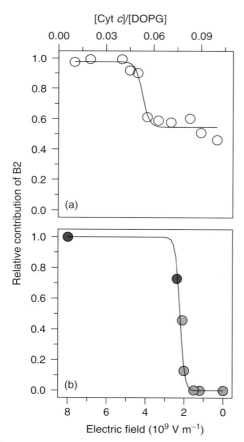

Figure 10.4 Relative contributions of the B2 states for (a) ferric Cyt c bound to DOPG as a function of the protein/lipid ratio, determined by RR spectroscopy [12] and (b) for Cyt c bound to coated Ag electrodes as a function of the electric field strength, determined by SERR spectroscopy [7]. The latter plot includes data from electrodes with C_x-COOH ($x = 15, 10, 5, 2, 1$) coatings (light grey), sulfate and C_{11}-PO_3 coatings (dark grey) [6]. The solid lines are included to guide the eyes.

spectrum displays qualitatively the same changes compared to the RR spectrum in solution as noted for Cyt c/DOPG complexes, indicating that also in this case the conformational state B2 is formed although the relative distribution among the various (sub-)states is different (Figure 10.4). Upon increasing the chain length under otherwise identical conditions, the local electric field at the SAM/protein interface is weakened and the SERR spectra reveal a decreasing relative contribution of the B2 state [7]. The same effect is observed upon increasing the surface concentration of the protein on the SAM.

Thus, we note that immobilization of Cyt c on negatively charged lipid vesicles and SAM surfaces has the same effect on the protein structure: with increasing field strength, a structural transition to the conformational state B2 is favoured, whereas at low electric fields the native structure of the protein is preserved. Consequently, we conclude that – as far as electrostatic interactions are concerned – SAM-coated electrodes exert a similar influence on peripherally bound proteins as phospholipids vesicles, that is, *model membranes*. In this respect, the electrode/SAM device may be considered a *membrane model*; moreover, it is a functional membrane model since it allows monitoring the ET processes of membrane-attached redox proteins.

On the basis of Equation 10.1, the B2/B1 distribution derived from the SERR experiments for SAMs of different thickness can be plotted as a function of the electric field. The corresponding plot in Figure 10.4 also includes data obtained from Cyt c immobilized on a phosphonyl-terminated mercaptane and a sulfate monolayer, which exhibits distinctly higher surface charge densities than layers with carboxyl head groups [6]. A similar estimate of the electric field strength for liposome is not as easy, but the comparison of the $K_{B2/B1}$ values determined in both systems indicates that the local electric field strength at the negatively charged DOPG model membrane is even higher than at Ag electrodes coated with carboxyl-terminated mercaptanes.

10.3
Methods for Probing Electron Transfer Processes of Cytochrome c at Coated Electrodes

ET transfer reactions of immobilized redox proteins on SAM-coated Ag and Au electrodes are usually studied by cyclic voltammetry (CV) [17]. This technique, which has been widely applied to Cyt c, monitors the current flow as a function of the electrode potential and thus only probes the processes of the redox-active proteins. Furthermore, it does not provide information about the molecular mechanism of the interfacial processes, as it can be obtained by the structure-sensitive SERR and surface-enhanced infrared absorption (SEIRA) spectroscopies. To probe the dynamics of the molecular structure changes during the redox process, these techniques have to be operated in a time-resolved (TR) manner, that is, by coupling the SERR and SEIRA detection with the potential jump technique [18, 19]. In this approach, a rapid potential jump by ΔE is applied to the working electrode, leading to a perturbation of the equilibrium at the initial potential E_i.

The subsequent relaxation processes restore thermodynamic equilibrium at the final potential E_f ($E_f = \Delta E + E_i$). These processes can now be probed by TR SERR and SEIRA spectroscopic techniques, which provide various types of information.

First, TR SERR spectroscopy using excitation lines in resonance with the strong electronic transition of the haem cofactor (i.e. 413 nm) monitors the time-dependent changes of the marker bands (see above) and thus the reduction or oxidation of the haem iron as well as possible coupled processes involving changes of the coordination state (Figure 10.5). Specifically, the strongest SERR band originating from the ν_4 mode at 1372 cm^{-1} in the oxidized form shifts down by about 10 cm^{-1} upon reduction of the haem [13]. This downshift is the result of force constant changes, since the additional negative charge increases the electron density in the anti-bonding orbitals of the porphyrin, thereby weakening the C–N bonds which represent the main internal coordinate of the ν_4 mode. This redistribution of electron density in the haem is much faster than the ET from or to Cyt c, implying that the temporal evolution of the ν_4 modes in the TR SERR experiments directly reflects the ET reaction.

Second, under Q-band excitation, that is, using excitation lines close to the weak Q-transition of the haem, the surface enhancement of totally symmetric (A_{1g}) and non-totally symmetric modes (e.g. B_{1g}) of the haem depends on its orientation relative to the surface (Figure 10.5) [16]. Thus, time-dependent changes of the relative intensities of B_{1g} versus A_{1g} modes reflect changes of the protein orientation during the relaxation process.

Third, TR SEIRA spectroscopy provides complementary information about redox-linked structural changes of the protein and orientational changes of individual peptide segments [19]. TR SEIRA spectroscopy is carried out by either employing the step scan or the rapid-scan technique for probing processes faster or slower than about 100 ms. To enhance the sensitivity, SEIRA experiments are carried out in the difference mode, that is, the spectra measured at a potential E_f are always related to a reference spectrum obtained at potential E_i, which, for instance, may be chosen such that the reference spectrum exclusively reflects the reduced form of Cyt c. Then the difference spectra ($E_f - E_i$) display only those bands that undergo potential-dependent changes. The characteristic marker bands for this technique are located in the amide I band region which is indicative of the various secondary structure elements of Cyt c (Figure 10.5). Among them are the bands at 1693 and 1673 cm^{-1} of the reduced and oxidized Cyt c, respectively. These bands are tentatively assigned to the amide I mode of the β-turn III segment 67–70. This band pair is also observed in spectroelectrochemical studies in solution, implying that they are characteristic features of the native protein structure in state B1. Furthermore, due to the similarities with the IR difference spectra in solution, the appearance of these bands in the SEIRA difference spectra indicate true secondary structure rather than orientational changes of the peptide segment, that is, tertiary structure changes. This conclusion might not hold for the band pair at 1666 and 1660 cm^{-1} since it does not directly coincide with analogous features in the solution difference spectrum [19].

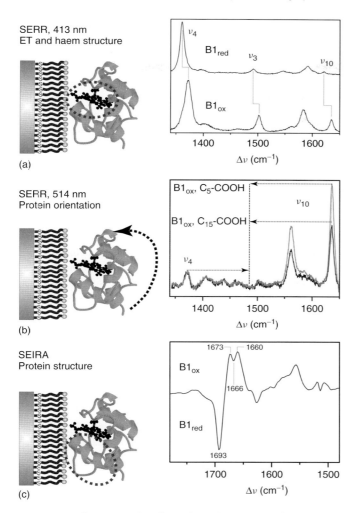

Figure 10.5 Illustration of surface-enhanced vibrational spectroscopic approaches for studying Cyt c. (a) SERR spectroscopy under rigorous resonance conditions for probing ET and haem structure changes based on the characteristic shifts of the marker bands. (b) SERR spectroscopy under Q-band excitation for probing protein orientational changes based on the intensity ratio changes of the ν_4 and ν_{10} modes. (c) SEIRA spectroscopy for probing protein structural and orientational changes based on the amide I band difference signals.

Although these spectroscopic methods together with CV promise to provide a comprehensive picture on the interfacial redox processes, the information obtained by these techniques do not exactly refer to the same conditions. Probing potential-dependent changes of the haem cofactor by SERR spectroscopy requires Ag electrodes, which upon electrochemical roughening affords a surface enhancement in the spectral region below 600 nm, necessary to combine the

SER enhancement with the molecular resonance and pre-resonance enhancement between 400–450 nm and 500–570 nm, respectively. SEIRA spectroscopy is carried out on thin Au films that are chemically deposited on silicon ATR crystals used for attenuated total reflection spectroscopy. CV measurements are usually performed on polished Au surfaces, even though extension to deposited Au films and even to electrochemically roughened Ag surfaces is possible albeit at the expense of a decreased signal-to-background ratio and, in the case of Ag, of a restricted potential scan range.

In this respect, progress in developing novel SER-active devices has been made, which promises to bridge the gap between the Ag- and Au-based experiments [20]. Such a device consists of an electrochemically roughened Ag electrode that is coated with an amino-terminated SAM ($x = 11$). After addition of tetrachloroaureate and subsequent electrochemical reduction, a thin Au film is deposited on the SAM with a thickness of about 15–20 nm, depending on the experimental conditions. Such a device exhibits electrochemical properties similar to bulk Au electrodes but provides a strong surface enhancement in the spectral region below 550 nm that is comparable to that of pure Ag surfaces. This remarkable phenomenon may be qualitatively explained by a non-radiative excitation energy transfer from the optically excited surface plasmons of Ag to those of the Au layer, given that the Au layer is thin enough and thus sufficiently transparent in this wavelength region. Note that the direct optical excitation of Au plasmons below 550 nm is impaired by the strong d → d transitions of the metal.

10.4
The Unusual Distance Dependence of the Interfacial Electron Transfer Process

We first focus on the TR SERR experiments with 413 nm excitation (Figure 10.5a) probing the haem oxidation and reduction. Designing the potential jump experiments such that the final potential is equal to the redox potential, that is, $E_f = E^0$, refers to conditions where the forward and backward ET rate constants are the same corresponding to the formal heterogeneous rate constant k^0_{ET}. Thus, in the absence of any other coupled processes, this quantity can be directly derived from the experimentally determined relaxation time τ_{relax}:

$$\tau^{-1}_{relax} = k^0_{ET,forward} + k^0_{ET,backward} = 2k^0_{ET} = k_{app} \tag{10.2}$$

Such TR SERR experiments with Soret-band excitation have been carried out with Cyt c immobilized on Ag electrodes coated with carboxyl-terminated SAMs of different chain lengths (C_x-COOH; $x = 15, 10, 5, 2, 1$) [18]. The spectra were measured following a potential jump from $E_i = +0.13$ V to $E_f = E^0$ after different delay times (Figure 10.6). The quantitative evaluation of the spectra is straightforward on the basis of a component analysis in which the complete spectra of the individual components are fitted to the experimental spectra. The relative amplitudes of the component spectra, which are linked to the relative concentration via the RR cross sections, are thus the only variables in the fitting process, which is therefore associated with a high accuracy, given that the assumed number of components is

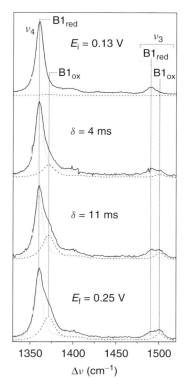

Figure 10.6 Stationary SERR spectra of Cyt c immobilized on Ag electrodes coated with C_1-COOH at the initial potential $E_i = 0.13$ V (top) and the final potential $E_f = 0.25$ V (redox potential, bottom) and TR SERR spectra obtained at different delay times δ following a potential jump from E_i to E_f. All spectra were measured with 413 nm excitation [18]. The dotted and dashed lines represent the component spectra of the reduced and oxidized state of B1.

correct. For shorter chain lengths ($x < 10$), contributions of the B2 states have to be included, specifically at low protein surface concentration, which, however, do not participate in the redox processes under the chosen experimental conditions. Correspondingly, the kinetic analysis is consistent with a one-step relaxation process, and for $x = 15$, τ_{relax} is determined to be 7.1 s, corresponding to $k_{app} = 0.14$ s^{-1} (Equation 10.2; Figures 10.7 and 10.8) [18]. At $x = 10$, k_{app} has increased substantially to a value of 83 s^{-1}. The data for $x = 15$ and $x = 10$ are consistent with a non-adiabatic electron tunnelling process, as concluded by comparison with rate constants determined for long-range ET in solution. The rate constants determined for shorter SAM lengths, however, do not follow the increase expected from the exponential distance dependence of electron tunnelling but instead level off to afford a largely constant value of about 250 s^{-1} for $x \leq 5$ (Figure 10.7). These findings imply that another process rather than electron tunnelling is rate-limiting. This conclusion is further supported by measuring the rate constant as a function of the overpotential ($\eta = |E_f - E^0|$ with $E_f \neq E^0$), which increases the driving force for the ET. At $x = 15$, $k_{app} = f(\eta)$ nicely follows the Marcus description of heterogeneous ET, affording a reorganization energy of about 0.25 eV [21]. This quantity is distinctly smaller than the value determined for Cyt c in solution, but can be rationalized mainly in view of the reduced contribution of the solvent reorganization for the immobilized protein. At SAMs with $x = 2$, however, k_{app} is

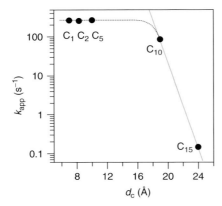

Figure 10.7 Distance dependence of the ET rate constants k_{app} determined for Cyt c immobilized on Ag electrodes coated with C_x-COOH. The data were determined from TR SERR experiments (413 nm excitation) as described elsewhere [18]. The dotted line represent the expected distance dependence for electron tunnelling. The dashed line is included to guide the eyes.

essentially independent of the overpotential, which is an additional evidence for a rate-limiting process other than electron tunnelling at short distances.

The non-exponential distance dependence of the ET process is not a unique property of Cyt c but has been observed for a variety of proteins using SERR spectroscopy and electrochemical methods [6]. However, the most powerful methods for elucidating the mechanism of the interfacial redox processes are

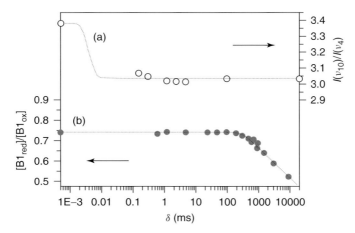

Figure 10.8 Kinetic data derived from TR SERR experiments of Cyt c immobilized on Ag electrodes coated with C_{15}-COOH, following a potential jump from $E_i = +0.25$ V to $E_f = +0.15$ V. (a) Intensity ratio changes of the ν_{10} and ν_4 modes obtained from experiments with 514 nm excitation (the dotted line is included to guide the eyes). (b) Concentration ratio of B1$_{red}$ and B1$_{ox}$ obtained from 413 nm experiments (the dotted line represents a monoexponential fit). Further details are described elsewhere [16].

surface-enhanced vibrational spectroscopies, specifically when applied to haem proteins. Using the methods described in the previous section, we may monitor the potential-jump-induced processes in terms of ET and haem structure changes, protein reorientation and protein structural changes in order to elucidate the origin of the unusual distance dependence of the interfacial ET reaction.

10.5
Electron Transfer and Protein Orientational Dynamics

Careful analyses of the TR SERR spectra probed under rigorous resonance conditions (413 nm excitation, see above) do not reveal any haem structural changes coupled to haem oxidation and reduction at any SAM thickness. This conclusion is derived from the entire spectral range of the porphyrin fundamentals, including those modes (200–500 cm^{-1}) that are known to respond sensitively to subtle changes of the protein–cofactor interactions. Evidently, the structural adaptation of the haem pocket occurs quasi-instantaneously with the ET. As discussed above, the conformational transition from the native state B1 to the state B2, which is associated with a ligand exchange and a substantial alteration of the tertiary structure [14], is distinctly slower than the ET process.

Next we analyse the rotational diffusion of the immobilized protein on the SAM surface. Such a process is associated with a change of the orientation of the haem with respect to the metal surface, which in turn is reflected by changes of the intensity ratio of modes of different symmetries under Q-band excitation (514 nm excitation). At C_{15}-COOH, this process is much faster than ET and at the limit of the time resolution of the experimental setup such that only a lower limit of 6000 s^{-1} can be defined (Figure 10.8) [16]. At C_{10}-COOH, this constant has decreased drastically (379 s^{-1}) but still remains larger than the relaxation constant of the ET process. When at this SAM thickness, the driving force for electron tunnelling is raised, k_{app} only increases up to the value of the reorientation constant, which in turn is overpotential-independent. These findings imply that, for C_{10}-COOH at zero driving force, k_{app} exclusively reflects electron tunnelling but rotational diffusion of the immobilized protein may constitute the rate-limiting event for the overall interfacial redox process when the tunnelling rate is increased. Consequently, it is very likely that at shorter SAMs the deviation from the non-exponential distance dependence of k_{app} is due to protein reorientation. In fact, for C_5-COOH the value for k_{app} was found to be the same as the rate constant derived from the intensity ratio changes (about 250 s^{-1}), implying that reorientation is further slowed down with decreasing distance and thus has become the rate-limiting event. This conclusion is further supported by measurements of the viscosity dependence, which reveal a decrease of the reorientation rate constant and thus of k_{app} by a factor of 2 when the viscosity in solution is increased by 20% via the addition of sucrose [16].

The relationship between rotational diffusion of the protein and ET can be understood on the basis of recent molecular dynamic (MD) simulations of Cyt c immobilized on an SAM surface [22]. The study revealed an orientational distribution due to the involvement of different Lys residues in electrostatic binding. Among them are a medium-affinity binding domain (BD2; Lys25, Lys27) and a high-affinity binding domain (BD3, Lys13, Lys72, Lys73, Lys79, Lys86, Lys87) (cf. Figure 10.2), which are associated with different average electronic coupling parameters as demonstrated by pathway analyses. It was found that for the preferred high-affinity binding domain BD3 the average coupling parameter is about 10 times lower, implying that electron tunnelling is approximately 100 times slower for the majority of the Cyt c species as compared to the minor fraction of Cyt c species bound via BD2.

If transitions between the various orientations are sufficiently fast, ET will predominantly occur via the BD2-bound species. This is evidently the case for Cyt c on C_{15}-COOH and C_{10}-COOH. At C_5-COOH, the distance-dependent increase of the electron tunnelling rate and the decrease of the reorientation rate lead to a rather complex kinetic behaviour due to the convolution of the distribution of various orientations associated with orientation-dependent electron tunnelling probabilities and the transitions between the various orientations, that is, rotational diffusion. Thus, one would expect a non-monoexponential ET kinetics, which, however, could neither be confirmed nor ruled out taking into account the accuracy of the TR SERR measurements. It should be noted that under these conditions the time-dependent changes of the orientational distribution as reflected by the TR SERR experiments may reflect rotational diffusion *and* the orientation-dependent ET.

10.6
Electric Field Effects on the Electron Transfer Dynamics

The results discussed so far show that the plateau region of the distance-dependent ET is due to interference with the orientational dynamics which becomes eventually rate-determining at short tunnelling distances (Figure 10.7). It is reasonable to assume that this distance dependence of the rotational diffusion of the immobilized protein is due to the interfacial electric field strength $\mathbf{E}_F$. The electric field dependence of the binding energy $\Delta G_i(E_F)$ for Cyt c in a specific orientation may be expressed by

$$(\Delta G_i(E_F) - \Delta G_i(0)) \propto |\mathbf{E}_F| \cdot |\mu_i| \cdot \cos\alpha \qquad (10.3)$$

where $\Delta G_i(0)$ is the binding energy in the absence of an electric field, μ_i is the molecular dipole moment of Cyt c and α is the angle of the dipole moment vector with the surface normal, that is, the vector of the electric field. One can easily see that the differences between the binding energies for different orientations will increase with the electric field, leading to different equilibrium distributions of the protein orientations. This conclusion is, in fact, confirmed by the Q-band-excited SERR spectra of Cyt c immobilized on SAMs of different thickness and thus under the influence of different electric fields [16]. For a given electrode potential, the intensity ratio $\nu_{10}(B_{1g})/\nu_4(A_{1g})$ decreases with the thickness of the SAM, tending to the value of the isotropic sample in solution at very long distances. The same tendency is observed for a given SAM upon decreasing the electrode potential such that $|E - E_{pzc}|$ decreases. Both observations indicate that higher electric fields tend to align more effectively the dipole moment of the adsorbed protein, leading to a more perpendicular average orientation of the haem plane with respect to the electrode surface. Accordingly, also the reorientation rate constant of the immobilized protein is expected to decrease with the chain length since the concomitant increase of the electric field strength raises the activation energies for the transitions between the individual orientations.

On the basis of this scheme, one may easily understand the behaviour of other redox proteins on SAM-coated electrodes. Particularly instructive is the comparison of the results obtained for iso-1 cytochrome c (iso-1 Cyt c) from yeast with those of Cyt c from horse heart (HH) which have been discussed so far [23]. Both proteins are highly homologous and possess very similar three-dimensional structures including the same Met-His ligation pattern of the haem. Also, the redox potentials and the RR spectra are nearly identical. The Lys residues involved in electrostatic binding of HH Cyt c are conserved in iso-1 Cyt c. Thus, it is quite surprising, at first sight, that the behaviour of iso-1 Cyt c in the immobilized state differs substantially from that of HH Cyt c. First, binding of iso-1 Cyt c to C_{10}-COOH SAMs leads to a nearly complete conversion to the conformational state B2, whereas for HH Cyt c essentially no state B2 could be detected under these conditions. Second, using mixed monolayers of C_{10}-COOH and C_{10}-OH stabilizes the native B1 state of iso-1 Cyt c but the interfacial ET is 10 times slower (5 s^{-1}) than for HH Cyt c. The same value is determined for the orientational changes of the immobilized iso-1 Cyt c, indicating a gated ET already at this SAM thickness. These discrepancies compared to HH Cyt c can be understood taking into account the different dipole moments. MD simulations combined with electrostatic calculations afforded a dipole moment for iso-1 Cyt c, which is about four times larger than for HH Cyt c. This remarkable difference, which results from subtle differences of the protein structure, leads to a substantially higher electric field–dependent energy term $|E_F| \cdot |\mu_i| \cdot \cos \alpha$ (see Equation 10.3), which evidently causes a dramatic shift of the conformational equilibrium towards the B2 state and an increase of the activation barrier for rotational diffusion of the protein.

Thus, the electric field control of the rotational diffusion appears to be a general phenomenon in the heterogeneous ET of immobilized proteins. This conclusion is further supported for other haem proteins at different types of monolayers on Ag electrodes using SERR spectroscopy [8, 24]. The same mechanism evidently holds for redox proteins on SAM-coated Au electrodes where electrochemical studies have revealed qualitatively similar results with a largely distance-independent ET rate constant for short monolayers (cf. [6]). However, the absolute values are higher by more than a factor of 2 for Au electrodes than for Ag electrodes when comparing the same protein and the same monolayer. Again, this effect can be interpreted in terms of the interfacial electric fields that are different for coated Au and Ag electrodes.

According to Equation 10.1, the electric field strength at the SAM/protein interface is approximately given by the sum of two terms: the charge density on the SAM surface σ_C and the quantity $|E - E_{pzc}|$. While σ_C is negative and presumably quite similar for SAM-coated Ag and Au electrodes, E_{pzc} is >0.0 and about -0.7 V for Au and Ag, respectively. Thus, for electrode potentials around the redox potential of Cyt, $(E - E_{pzc})$ is negative for Ag but positive or close to zero for Au, such that the electric field strength is expected to be distinctly larger for Ag than for Au electrodes. As a consequence, the electric field–dependent increase of the activation barrier for rotational diffusion is less pronounced for Au and the limiting rate constant is higher.

10.7
Electron Transfer and Protein Structural Changes

So far we have only focused on the ET reaction and possible structural changes of the haem as well as on the rotational diffusion of the protein. Now we turn to the TR SEIRA spectroscopic analysis of the redox-linked protein structural changes on the basis of the amide I bands. These experiments were carried out with HH Cyt c on coated electrodes. As in the SERR experiments, a potential jump to the redox potential was employed and the subsequent band intensity changes were probed by rapid-scan SEIRA for C_{15}-COOH and step-scan SEIRA measurements for C_{10}-COOH and C_5-COOH (Figure 10.9) [19]. As expected, the overall distance

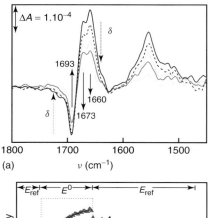

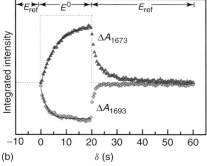

Figure 10.9 (a) Rapid-scan SEIRA spectra of Cyt c immobilized on Au electrodes coated with C_{15}-COOH using the spectrum measured at $+0.1$ V as a reference. The individual traces represent the difference spectra obtained at 1 s (grey, dotted), 3 s (grey, solid), 5 s (black, dotted), 10 s (black, dashed) and 20 s (black, solid) after the potential jump from $E_{ref} = 0.1$ V to the redox potential ($E^0 = 0.24$ V). (b) Kinetic traces of the intensity changes of the SEIRA bands at 1693 cm^{-1} (grey squares, reduced) and 1673 cm^{-1} (grey triangles, oxidized), obtained from the above TR SEIRA spectra.

dependence strongly deviates from an exponential behaviour. In general, the relaxation rate constants derived from the intensities of the amide I bands at 1693 and 1673 cm^{-1} (β-turn III segment 67–70 of the reduced and oxidized form, respectively) are the same for each SAM, and for C_{10}-COOH coatings also nicely agree with k_{app} of the ET reaction on Ag determined by TR SERR experiments (413 nm). We may, therefore, safely conclude that structural changes of β-turn III segment 67–70 occur concomitant with the ET. In the gating region, that is, at C_5-COOH, the relaxation constants determined by SEIRA spectroscopy for Cyt c on Au are about three times larger than those derived from the SERR measurements for the Ag electrode, which, however, is consistent with previous electrochemical data as discussed above. Also at very long distances, that is, at C_{15}-COOH, the relaxation constants for the SEIRA intensities at 1693 and 1673 cm^{-1} on Au is larger by a factor of about 2 than that of the SERR intensities of the oxidation state marker bands, indicating faster electron tunnelling on Au electrodes at very long distances. This acceleration was suggested to be due to the different SAM structures on both metals which result from a larger tilt angle of the Au–S compared to the Ag–S bond and thus may lead to a slightly shorter protein-to-electrode distance [5]. The effect of the larger tilt angle is less pronounced for C_{10}-COOH, for which the same tunnelling rate is determined for Ag and Au electrodes.

The fact that SEIRA and SERR experiments are usually carried out on different metals leaves some uncertainty about the comparability of the results, despite the explanations discussed above. With the development of Au–Ag hybrid electrodes as described above, SERR spectra could also be measured for Cyt c immobilized on coated Au surfaces. In the stationary spectra (413 nm), no differences were observed for Au–Ag hybrid devices with C_{10}-COOH and C_{15}-COOH coatings as compared to the respective coated Ag electrodes, implying that in each case the immobilized Cyt c is exclusively in the native conformational state B1. On the other hand, the ET kinetics for a C_{15}-COOH coating affords an apparent rate constant of 0.8 s^{-1}, which is distinctly larger than those determined for pure Ag (0.14 s^{-1}; SERR) and Au electrodes (0.4 s^{-1}; SEIRA). Conversely, the same rate constants are determined for all three types of electrodes upon coating with C_{10}-COOH. In view of these results, it seems to be questionable that different SAM structures on Au and Ag surfaces are the origin for the quite substantial variation of the electron tunnelling rates at long distances on Ag, Au and Au–Ag electrodes.

10.8
Overall Description of the Mechanism and Dynamics of the Interfacial Processes

We may now combine the various kinetic data measured by the three techniques to obtain an overall description of the dynamics of the interfacial processes of Cyt c as shown in Figure 10.10. The rate constants are related to the electric field strength at the SAM/protein interface as estimated according to Equation 10.1. It can be seen that for Cyt c the transition between the electron tunnelling

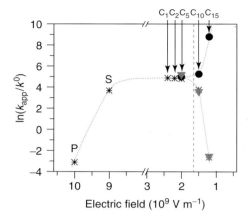

Figure 10.10 Electric field dependence of the rate constants for protein reorientations (circles), redox-linked protein structural changes (triangles) and ET (stars), which were obtained by TR SERR spectroscopy with 514 nm excitation, TR SEIRA spectroscopy and TR SERR spectroscopy with 413 nm excitation, respectively. The data include the rate constants obtained from electrodes covered with carboxyl-terminated SAMs (C_x), sulfate (S) and C_{11}-PO_3 (P) and are related to the electric field strength according to Equation (10.1) [6]. The dashed line separates the region of electron tunnelling (right) and the gated region (left). The dotted lines are included to guide the eyes. Further information is given in the text.

and the gated region occurs above a field strength of 1.5×10^9 V m^{-1}. Up to this threshold, rotational diffusion of the protein is sufficiently fast to ensure ET via the protein orientation that is most efficient for electron tunnelling. Because of the field-induced decrease of the reorientation rate, the mobility of the protein on the surface is increasingly reduced, resulting in a complex interplay between orientation-dependent electron tunnelling and rotational diffusion. This scenario can account for the largely distance-independent rate constants between 2.0 and 2.5×10^9 V m^{-1}. Both in the tunnelling and in the gated region, the redox-linked protein structural changes as reflected by the amide I bands of the β-turn III segment 67–70 occur instantaneously with the ET. The only exception refers to the yet unassigned amide I band at 1660 cm^{-1} of the oxidized form, for which a retardation of the kinetics by a factor of 2 has been observed [19].

Figure 10.10 also includes the apparent ET rate constants determined from TR SERR experiments of Cyt c bound to Ag electrodes covered with sulfate ions and phosphonyl-terminated SAMs (C_{11}-PO_3). In these systems, the surface charge density is significantly higher than for carboxyl-terminated SAMs such that Equation 10.1 predicts local electric fields higher than 8×10^9 V m^{-1}. Although the estimates of the field strengths are associated with a considerable uncertainty and the determination of the respective rate constants is aggravated by interference with the conformational transitions between the states B1 and B2, it is justified to conclude that at sufficiently high electric fields k_{app} decreases.

10.9
Interfacial Electric Fields and the Biological Functions of Cytochrome c

In the previous sections we have demonstrated that SERR and SEIRA spectroscopic techniques can provide detailed insight into the mechanism and dynamics of the potential-dependent processes of Cyt c on coated electrodes. Now we discuss whether these finding are also relevant for the processes of Cyt c in vivo.

Cyt c exerts two qualitatively different physiological functions related to life and death of cells. On the one hand, Cyt c serves as an electron transporter in the respiratory chain shuttling electrons from the membrane-bound enzyme complex CcR to CcO where oxygen is reduced to water [1]. Whereas this function has been known for decades, more recently Cyt c has been identified as being central for apoptotic processes [25]. In one case, Cyt c binds to Apaf-1 protein, thereby causing its oligomerization and the subsequent activation of caspasis. Prior to this reaction sequence, however, Cyt c has to be transferred through the mitochondrial membrane to the cytosol, a process that is most likely related to the second function of Cyt c in apoptosis, the catalysis of H_2O_2-dependent peroxidation of the mitochondria-specific lipid component cardiolipin [10]. Cardiolipin peroxidation causes an increased permeability of the mitochondrial membrane such that Cyt c can be released to the cytosol. Thus, the peroxidation and the 'normal' redox function take place at the inner mitochondrial membrane and thus under the influence of interfacial electric fields.

We first consider the redox function, that is, the transfer of electrons to CcO, on the basis of the overall reaction in Figure 10.11. Prior to the inter-protein ET, a transient complex is formed most likely via the high-affinity binding domain of Cyt c (BD3 – see above) and the complementarily charged docking site on CcO close to the Cu_A centre. Inter-protein ET rate constants between 10^4 and 10^5 s^{-1} have been estimated from studies of the Cyt c/CcO complexes in solution [26], that is, under conditions that may well be compared with the low-field regime of the electrochemical system studied here. In fact, then rotational diffusion may occur on the microsecond time scale and there is no indication for rate-limiting protein structural changes. Thus, electron tunnelling via the most favourable orientation of Cyt c is rate-limiting. Based on a structural model of the Cyt c/CcO complex, a tunnelling rate of 10^5 s^{-1} is indeed consistent with the distance between haem c and the Cu_A site.

Earlier studies on CcO reconstituted in liposomes have shown that an increase of the transmembrane potential may drastically slow down ET from Cyt c to CcO as well as intra-protein ET in CcO [27–29]. In view of the results obtained from the electrochemical model systems, it is tempting to assume that the increase of the transmembrane potential may cause the transition to the high-field regime. Consequently, the reorientation rate may decrease such that it becomes rate-limiting for the overall redox process. It has been suggested that the modulation of transmembrane potential reflects the interplay between the CcO-driven generation of a proton gradient and the proton consumption by ATPase [30]. In this sense, the

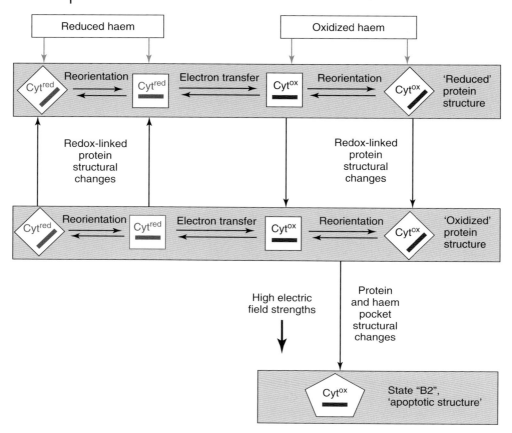

Figure 10.11 Schematic representation of the overall interfacial processes of Cyt c at negatively charged surfaces.

electric field control of the reduction of CcO by Cyt c might represent a regulatory feedback mechanism to avoid unproductive oxygen consumption.

Higher electric fields also cause larger binding energies and thus a longer lifetime of the Cyt c/CcO complex. As a result, the probability increases that the bound ferric Cyt c is converted to the conformational state B2. In fact, this conformational state has been detected in the fully oxidized complex of Cyt c and CcO in solution stabilized at low ionic strength. Durable formation of the conformational state B2 would block the ET chain between CcR and CcO, since, due to the very negative redox potential (about -0.15 V) [30], the ferric state B2 cannot accept electrons from CcR.

The formation of the state B2, which is prohibitive for the redox function of Cyt c, may also occur upon interactions with the mitochondrial membrane itself. The high content of negatively charged cardiolipin plus an increase of the transmembrane potential may constitute a high electric field regime which stabilizes state B2 (see Figure 10.4). This may be the initial step of Cyt c for switching from the 'normal' redox function to the apoptotic function (Figure 10.11).

Although the hypotheses outlined in this section have to be checked by further experiments, one may conclude that local electric fields play a critical role for the biological functions of Cyt c. In this respect, studies on electrochemical model systems by SERR and SEIRA spectroscopy can provide deep insights into elementary reaction steps of biological processes at interfaces which are not accessible by other techniques.

References

1. Scott, R.A. and Mauk, A.G. (eds) (1995) *Cytochrome c-A Multidisciplinary Approach*, University Science Books, Sausalito.
2. Archer, M. D. and Barber, J. (eds) (2004) *Molecular to Global Photosynthesis*, Series on Photoconversion of Solar Energy, vol. 2, Imperial College Press, London.
3. Clarke, R.J. (2001) The dipole potential of phospholipid membranes and methods for its detection. *Adv. Colloid Interface Sci.*, **89**, 263–281.
4. Cotton, T.M., Schultz, S.G. and Van Duyne, R.P. (1980) Surface-enhanced resonance Raman-scattering from cytochrome-C and myoglobin adsorbed on a silver electrode. *J. Am. Chem. Soc.*, **102**, 7960–7962.
5. Love, J.C., Estroff, L.A., Kriebel, J.K., Nuzzo, R.G. and Whitesides, G.M. (2005) Self-assembled monolayers of thiolates on metals as a form of nanotechnology. *Chem. Rev.*, **105**, 1103–1169.
6. Murgida, D.H. and Hildebrandt, P. (2005) Redox and redox-coupled processes of heme proteins and enzymes at electrochemical interfaces. *Phys. Chem. Chem. Phys.*, **7**, 3773–3784.
7. Murgida, D.H. and Hildebrandt, P. (2001) The heterogeneous electron transfer of cytochrome c adsorbed on coated silver electrodes. Electric field effects on structure and redox potential. *J. Phys. Chem. B*, **105**, 1578–1586.
8. Zuo, P., Albrecht, T., Barker, P.D., Murgida, D.H. and Hildebrandt, P. (2009) Interfacial redox processes of cytochrome b_{562}. *Phys. Chem. Chem. Phys.* doi: 10.1039/b904926f.
9. Smith, C.P. and White, H.S. (1992) Theory of the interfacial potential distribution and reversible voltametric response of electrodes coated with electroactive molecular films. *Anal. Chem.*, **64**, 2398–2405.
10. Basova, L.V., Kurnikov, I.V., Wang, L., Ritov, V.B., Belikova, N.A., Vlasova, I.I., Pacheco, A.A., Winnica, D.E., Peterson, J., Bayir, H., Waldeck, D.H. and Kagan, V.E. (2007) Cardiolipin switch in mitochondria: shutting off the reduction of cytochrome c and turning on the peroxidase activity. *Biochemistry*, **46**, 3423.
11. Berghuis, A.M. and Brayer, G.D. (1992) Oxidation state-dependent conformational-changes in cytochrome-C. *J. Mol. Biol.*, **223**, 959–976.
12. Oellerich, S., Lecomte, S., Paternostre, M., Heimburg, T. and Hildebrandt, P. (2004) Peripheral and integral binding of cytochrome c to phospholipid vesicles. *J. Phys. Chem. B*, **108**, 3871–3878.
13. Siebert, F. and Hildebrandt, P. (2007) *Vibrational Spectroscopy in Life Science*, Wiley-VCH Verlag GmbH, Weinheim.
14. Oellerich, S., Wackerbarth, H. and Hildebrandt, P. (2002) Spectroscopic characterization of non-native states of cytochrome c. *J. Phys. Chem. B*, **106**, 6566–6580.
15. Yeh, S.R., Han, S.W. and Rousseau, D.L. (1998) Cytochrome c folding and unfolding: a biphasic mechanism. *Acc. Chem. Res.*, **31**, 727–736.
16. Kranich, A., Ly, K.H., Hildebrandt, P. and Murgida, D.H. (2008) Direct observation of the gating step in protein electron transfer: electric field controlled

protein dynamics. *J. Am. Chem. Soc.*, **130**, 9844–9848.
17. Armstrong, F.A. (2005) Recent developments in dynamic electrochemical studies of adsorbed enzymes and their active sites. *Curr. Opin. Chem. Biol.*, **9**, 110–117.
18. Murgida, D.H. and Hildebrandt, P. (2000) Proton coupled electron transfer in cytochrome c. *J. Am. Chem. Soc.*, **123**, 4062–4068.
19. Wisitruangsakul, N., Zebger, I., Ly, K.H., Murgida, D.H. and Hildebrandt, P. (2008) Redox-linked protein dynamics probed by time-resolved surface enhanced infrared absorption spectroscopy. *Phys. Chem. Chem. Phys*, **10**, 5287–5297.
20. Feng, J.J., Gernert, U., Sezer, M., Kuhlmann, U., Murgida, D.H., David, C., Richter, M., Knorr, A., Hildebrandt, P. and Weidinger, I. (2009) A novel Au-Ag hybrid device for surface enhanced (resonance) Raman spectroscopy. *Nano Lett.*, **9**, 298–303.
21. Murgida, D.H. and Hildebrandt, P. (2002) Electrostatic-field dependent activation energies control biological electron transfer. *J. Phys. Chem. B*, **106**, 12814–12819.
22. Paggi, D.A., Martín, D.F., Kranich, A., Hildebrandt, P., Martí, M. and Murgida, D.H. (2009) Computer simulation and SERR detection of cytochrome c dynamics at SAM-coated electrodes. *Electrochim. Acta*, **54**, 4963–4970.
23. Feng, J.J., Murgida, D.H., Utesch, T., Mroginski, M.A., Hildebrandt, P. and Weidinger, I. (2008) Gated electron transfer of yeast iso-1 cytochrome c on SAM-coated electrodes. *J. Phys. Chem. B*, **112**, 15202–15211.
24. Kranich, A., Naumann, H., Molina-Heredia, F.P., Moore, H.J., Lee, T.R., Lecomte, S., de la Rosa, M.A., Hildebrandt, P. and Murgida, D.H. (2009) Gated electron transfer of cytochrome c_6 at biomimetic interfaces: a time-resolved SERR study. *Phys. Chem. Chem. Phys.* doi: 10.1039/b904434e.
25. Jiang, X. and Wang, X. (2004) Cytochrome c-mediated apoptosis. *Annu. Rev. Biochem.*, **73**, 87–106.
26. Geren, L.M., Beasley, J.R., Fine, B.R., Saunders, A.J., Hibdon, S., Pielak, G.J., Durham, B. and Millett, F. (1995) Design of a ruthenium cytochrome-C derivative to measure electron-transfer to the initial acceptor in cytochrome-C-oxidase. *J. Biol. Chem.*, **270**, 2466–2472.
27. Gregory, L. and Fergusonmiller, S. (1989) Independent control of respiration in cytochrome-C oxidase vesicles by Ph and electrical gradients. *Biochemistry*, **28**, 2655–2662.
28. Sarti, P., Antonini, G., Malatesta, F. and Brunori, M. (1992) Respiratory control in cytochrome-oxidase vesicles is correlated with the rate of internal electron-transfer. *Biochem. J.*, **284**, 123–127.
29. Nicholls, P. and Butko, P. (1993) Protons, pumps, and potentials - control of cytochrome-oxidase. *J. Bioenerg. Biomembr.*, **25**, 137–143.
30. Murgida, D.H. and Hildebrandt, P. (2004) Electron transfer processes of cytochrome c at interfaces. New insights by surface-enhanced resonance Raman spectroscopy. *Acc. Chem. Res.*, **37**, 854–861.

11
Quantitative DNA Analysis Using Surface-Enhanced Resonance Raman Scattering

Ross Stevenson, Karen Faulds, and Duncan Graham

11.1
Introduction

The ability to detect specific DNA sequences is of fundamental importance to modern molecular biology and molecular diagnostics [1]. Detection and identification of unique strands of DNA can play important roles in the early diagnosis of disease states or in the forensic identification of criminal suspects [2, 3]. Although current detection procedures are adequate for most applications, development of new DNA detection techniques capable of detecting multiple sequences in a short time frame could be a major advantage. Doing so could lead to the earlier diagnosis of many diseases, including cancer, from which improved treatment opportunities would be possible. A more sensitive analysis could be used to detect disease progression, giving rise to more tailored therapies. Faster and more sensitive DNA detection approaches could also lead to improved screens for bacterial infections, particularly relevant at a time when there are so many issues relating to hospital-acquired infections (HAIs) and the associated human and financial costs. The current practice for diagnosing HAIs are based on time-consuming cell culture techniques, where typically the sample must undergo 48 h incubation on selected antibiotic-treated agar plates for identification.

The majority of techniques for quantitative detection of specific DNA sequences rely on the detection on fluorescent spectra [4–7]. Fluorescence provides popular and robust detection methodologies; fluorescently labelled biological components are well established and have been common practice in a wide range of procedures including microarrays, ELISA, separation science and, importantly for molecular diagnostics, in real-time polymerase chain reaction (rt-PCR). PCR is a fundamental process in molecular biology that amplifies specific sequences of DNA using polymerase enzymes contained within a nucleotide and salt broth. PCR is the current gold standard technique in DNA analysis and gives rise to an extremely sensitive DNA detection technique. A further benefit with the approach is that it uses a closed-tube format, ensuring contamination is kept minimal. Advances in the basic technique led to rt-PCR offering researchers the opportunity to quantify DNA levels cycle by cycle by utilizing fluorescent probes. Fluorescent probes for

DNA provide many advantages in terms of ease of use and sensitivity. They can either be sequence-specific probes or general intercalators such as SyBr Green. Other fluorescent approaches for DNA detection that use closed-tube formats include molecular beacons, Taqman and Scorpions [5, 7, 8]. Each illustrates an application where specific DNA sequences can be resolved in a short time frame using fluorescent labels. The main problems associated with working with fluorophores are generally linked to the overlapping spectral features and non-uniform photobleaching rates [9]. Fluorescent profiles are typically broad (100 nm) and contain little specific molecular information. Using the standard closed-tube system, it is possible to multiplex four fluorophores with a single excitation wavelength, but if an internal standard is required, then that reduces to a maximum of three sequences that can be probed simultaneously. A further disadvantage of using a fluorescence-based approach is photobleaching, and although many of the newer commercially available dyes are more photostable than the classically used dyes, bleaching does still occur. The problem becomes significantly more challenging if the experiment requires the use of multiple fluorophores, each with a differing photobleaching rate, making quantitative analysis difficult. Fluorescent PCR assays have been developed for the detection of specific RNA sequences associated with a number of HAIs, including the *mecA* gene linked to methicillin-resistant *Staphylococcus aureus* (MRSA) [10]. This approach has led to a clinically relevant, fast screen that can target a single key gene. There remains a need for an assay that can simultaneously screen multiple RNA targets.

Surface-enhanced resonance Raman scattering (SERRS) for DNA analysis has a number of advantages over the currently employed molecular diagnostic techniques. Studies have reported that SERRS is capable of almost unparalleled sensitivity, with single-molecule detection possible [11, 12]. SERRS affords multiplexing capabilities that exceed those possible with fluorescence. Molecule-specific Raman fingerprints contain numerous unique, sharp vibrational bands from which highly multivariant analysis can be observed [13]. A further advantage of most SERRS applications is that they require no separation step, leading to faster, higher throughput result. The sensitivity of SERRS makes it a very attractive method in its own right, but when coupled with speed and the ability to resolve multiple signals from a mixture, it becomes an incredibly powerful tool. Limitations with the SERRS technique, often linked to the surface from which enhancement of the Raman signal occurs, are described later in the chapter. It is hoped that the use of SERRS in DNA detection will lead to an increase in the amount of real-time data observed per experiment, meaning fewer experiments required for complex DNA analysis.

11.2
SERRS Surfaces

Metal nanoparticles are used as enhancing agents for the Raman signal from an associated label. When roughened metallic nanostructures are illuminated by a specially selected wavelength of light, electrons (also called *surface plasmons*; see also Chapter 1) oscillate synchronously with the incident frequency, generating a secondary electromagnetic field above that of the incident field [14]. The resultant resonance response gives rise to improved Raman scattering due to the enhanced local electromagnetic field intensity, with enhancements of anything from 10^6 to 10^{15} possible for molecules close to the metallic surface [15]. The distance the reporter is from the metallic surface is of key importance. Theoretical studies have shown that SERS enhancement (G) diminishes as $G = [r/(r+d)]^{12}$ for single analyte molecules located at a distance (d) from the surface of a metal nanoparticle with radius (r) [16]. Care must therefore be taken in designing Raman probes to ensure that the label lays on the surface, or as close to the surface as possible.

A number of metallic surfaces have been analysed for their abilities to facilitate SERRS [17]. The earliest SERS results were obtained from pyridine adsorbed on a silver electrode surface by Fleischmann *et al.* [18]. Although such electrodes roughened by the oxidation–reduction cycle offer an effective SERRS template, current work generally utilizes colloidal suspensions or surfaces with a nanoengineered roughness [19]. Other general examples of suitable SERRS surfaces not described in this chapter include hollow gold nanospheres [20], gold and silver rods [21] and silica-seeded nanoparticles [22]. Coinage metal colloidal solutions (typically gold or silver) are the most commonly used SERRS surface due to their cost and ease of synthesis (see also Chapter 2). Gold or silver nanoparticles are prepared by reducing the appropriate metal salt, typically citrate, borohydride, hydroxylamine or EDTA. The majority of DNA nanoparticle applications use citrate-reduced silver nanoparticles containing a silver core encased within a physisorbed anionic citrate layer [23].

A significant disadvantage in nanoparticle use is related to reproducibility that is inherent from the batch-to-batch variability observed during colloid synthesis. It is essential that particle size, shape and size distribution remain as uniform as possible and, although careful choice of reducing agent conditions can lead to significant improvements, batch screening is still necessary. An alternative method to control stock variability is to blend batches into a much larger, more reproducible solution. Colloid solutions can also contain 'hot nanoparticles': that is, a small percentage of nanoparticles that give rise to significantly higher Raman responses than the general population [24]. For the majority of colloidal suspension research, the laser beam interrogation area covers a large number of nanoparticles, so random hot spot characteristics are averaged out; however, when using microfluidic/nanofluidic analysis of almost single nanoparticles, such effects can be the cause of rogue results [25].

The alternative approach to the use of nanoparticles is use of an engineered roughened surface. This can take the form of a metal electrode [18], thin vapour-deposited metal films [26] or specifically designed structured surfaces formed by lithography or deposition [19]. Thin metal films are advantageous as they can be 'tuned' to a specific wavelength by varying the thickness of the metal surface; however, the enhancement factors obtained are generally 10^4–10^5 smaller than other SERRS substrates [27]. An example of their use with a DNA SERRS application is where Vo-Dinh *et al.* created a 9 nm thick silver island, coined a 'gene-chip', to detect the oncogene *BCL-2* with a dye-labelled oligonucleotide probe [28]. Another unique SERRS platform comes in the form of the nano-scaffold Klarite, fabricated with a particular plasmon frequency owing to its inverted pyramid structure. Klarite has been used by Graham *et al.* to detect a hybridized DNA probe on the surface [29].

Effective adsorption of the target molecule onto the surface is of critical importance for good Raman enhancement [30], and two techniques are commonly used. The surface can be (i) modified to make it seek the analyte, as pioneered by Van Duyne in the SERS detection of glucose [31], and in use of the electrostatic layering technique [32] or (ii) the analyte can be modified to seek the surface. Effective adsorption of DNA is challenging due to the overall negative charge imparted by the phosphate backbone. Interaction with the negatively charged colloid surface is electrostatically unfavourable, making a strategy necessary to adsorb DNA onto colloid. It is possible to overcome the charge repulsion by utilizing either an electrostatic layering technique (as shown in Figure 11.1) or by modification of the DNA to include a positively charged species. The basic layering theory has a positively charged species introduced into the system that forms a 'sandwich' or 'bridge' between the negatively charged surface and the negatively charged DNA. Poly-(L-lysine) and spermine have both been analysed for their ability to effectively adsorb oligonucleotides, with spermine giving the most sensitive and reproducible results [33]. Interestingly, the spermine bridges between the negatively charged species but it does not replace the negatively charged species on the colloid surface: it merely enhances analyte adsorption. Spermine can also carry out a dual role in the sensitive detection of oligonucleotides as it has the ability to aggregate nanoparticles, from which larger Raman signals can be observed as a result of the coupling of surface plasmons when surfaces come into close proximity [34].

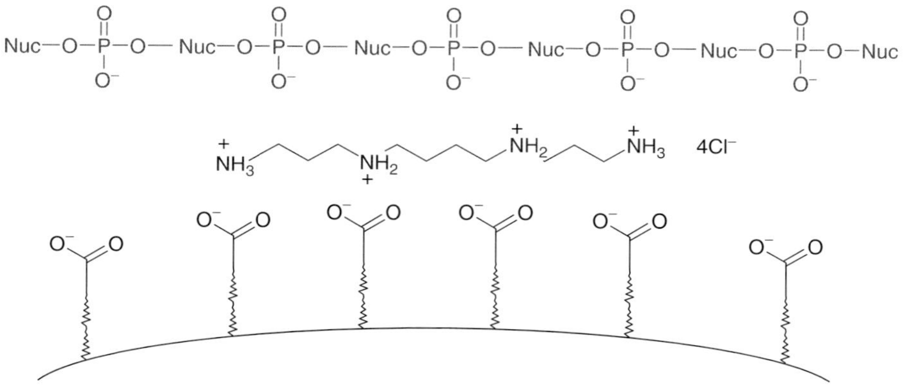

Figure 11.1 Spermine acting as a bridge between the electrostatically repulsed colloid citrate layer and the phosphate backbone of DNA. Spermine can also be used to aggregate nanoparticles to achieve maximum SERRS signals [33]. (Reproduced with permission from The Royal Society of Chemistry.)

The approach of target modification, where the analyte is modified to include a surface-seeking species, lends itself for use in DNA-targeted SERRS. Modification not only allows the addition of a surface-seeking group, it can also be used to introduce an intrinsically strong Raman scattering molecule into the system. Most commonly used SERRS labels employ thiols, disulfides, dithiolanes or benzotriazole moieties for chemisorption to the metal surface [35]. Such compounds form densely packed monolayers to which the oligonucleotides can be attached. Benzotriazoles have been used for generations as anti-tarnish reagents for silver. The nitrogen groups are believed to complex to multiple silver atoms to form surface clusters [30]. Benzotriazole modification of DNA is discussed later in the chapter.

Thus, to facilitate the highly sensitive SERRS experiments required to improve upon the current molecular biodiagnostic techniques, it is necessary to introduce (i) a method for preparing well-aligned monolayers of reporter onto a metallic surface and (ii) ensure that the chromophore introduced is in resonance with (has an adsorption maxima close to) the wavelength of the incident laser. For applications where the target molecules are poor Raman scatterers, it may be necessary to introduce a dye-type molecule/Raman reporter into the system. DNA is a relatively poor Raman material and modifications are necessary for sensitive SERRS experiments.

11.3
Raman Reporters

Raman reporter molecules are inherently strong light scatters and can be split into two general categories; dyes that do or that do not fluoresce. Some of best Raman scatterers are fluorophores and the large number, availability and the simple chemistry involved in their coupling to DNA or other biomolecules make the commercially available probes a popular starting point for SERRS [36]. Metal surfaces are also particularly useful with regard to the use of fluorophores, as they not only act as excellent scatterers of light but are also efficient fluorescence quenchers. In SERRS applications, it is important to ensure that all unconjugated dye is washed from the vessel, or an unwanted fluorescence emission can mask some, if not all, of any potential Raman fingerprint.

To analyse the capability of SERRS to examine popular, commercially available DNA dyes, spectra for nine dye-labelled oligonucleotides were recorded using 514.5 nm laser excitation (shown in Figure 11.2). There are distinct differences between the spectra of all labels, allowing easy identification of each dye and thus easy identification of the DNA sequence present. The study is particularly interesting, as FAM, TET and HEX have very similar chemical structures (Table 11.1), differing only in the number of chlorine atoms on the ring system. This is an interesting example that not only demonstrates the specificity achievable from SERRS but also highlights the multiplexing capability.

Most commonly used fluorescent labelling probes are those that are amine-reactive. Amine modifications are widely used, as the covalent bond

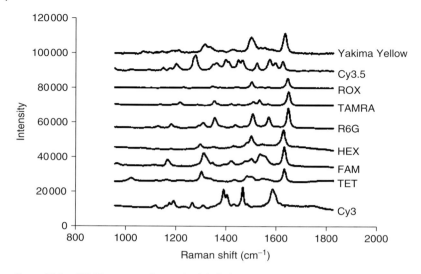

Figure 11.2 SERRS spectra of nine dye-labelled oligonucleotides. All spectra were obtained using silver nanoparticles and 514.5 nm laser excitation. (Reprinted with permission from [40]. © 2010 American Chemical Society.)

between dye and biomolecule leads to stable conjugates, capable of withstanding rigorous incubation, hybridization and washing steps. If the dye has a net negative charge, electrostatic layering can be used to ensure surface adsorption, improving the signal-to-noise ratio and enhancing the limits of detection (LoDs).

Non-fluorescent labels can be further split in two distinct categories: dyes that label the DNA, or cross-linking dyes designed to covalently bind oligonucleotide and attach to the surface. Labels such as DABCYL, phthalocyanines and black hole quenchers (BHQs) are examples of non-fluorescent labels that adsorb onto the surface of the nanoparticle (Table 11.2).

Dyes designed for complexing to the metal surface through formation of bonds between the metal and a surface-seeking group are generally favoured over those dyes that are electrostatically attracted to the surface. The most successful of the covalently attached dyes often utilize azo dyes, containing a benzotriazole group that effectively complexes to the silver surface. Two approaches are available for the addition of benzotriazole azo dyes to DNA sequences (Figure 11.3). In one, the dye can be synthesized as a phosphoramidite that can be incorporated onto the 5′end of the sequence during the solid-phase synthesis of the probe [37]. The alternate approach is to post-synthetically couple a reactive benzotriazole azo dye with a functional group, such as amine or thiol, on an oligonucleotide probe. A maleimide dye can be coupled with diene-modified oligonucleotide via the Diels Alder cycloaddition. Such strategies produce 5′-labelled sequences ideally placed for direct complexation to metal surfaces.

Table 11.1 Structures and charge of many commercially available dye labels used in DNA SERRS analysis [33].

Dye label	Structure	Charge
R, R1 = H = FAM R = Cl, R1 = H = TET R, R1 = Cl = HEX		Negative
R, R3 = H, R1 = $(CH_3)_2N$, R2 = $N^+(CH_3)_2$ = TAMRA R, R3 = CH_3, R1 = CH_3CH_2NH, R2 = $N^+HCH_2CH_3$ = R6 G		Positive
ROX		Positive
Yakima Yellow		Negative
BODIPY TR-X		Negative

(*continued overleaf*)

Table 11.1 (Continued)

Dye label	Structure	Charge
n = 1 = Cy3 n = 2 = Cy5 n = 3 = Cy7		Positive
n = 1 = Cy3.5 n = 2 = Cy5.5		Positive

Reproduced with permission from The Royal Society of Chemistry.

Numerous dye approaches are available to exploit the different chemistries of surface adsorption. The vast majority of labels used to date modify DNA at the 5′ terminus; however, it should be noted that 3′ dye modification techniques exist [38] and, as it is the dye that is observed as opposed to the oligonucleotide sequence, it is possible that there are applications that would benefit from alternative site labelling of DNA.

11.4
SERRS DNA Probes

As discussed, SERRS requires a visible chromophore capable of effective adsorption onto a roughened metal surface. Intrinsically, DNA is neither a visible chromophore nor is it good at adsorbing to colloid surfaces. The electrostatic repulsion of the negativity charged phosphate backbone of DNA with the negatively charged colloid surface gives rise to poor surface coverage. For most DNA-based SERRS applications, the introduction of single or multiple modifications are necessary to achieve spectroscopic detection.

Bell *et al.* observed the spectral fingerprints of the individual mononucleosides [39]. Of particular note is that single-molecule detection could be achieved when the spectra of adenine on colloidal silver was observed. In this application, the authors overcame electrostatic repulsion and aggregated the nanoparticles with the introduction of spermine hydrochloride. Observing the spectral fingerprint of individual mononucleotides is clearly possible in a simple colloid system, but advancing the technique to DNA sequences containing multiple, mixed bases

Table 11.2 Non-fluorescent dyes used in DNA SERRS analysis [33].

Dye label	Structure	Charge
DABCYL		Positive
BHQ2		Positive
X = Co = PtcCo X = Al = PtcAl X = Zn = PtcZn		Positive

Reproduced with permission from The Royal Society of Chemistry.

is problematic. Sequence spectra are highly convoluted and not easily resolved. Typically, the large cross section of adenine means that its signal overrides the spectra from each of the other bases, making specific sequence detection at best extremely difficult if not impossible.

There are currently two fluorescence readout approaches commonly used in DNA assays: (i) fluorescent dyes that covalently attach to DNA, as used in

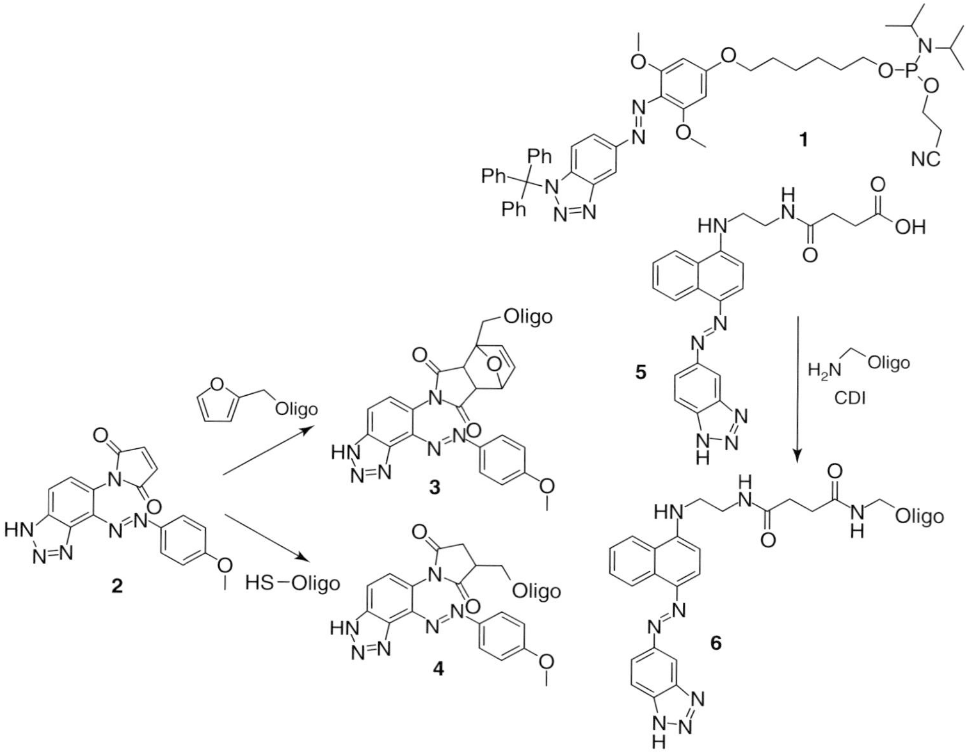

Figure 11.3 Approaches used to attach benzotriazole and benzotriazole azo dyes to oligonucleotides [33]. (Reproduced with permission from The Royal Society of Chemistry.)

rt-PCR and microarray-type formats and (ii) dyes that intercalate the double helix, common in agarose gel stains. Although the latter approach can sensitively detect double-stranded DNA (dsDNA), the intercalation is indiscriminate and lacks the ability to detect specific sequences, making the former preferable for multiplexing analysis. The Taqman approach is an example of a covalently attached DNA probe commonly used in rt-PCR. It utilizes a sequence-specific probe modified with a fluorophore at one terminus and a quencher at the other. If the target sequence is present, the Taqman probe anneals and no fluorescence is observed because of the proximity of fluorophore and quencher. During PCR, a Taq polymerase enzyme extends along the primer, synthesizing the nascent strand and displacing fluorophore, giving rise to a quantifiable fluorescent readout.

Similar to the surface coverage problems faced by DNA, fluorophores are charged species, and therefore consideration must be given to ensure that electrostatically repulsed dyes are adsorbed effectively onto the colloid surface. Again, the electrostatic layering approach has been used with success to prepare fluorescent nanoparticles. Faulds *et al.* carried out a study with eight commercially available

11.4 SERRS DNA Probes

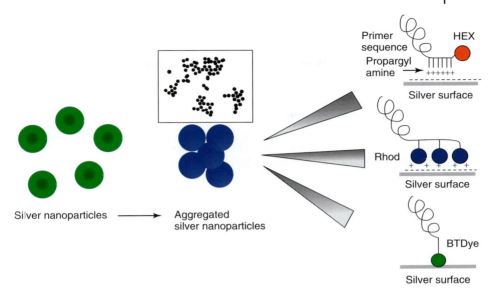

Figure 11.4 Schematic representation of the aggregation process required to achieve maximum detection of labelled DNA by SERRS. Shown are routes to achieve optimum surface coverage using three commonly used but differently charged molecules. Negatively charged label HEX requires propargylamine modification to confer a positive charge on to the molecule and allow surface adsorption to occur. The R6G label is positively charged; therefore, no surface modification is required. The alternative route is to use a specially synthesized dye containing the benzotriazole group which attaches strongly and essentially irreversibly to silver surfaces [41]. (Reproduced with permission from The Royal Society of Chemistry.)

dye labels used routinely in the fluorescence detection of oligonucleotides [40]. The dyes chosen (ROX, rhodamine 6G, HEX, FAM, TET, Cy3, Cy5, TAMRA) covered those that, when in aqueous solution, are positively charged (R6G, ROX, TAMRA) and those that are negatively charged (HEX, FAM). The positively charged dyes required no further modification to attach to the negatively charged colloid surface (silver in this study, see Figure 11.4); however, effective adsorption of the latter two dyes required surface modification.

An alternative route to enhance surface adsorption is the incorporation of surface-seeking propargylamine–deoxyuridines into the DNA sequences. At physiological pH, the modified bases contain a protonated aliphatic amine, providing a site for electrostatic interaction between DNA and the colloid surface. Best SERRS results are obtained if the fluorophore is placed as close to the modified bases as possible, ensuring the largest Raman signal and complete quenching due to the proximity to the nanoparticle surface. It was demonstrated that addition of two propargylamine-modified nucleotides at the 5′ end gave the most sensitive results [42]. Comparison of a positively charged dye and a negatively charged dye containing the propargylamine-modified bases was made (with added spermine).

Good adsorption of DNA was observed and hence successful SERRS was obtained from both.

11.5 Sensitivity

There are numerous reports that describe the quantitative and sensitive detection of labelled oligonucleotides [43, 44]. Linear responses can be achieved under optimized conditions if the concentration of labelled oligonucleotide is below monolayer surface coverage, from which observation of the SERRS detection limit is possible. When levels of DNA are less tightly controlled, giving more than monolayer coverage, the calibration graph curves off towards steady state. This is a result of the competition between the formation of multi-layers and adsorption onto the walls of the vessel due to the high concentration of analyte present. Stokes *et al.* carried out a study to analyse the LoDs for many of the most routinely used fluorescent labels for DNA, and compared the limits achieved by fluorescence

Table 11.3 Raman and fluorescent limits of detection for commercially available dyes [45].

Dye label	λ_{max} (nm)	λ_{ex}: 514.5 nm	λ_{ex}: 632.8 nm		λ_{ex}: 785 nm	Fluorescence LoD
		Silver (mol dm^{-3})	Silver (mol dm^{-3})	Gold (mol dm^{-3})	Silver (mol dm^{-3})	(mol dm^{-3})
FAM	492	2.7×10^{-12}	2.0×10^{-9}	–	–	6.5×10^{-8}
TET	521	1.6×10^{-11}	2.0×10^{-9}	–	–	2.6×10^{-8}
R6G	524	1.2×10^{-12}	1.1×10^{-10}	–	–	3.5×10^{-8}
Yakima Yellow	526	1.7×10^{-11}	–	–	–	N.M.
HEX	535	7.8×10^{-12}	1.2×10^{-9}	–	–	1.1×10^{-8}
Cy3	552	2.6×10^{-10}	1.5×10^{-10}	–	–	4.6×10^{-9}
TAMRA	565	3.5×10^{-12}	1.8×10^{-10}	–	–	1.1×10^{-8}
Cy3.5	581	2.5×10^{-11}	7.5×10^{-13}	2.5×10^{-10}	–	N.M.
ROX	585	8.1×10^{-11}	3.3×10^{-11}	1.1×10^{-9}	–	2.3×10^{-8}
BODIPY TR-X	588	1.3×10^{-10}	7.9×10^{12}	4.9×10^{-10}	–	N.M.
PtcCo	625	–	3.2×10^{-11}	N.M.	–	Not fluorescent
PtcAl	640	–	2.8×10^{-11}	N.M.	–	Not fluorescent
Cy5	643	–	8.3×10^{-11}	1.7×10^{-9}	–	3.1×10^{-9}
PtcZn	680	1.4×10^{-10}	3.2×10^{-11}	N.M.	–	Not fluorescent
Cy5.5	683	–	5.2×10^{-12}	7.3×10^{-11}	–	N.M.
Cy7	748	–	–	–	5.8×10^{-11}	N.M.

Copyright Wiley-VCH Verlag GmbH & Co. KGaA. Reproduced with permission.

with the detection limits from SERRS, shown in Table 11.3 [45]. The study was particularly interesting, as it served to prove that Raman spectroscopy can be more sensitive than fluorescence. Typically, a 100 × greater sensitivity was achieved by SERRS, although special note should go to the TAMRA dye, where SERRS gives a 10 000× lower LoD than fluorescence. The data also illustrates the advantage of SERRS over the non-resonant form, SERS. There is a clear correlation between the λ_{max} of the dye and the excitation wavelength used, showing that the resonance contribution of the system can improve sensitivity limits 100–1000 fold. From the table, the most sensitive excitation wavelength/dye combination was achieved using silver nanoparticles, 632.8 nm excitation and a Cy3.5 dye, with which levels as low as 7.5×10^{-13} M could be detected.

A final observation from the table is that gold and silver nanoparticles have differing LoDs, with silver 10–100 times more sensitive than the gold counterpart in most cases. Even though gold lacks the detection limits of silver, it is still commonly used, as silver suffers from difficulties in reproducible synthesis and is more prone to unwanted aggregation upon addition of analyte [46].

11.6
Multiplexing

Our definition of SERRS multiplexing is the ability to detect multiple labels attached to different DNA probes at the same time and without any additional separation procedure. It is possible to observe multiple DNA sequences at the same time using a microarray format. Mirkin and co-workers developed a multiplexed DNA/RNA detection system based on a microarray-type format [47]. The strategy used six dissimilar DNA targets with six Raman labelled probes and six specific sites containing capture DNA strands. Addition of target produced a nano-scaffold, with the capture strand leaving sufficient bases to allow the Raman-tagged DNA probe to hybridize onto the scaffold. In the application, the authors introduced a post-binding silver reduction step to enhance the SERS readout. The strategy provides a high-sensitivity (unoptimized detection limit is 20 fM) and high-selectivity detection technique. As spatial separation is used to resolve multiple signals, such arrays do not fall into our multiplex definition. In this chapter, we aim to review the multiplexing capabilities of SERRS from a solution; therefore, the tremendously powerful array format is not discussed in this review.

As alluded to in the introduction, the multiplexing capabilities of SERRS have been a driving force in the resurgence of Raman detection techniques over the past decade. The sharp vibrational fingerprint spectra observed are ideally placed to allow the separation of complex component mixtures with a non-invasive spectroscopic technique. Clearly, the easiest and predictably earliest example of a multiplexed SERRS response was where two analytes were present. Between 2001 and 2003, the first examples of a SERRS DNA 2-plex, using rhodamine- and HEX-labelled oligonucleotides [48], and a 3-plex, using FAM, TET and Cy3, were developed by Graham and co-workers [49].

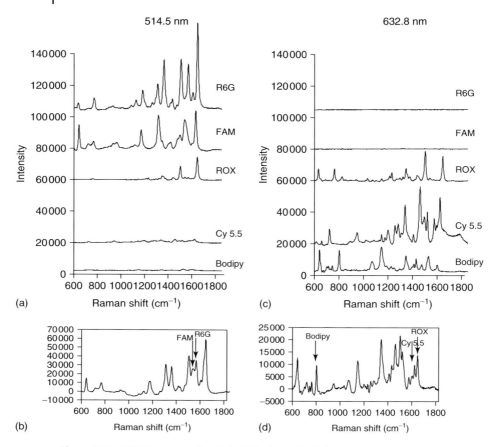

Figure 11.5 SERRS spectra of each individual dye-labelled oligonucleotide using laser excitation at (a) 514.5 nm and (c) 632.8 nm. The multiplexed SERRS spectra of the 5-plex using (b) 514.5 nm and (d) 632.8 nm excitation are also shown [50]. (Copyright Wiley-VCH Verlag GmbH & Co. KGaA. Reproduced with permission.)

Faulds et al. demonstrated the first 5-plex system where FAM-, Cy5.5-, BODIPY-, R6G- and ROX-labelled oligonucleotides were detected in situ as shown in Figure 11.5 [50]. In the study, the oligonucleotide mixture was maintained at a concentration of 1.82×10^{-9} mol dm^{-3} throughout. Each label gives a distinctive spectrum, allowing specific oligonucleotide sequences to be distinguished from one another in a simple and fast manner. This work is especially interesting, as it opened the possibility of using multiple excitation wavelengths as a selection method by exploiting a mixture of dyes resonantly enhanced at different frequencies. Each dye used in the experiment has different absorbance maxima and thus not all the dyes are in resonance at the same laser excitation frequency. Three of the dyes used (R6G, FAM and ROX) gave excellent LoDs when excited at 514.5 nm, as they were in resonance with that wavelength. The other two were best observed at 632.8 nm, although it should be noted that ROX was unique in that it could be observed at either wavelength. ROX has an absorbance maximum at 585 nm, but a second significant absorbance peak at 530 nm makes the molecule more in resonance at 514.5 nm, and thus more sensitive detection could be observed with excitation at this wavelength. The LoD work from this paper is also significant, as it allows the direct comparison of the LoD of single analytes as opposed to those in a complex mixture, and concluding that there was very little difference.

The 5-plex system allows the observer to note by eye the presence or absence of specific signals; however, exactly how many dyes can be picked out casually is dependent on the quality of dyes available and the ability of those dyes to work together. To exploit the maximum multiplicity that SERRS can offer, it is necessary to replace visual observation by more complex chemometric algorithms [13]. In such an approach, rather than detect specific discriminatory Raman bands, a multivariate analysis (MVA) approach is used, similar to those used in earlier vibrational spectroscopy studies. MVA considers the complete SERRS spectrum, with each Raman reporter ascribed a unique dimension, such that is there are n variables (Raman scatterers) and each object (analyte measured) may be said to reside at a unique position in an abstract entity referred to as *n-dimensional hyperspace*. A SERRS spectrum will contain hundreds of variables giving hundreds of dimensions, making the hyperspace difficult to visualize or to use in predictive modelling. MVA is therefore a tool that simplifies the dimensional reduction, either by the use of an unsupervised algorithm to summarize the natural variance in the data, or by using supervised learning via partial least squares (PLSs) regression of artificial neural networks based upon the observer's *a priori* knowledge of the samples being studied. The natural variance mentioned in the first approach, due to experimental error, can mask the interesting artifacts related to the studied hypothesis. If so, then it is necessary to introduce supervised methods, used with suitable model validation steps, which offers a more targeted approach to identifying the spectral variations to correlate with the anticipated patterns in the data. Fittingly, MVA is a potent tool, as it makes possible the ability to manipulate the resonance contribution to our advantage; this is not, however, always possible, especially when it is necessary to increase the complexing capability, and it then becomes necessary to apply chemometric methods to resolve the separate components of a multiplex. Figure 11.6 is a schematic representation of the concept. To date, the presence or absence of six uniquely labelled oligonucleotides has been established using such methods. In the study, six commercially available DNA dye labels (ROX, HEX, FAM, TET, Cy3 and TAMRA) were used to detect six different DNA sequences corresponding to different strains of *E. coli* with MVA. This was the first example of the use of chemometrics in the multivariant analysis of DNA sequences by SERRS, and this particular example applied discriminant analysis, supervised learning by PLS regression. The ability to discriminate whether a particular label was present or absent in a mixture was achieved with very high sensitivity (0.98–1), specificity (0.98–1), accuracy (range 0.99–1) and precision (0.98–1).

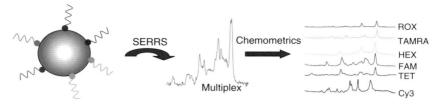

Figure 11.6 A representation of how chemometrics may be used to deconvolute a multiplexed spectrum of six dye-labelled oligonucleotides [13]. (Reproduced with permission from The Royal Society of Chemistry.)

11.7
Assays

The development of a useful assay capable of detecting multiple, specific DNA sequences from a real biological sample is the ultimate aim of all SERRS-based DNA analysis. To appreciate the advances made in SERRS DNA nanotechnology, it is important to build a foundation on other nanoparticle DNA-based work.

Non-SERRS DNA-functionalized nanoparticles have been used in a range of non-sequencing applications. DNA nanoparticle probes have been used as non-viral vectors for drug delivery [51, 52], for intracellular gene regulation [53], in the controlled hybridization of DNA probes with electronic control [54], and in Bio-Barcodes developed by Mirkin et al. [55]. A noteworthy DNA SERS nanoparticle used not in the detection of DNA is the aptasensor developed by Wang and co-workers [56]. The authors developed an aptamer SERS platform capable of detecting α-thrombin.

Also, detection of specific DNA sequences using functionalized nanoparticles has been around for a number of years. The earliest and crudest experiments relied on the observable changes in surface plasmon when two or more nanoparticles are brought into close proximity [57], using a sandwich/split probe-type assay that aggregates nanoparticles. The complementary sequence to a target strand of DNA/RNA is split, and each 'fragment' is conjugated to its own nanoparticle. Upon hybridization of the target sequence to each fragment, the nanoparticles are brought close enough together in a nanoparticle assembly to cause a plasmon shift measurable by UV spectroscopy [58]. The split-probe format was advanced by McKenzie et al. with the introduction of locked nucleic acid (LNA) in place of the standard DNA probes [59]. LNA enhances base stacking and backbone preorganization, significantly increasing the melting temperature (T_m), thereby improving the hybridization specificity and the mismatch discrimination of DNA bases. This results in a more sensitive and discriminatory sequence-specific probe system [60]. Although neither is a Raman experiment, both are fundamentally important to DNA detection with nanoparticles. By introducing a Raman dye(s) into the system, the same techniques give rise to more sensitive and quantifiable detection systems. Graham et al. used the split-probe format to show the controlled aggregation of silver DNA nanoparticles, and used the system to show that it was possible to 'turn on' the SERRS signal in a controlled manner. The approach differs from earlier published work, as the signal only appears when the recognition event of DNA hybridization occurs [61].

As discussed in the multiplexing section, Graham used a 2-plex system to identify different gene sequences from human patient samples. In the study, the authors detected the mutational status of the cystic fibrosis transmembrane conductance regulator gene with an amplification refractory mutation system (ARMS) by SERRS. The gene can be found in any one of three genetic variants, the wild type (both alleles normal), the heterozygote (one allele normal, one mutant) and homozygote (both alleles mutant). The predominant mutation is a three-base deletion in the sequence that codes for the amino acid tryptophan in the wild-type protein. Specific primers were designed to amplify the region where the deletion occurs and were used in a multiplexed PCR assay where the PCR product was identified by SERRS upon removal of primers [48].

The first multiplexed SERRS detection using labelled oligonucleotides in a microfluidic system was developed by Docherty *et al.* [49]. The study detected three specific *E. coli* genes simultaneously; however, it is the SERRS detection from a microfluidic channel (see also Chapter 8) that makes the paper particularly interesting, potentially opening up new detection platforms.

Vo-Dinh and co-workers developed a molecular beacon type assay, coined the 'molecular sentinel', for SERS detection of human immunodeficiency virus type-1 (HIV-1) DNA sequence [62]. The technique used silver nanoparticles with 5'-thiolated DNA probes each with a different 3' Raman tag. The sequences are covalently held to the nanoparticle surface via a sulfide linkage and are engineered with short complementary sequences at the 5' and 3' terminals to ensure that a hairpin is formed in the native state, holding the dye in close proximity to the nanoparticle surface. Upon addition of target DNA, the hairpin is forced open and the linear conformation adopted lifts the dye from the surface, diminishing the Raman output. An advanced study using the same principles led to a multiplexed detection of specific target gene targets, *erbB-2* and *ki-67*, both of which are critical biomarkers for breast cancer [16]. Faulds *et al.* developed a dual-readout system

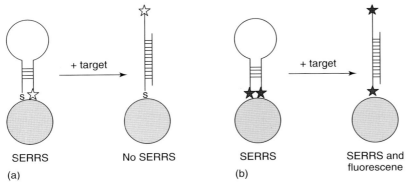

Figure 11.7 The molecular sentinel/SERRS beacon DNA detection techniques as developed by (a) Vo-Dinh *et al.* and (b) Faulds *et al.*

similar to the molecular sentinel (Figure 11.7). In this approach, 'hairpin' DNA probes containing a 5' fluorescent Raman dye and a 3' benzotriazole group for anchoring the DNA to the colloid are used [38]. The SERRS spectrum of the 'native' probe contains spectra of both 5' and 3' dyes and fluorescence is quenched. Addition of the complimentary DNA sequence opens the DNA, and the 5' fluorophore is removed from the surface, leading to a dual-readout system where the change in SERRS can be monitored as well as the introduction of fluorescence.

MacAskill and co-workers developed an SERRS approach that can be applied to detection of HAIs, monitoring three DNA sequences from the genes of different bacteria simultaneously [63]. The research lends from the observation that single-stranded DNA (ssDNA) adsorbs more readily to a colloid surface than dsDNA [64]. The researchers used LNA for improved discrimination and spermine to overcome the electrostatic repulsion. This work was a breakthrough in the SERRS detection: the first report of clinically relevant oligonucleotide sequences were detected with a closed-tube assay that could be potentially applied to clinical environments (Figure 11.8).

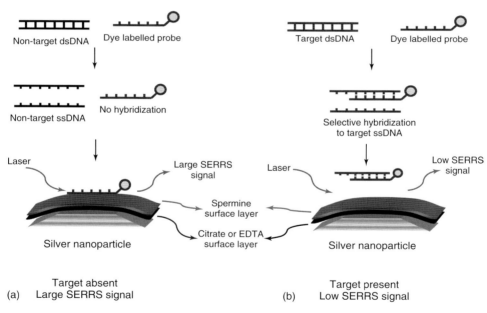

Figure 11.8 Schematic representation of the SERRS MRSA detection assay as developed by Macaskill *et al*. (a) if non-target DNA is present the dye-labelled DNA or LNA probe is free to adsorb on the surface of the silver nanoparticles resulting in an intense SERRS signal. In (b), the presence of target DNA allows the probe to hybridize to its complement, and in this case the double-stranded DNA (dsDNA) is not adsorbed onto the surface of the silver nanoparticles, resulting in a reduction in the SERRS signal. (Reprinted with permission from [63]. © 2010 American Chemical Society.)

An assay for alternative gene splicing using non-fluorescent Raman labels with sensitivity approaching 1 fM has been demonstrated by Sun et al. [65]. The authors detected alternative sequences in the gene splicing profile of *BCRA1* gene, linked to the malignant transformation of tumours in breast cancer. The work utilized a sandwich/split-probe approach, carried out on a surface and with a gold Raman probe. It is hoped that such a technique could readily replace the time-consuming and expensive sequencing of full-length cDNA, which is the current most reliable technique for an alternative splicing study.

11.8
Conclusion

This review focuses on how SERRS-active oligonucleotide probes can be designed and synthesized to provide a quantitative, low-level detection method for DNA sequences. With careful choice of molecular label and excitation source, SERRS offers an almost unparalleled multiplexing capability and an ultra-low LOD. SERRS probes offer tremendous scope for specific DNA sequence detection from a complex mixture, making them uniquely placed to observe and quantify multiple gene targets in a fast, high-throughput method. SERRS researchers can easily build on the foundations of the fluorescent DNA labelling technologies and can exploit the vast number and availability of many Raman dyes/fluorescent dyes.

The use of SERRS detection as a genuine molecular diagnostic technique is still in its infancy but the current data available shows a technique capable of rivalling, if not surpassing, the fluorescent equivalent in many applications. A number of groups are exploring the use of SERRS as a sensitive, high-information technique for DNA analysis, hopefully leading to the development of new SERRS-based assays to be used in a clinical environment.

References

1. Yang, S. and Rothman, R.E. (2004) PCR-based diagnostics for infectious diseases: uses, limitations, and future applications in acute-care settings. *Lancet Infect. Dis.*, **4**, 337–348.
2. Bissonnette, L. and Bergeron, M.G. (2006) Next revolution in the molecular theranostics of infectious diseases: microfabricated systems for personalized medicine. *Expert Rev. Mol. Diagn.*, **6**, 433–450.
3. Linacre, A. and Graham, D. (2002) Role of molecular diagnostics in forensic science. *Expert Rev. Mol. Diagn.*, **2**, 346–353.
4. Holland, P.M., Abramson, R.D., Watson, R. and Gelfand, D.H. (1991) Detection of specific polymerase chain reaction product by utilizing the 5′–3′ exonuclease activity of *Thermus aquaticus* DNA polymerase. *Proc. Natl. Acad. Sci.*, **88**, 7276–7280.
5. Tyagi, S. and Kramer, F.R. (1996) Molecular beacons: probes that fluoresce upon hybridization. *Nat. Biotechnol.*, **14**, 303–308.
6. Marras, S.A., Gold, B., Kramer, F.R., Smith, I. and Tyagi, S. (2004) Real-time measurement of in vitro transcription. *Nucleic Acids Res.*, **32**, e72.

7. Whitcombe, D., Theaker, J., Guy, S.P., Brown, T. and Little, S. (1999) Detection of PCR products using self-probing amplicons and fluorescence. *Nat. Biotechnol.*, **17**, 804–807.
8. Heid, C.A., Stevens, J., Livak, K.J. and Williams, P.M. (1996) Real time quantitative PCR. *Genome Res.*, **6**, 986–994.
9. Kneipp, K., Kneipp, H., Itzkan, I., Dasari, R.R. and Feld, M.S. (1999) Ultrasensitive chemical analysis by Raman spectroscopy. *Chem. Rev.*, **99**, 2957–2976.
10. Francois, P., Pittet, D., Bento, M., Pepey, B., Vaudaux, P., Lew, D. and Schrenzel, J. (2003) Rapid detection of methicillin-resistant Staphylococcus aureus directly from sterile or nonsterile clinical samples by a new molecular assay. *J. Clin. Microbiol.*, **41**, 254–260.
11. Nie, S. and Emory, S.R. (1997) Probing single molecules and single nanoparticles by surface-enhanced Raman scattering. *Science*, **275**, 1102–1106.
12. Kneipp, K. and Kneipp, H. (2006) Single molecule Raman scattering. *Appl. Spectrosc.*, **60**, 322A–334A.
13. Faulds, K., Jarvis, R., Smith, W.E., Graham, D. and Goodacre, R. (2008) Multiplexed detection of six labelled oligonucleotides using surface enhanced resonance Raman scattering (SERRS). *Analyst*, **133**, 1505–1512.
14. Schatz, G.C. and Van Duyne, R.P. (2002) in *Handbook of Vibrational Spectroscopy* (eds J.M. Chalmers and P.R. Griffiths), John Wiley & Sons, Inc., New York, pp. 759–774.
15. Kneipp, K., Wang, Y., Kneipp, H., Perelman, L.T., Itzkan, I., Dasari, R.R. and Feld, M.S. (1997) Single molecule detection using surface-enhanced Raman scattering (SERS). *Phys. Rev. Lett.*, **78**, 1667.
16. Wang, H.N. and Vo-Dinh, T. (2009) Multiplex detection of breast cancer biomarkers using plasmonic molecular sentinel nanoprobes. *Nanotechnology*, **20**, 65101.
17. Tian, Z.-Q., Yang, Z.-L., Ren, B. and Wu, D.-Y. (2006) in *Surface-Enhanced Raman Scattering: Physics and Applications*, Topics in Applied Physics, vol. 103 (eds K. Kneipp, M. Moskovits and H. Kneipp), Springer, Berlin, pp. 125–146.
18. Fleischmann, M., Hendra, P.J. and McQuillan, A.J. (1974) Raman spectra of pyridine adsorbed at a silver electrode. *Chem. Phys. Lett.*, **26**, 163–166.
19. Perney, N.M., Baumberg, J.J., Zoorob, M.E., Charlton, M.D., Mahnkopf, S. and Netti, C.M. (2006) Tuning localized plasmons in nanostructured substrates for surface-enhanced Raman scattering. *Opt. Express*, **14**, 847–857.
20. Levin, C.S., Bishnoi, S.W., Grady, N.K. and Halas, N.J. (2006) Determining the conformation of thiolated poly(ethylene glycol) on Au nanoshells by surface-enhanced Raman scattering spectroscopic assay. *Anal. Chem.*, **78**, 3277–3281.
21. Sha, M.Y., Walton, I.D., Norton, S.M., Taylor, M., Yamanaka, M., Natan, M.J., Xu, C., Drmanac, S., Huang, S., Borcherding, A., Drmanac, R. and Penn, S.G. (2006) Multiplexed SNP genotyping using nanobarcode particle technology. *Anal. Bioanal. Chem.*, **384**, 658–666.
22. Kim, J.H., Bryan, W.W. and Lee, T.R. (2008) Preparation, characterization, and optical properties of gold, silver, and gold-silver alloy nanoshells having silica cores. *Langmuir*, **24**, 11147–11152.
23. Lee, P.C. and Meisel, D. (1982) Adsorption and surface-enhanced Raman of dyes on silver and gold sols. *J. Phys. Chem.*, **86**, 3391–3395.
24. Lee, S.J., Morrill, A.R. and Moskovits, M. (2006) Hot spots in silver nanowire bundles for surface-enhanced Raman spectroscopy. *J. Am. Chem. Soc.*, **128**, 2200–2201.
25. Doering, W.E. and Nie, S. (2003) Spectroscopic tags using dye-embedded nanoparticles and surface-enhanced Raman scattering. *Anal. Chem.*, **75**, 6171–6176.
26. Constantino, C.J., Lemma, T., Antunes, P.A. and Aroca, R. (2001) Single-molecule detection using surface-enhanced resonance Raman scattering and Langmuir-Blodgett monolayers. *Anal. Chem.*, **73**, 3674–3678.
27. Haynes, C.L., McFarland, A.D. and Duyne, R.P.V. (2005) Surface-enhanced

Raman spectroscopy. *Anal. Chem.*, **77**, 338A–346A.
28. Culha, M., Stokes, D. and Vo-Dinh, T. (2003) Surface-enhanced Raman scattering for cancer diagnostics: detection of the BCL2 gene. *Expert Rev. Mol. Diagn.*, **3**, 669–675.
29. Stokes, R.J., Macaskill, A., Dougan, J.A., Hargreaves, P.G., Stanford, H.M., Smith, W.E., Faulds, K. and Graham, D. (2007) Highly sensitive detection of dye-labelled DNA using nanostructured gold surfaces. *Chem. Commun.*, **27**, 2811–2813.
30. Smith, W.E. (2008) Practical understanding and use of surface enhanced Raman scattering/surface enhanced resonance Raman scattering in chemical and biological analysis. *Chem. Soc. Rev.*, **37**, 955–964.
31. Stuart, D.A., Yuen, J.M., Shah, N., Lyandres, O., Yonzon, C.R., Glucksberg, M.R., Walsh, J.T. and Van Duyne, R.P. (2006) In vivo glucose measurement by surface-enhanced Raman spectroscopy. *Anal. Chem.*, **78**, 7211–7215.
32. Zhou, L., Yang, J., Estavillo, C., Stuart, J.D., Schenkman, J.B. and Rusling, J.F. (2003) Toxicity screening by electrochemical detection of DNA damage by metabolites generated in situ in ultrathin DNA-enzyme films. *J. Am. Chem. Soc.*, **125**, 1431–1436.
33. Graham, D. and Faulds, K. (2008) Quantitative SERRS for DNA sequence analysis. *Chem. Soc. Rev.*, **37**, 1042–1051.
34. Graham, D., Smith, W.E., Linacre, A.M.T., Munro, C.H., Watson, N.D. and White, P.C. (1997) Selective detection of deoxyribonucleic acid at ultralow concentrations by SERRS. *Anal. Chem.*, **69**, 4703–4707.
35. McAnally, G., McLaughlin, C., Brown, R., Robson, D.C., Faulds, K., Tackley, D.R., Smith, W.E. and Graham, D. (2002) SERRS dyes. Part I. Synthesis of benzotriazole monoazo dyes as model analytes for surface enhanced resonance Raman scattering. *Analyst*, **127**, 838–841.
36. Graham, D., Mallinder, B.J. and Smith, W.E. (2000) Surface-enhanced resonance Raman scattering as a novel method of DNA discrimination. *Angew. Chem. Int. Ed.*, **39**, 1061–1063.
37. Brown, R., Smith, W.E. and Graham, D. (2003) Synthesis of a benzotriazole azo dye phosphoramidite for labelling of oligonucleotides. *Tetrahedron Lett.*, **44**, 1339–1342.
38. Faulds, K., Fruk, L., Robson, D.C., Thompson, D.G., Enright, A., Smith, W.E. and Graham, D. (2006) A new approach for DNA detection by SERRS. *Faraday Discuss.*, **132**, 261–268.
39. Bell, S.E. and Sirimuthu, N.M. (2006) Surface-enhanced Raman spectroscopy (SERS) for sub-micromolar detection of DNA/RNA mononucleotides. *J. Am. Chem. Soc.*, **128**, 15580–15581.
40. Faulds, K., Smith, W.E. and Graham, D. (2004) Evaluation of surface-enhanced resonance Raman scattering for quantitative DNA analysis. *Anal. Chem.*, **76**, 412–417.
41. Graham, D., Faulds, K. and Smith, W.E. (2006) Biosensing using silver nanoparticles and surface enhanced resonance Raman scattering. *Chem. Commun.*, 4363–4371.
42. Faulds, K., McKenzie, F. and Graham, D. (2007) Evaluation of the number of modified bases required for quantitative SERRS from labelled DNA. *Analyst*, **132**, 1100–1102.
43. Faulds, K., Barbagallo, R.P., Keer, J.T., Smith, W.E. and Graham, D. (2004) SERRS as a more sensitive technique for the detection of labelled oligonucleotides compared to fluorescence. *Analyst*, **129**, 567–568.
44. Barhoumi, A., Zhang, D., Tam, F. and Halas, N.J. (2008) Surface-enhanced Raman spectroscopy of DNA. *J. Am. Chem. Soc.*, **130**, 5523–5529.
45. Stokes, R.J., Macaskill, A., Lundahl, P.J., Smith, W.E., Faulds, K. and Graham, D. (2007) Quantitative enhanced Raman scattering of labeled DNA from gold and silver nanoparticles. *Small*, **3**, 1593–1601.
46. Dougan, J.A., Karlsson, C., Smith, W.E. and Graham, D. (2007) Enhanced oligonucleotide-nanoparticle conjugate stability using thioctic acid modified oligonucleotides. *Nucleic Acids Res.*, **35**, 3668–3675.

47. Cao, Y.C., Jin, R. and Mirkin, C.A. (2002) Nanoparticles with Raman spectroscopic fingerprints for DNA and RNA detection. *Science*, **297**, 1536–1540.
48. Graham, D., Mallinder, B.J., Whitcombe, D. and Smith, W.E. (2001) Surface enhanced resonance Raman scattering (SERRS)-a first example of its use in multiplex genotyping. *ChemPhysChem*, **2**, 746–748.
49. Docherty, F.T., Monaghan, P.B., Keir, R., Graham, D., Smith, W.E. and Cooper, J.M. (2004) The first SERRS multiplexing from labelled oligonucleotides in a microfluidics lab-on-a-chip. *Chem. Commun.*, **1**, 118–119.
50. Faulds, K., McKenzie, F., Smith, W.E. and Graham, D. (2007) Quantitative simultaneous multianalyte detection of DNA by dual-wavelength surface-enhanced resonance Raman scattering. *Angew. Chem. Int. Ed.*, **46**, 1829–1831.
51. Thomas, M. and Klibanov, A.M. (2003) Conjugation to gold nanoparticles enhances polyethylenimine's transfer of plasmid DNA into mammalian cells. *Proc. Natl. Acad. Sci.*, **100**, 9138–9143.
52. Ow Sullivan, M.M., Green, J.J. and Przybycien, T.M. (2003) Development of a novel gene delivery scaffold utilizing colloidal gold-polyethylenimine conjugates for DNA condensation. *Gene Ther.*, **10**, 1882–1890.
53. Rosi, N.L., Giljohann, D.A., Thaxton, C.S., Lytton-Jean, A.K., Han, M.S. and Mirkin, C.A. (2006) Oligonucleotide-modified gold nanoparticles for intracellular gene regulation. *Science*, **312**, 1027–1030.
54. Hamad-Schifferli, K., Schwartz, J.J., Santos, A.T., Zhang, S. and Jacobson, J.M. (2002) Remote electronic control of DNA hybridization through inductive coupling to an attached metal nanocrystal antenna. *Nature*, **415**, 152–155.
55. Nam, J.-M., Park, S.-J. and Mirkin, C.A. (2002) Bio-barcodes based on oligonucleotide-modified nanoparticles. *J. Am. Chem. Soc.*, **124**, 3820–3821.
56. Wang, Y., Wei, H., Li, B., Ren, W., Guo, S., Dong, S. and Wang, E. (2007) SERS opens a new way in aptasensor for protein recognition with high sensitivity and selectivity. *Chem. Commun.*, **48**, 5220–5222.
57. Mirkin, C.A., Letsinger, R.L., Mucic, R.C. and Storhoff, J.J. (1996) A DNA-based method for rationally assembling nanoparticles into macroscopic materials. *Nature*, **382**, 607–609.
58. Jin, R., Wu, G., Li, Z., Mirkin, C.A. and Schatz, G.C. (2003) What controls the melting properties of DNA-linked gold nanoparticle assemblies? *J. Am. Chem. Soc.*, **125**, 1643–1654.
59. McKenzie, F., Faulds, K. and Graham, D. (2008) LNA functionalized gold nanoparticles as probes for double stranded DNA through triplex formation. *Chem. Commun.*, **20**, 2367–2369.
60. You, Y., Moreira, B.G., Behlke, M.A. and Owczarzy, R. (2006) Design of LNA probes that improve mismatch discrimination. *Nucleic Acids Res.*, **34**, e60.
61. Graham, D., Thompson, D.G., Smith, W.E. and Faulds, K. (2008) Control of enhanced Raman scattering using a DNA-based assembly process of dye-coded nanoparticles. *Nat. Nanotechnol.*, **3**, 548–551.
62. Wabuyele, M.B. and Vo-Dinh, T. (2005) Detection of human immunodeficiency virus type 1 DNA sequence using plasmonics nanoprobes. *Anal. Chem.*, **77**, 7810–7815.
63. Macaskill, A., Crawford, D., Graham, D. and Faulds, K. (2009) DNA sequence detection using surface-enhanced resonance Raman spectroscopy in a homogeneous multiplexed assay. *Anal. Chem.*, **81**, 8134–8140.
64. Li, H. and Rothberg, L. (2004) Colorimetric detection of DNA sequences based on electrostatic interactions with unmodified gold nanoparticles. *Proc. Natl. Acad. Sci.*, **101**, 14036–14039.
65. Sun, L., Yu, C. and Irudayaraj, J. (2008) Raman multiplexers for alternative gene splicing. *Anal. Chem.*, **80**, 3342–3349.

12
SERS Microscopy: Nanoparticle Probes and Biomedical Applications
Sebastian Schlücker

12.1
Introduction

Metal nanostructures including nanoparticles play an important role in plasmon-assisted vibrational spectroscopies such as SERS (see also Chapters 1 and 2) [1–4]. For quantitative SERS, various issues – ranging from colloid stability and reproducibility to internal standards – should be considered (see also Chapter 3). SERS benefits from very high Raman signal intensities: the sensitivity is sufficient for trace and even single-molecule detection, and numerous analytes have been identified by their characteristic molecular fingerprint (see also Chapters 4 and 5, together with Chapter 6). Efficient label-free detection by SERS is usually based on the chemisorption or at least physisorption of the analyte on or near the surface of the plasmonic nanoparticle (Figure 12.1a); that is, attractive forces such as strong Coulomb interactions between the molecule in solution and the particle surface are often a prerequisite (see also Chapters 6 and 11). This has direct implications regarding the classes of molecules that can be detected. In certain cases, it may not be trivial to achieve adsorption of the analyte from solution phase onto the particle surface.

A further important aspect in SERS is selectivity. Intelligent strategies have been developed to achieve selectivity via surface modifications (see also Chapter 5). Particularly promising are approaches based on supramolecular chemistry, exploring the opportunities arising from multiple, weak intermolecular interactions such as hydrogen bonds or hydrophobic effects and the concept of multi-valency.

In addition to the label-free detection of analytes, which till today represents the vast majority of SERS publications, a second class or domain of applications is rapidly emerging: the target-specific detection using so-called SERS labels or nanotags. In this approach, SERS is used as a readout method for the selective and sensitive detection of target molecules such as proteins and oligonucleotides (see also Chapter 5) [5–8]. In SERS nanoparticle probes [7–27], Raman labels or reporter molecules are permanently adsorbed to the metal surface (Figure 12.1b). A protective shell is optional but highly desired for particle stabilization, especially for preventing aggregation. Laser excitation – in biological applications preferably

Surface Enhanced Raman Spectroscopy: Analytical, Biophysical and Life Science Applications. Edited by Sebastian Schlücker
Copyright © 2011 WILEY-VCH Verlag GmbH & Co. KGaA, Weinheim
ISBN: 978-3-527-32567-2

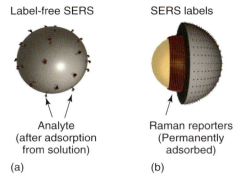

Figure 12.1 Two different modalities of using metal nanoparticles as colloidal SERS substrates. (a) The label-free detection of analytes is based on the characteristic SERS spectrum of the analyte. In most cases, the analyte is adsorbed from solution phase onto the particle surface. (b) In SERS labels as many Raman reporter molecules as possible, for example a complete self-assembled monolayer (SAM) with maximum surface coverage, are permanently adsorbed onto the particle surface. The characteristic SERS spectrum of the Raman reporters is used for particle identification. For target recognition, SERS probes must be conjugated to target-specific binding molecules such as antibodies.

in the red to near-infrared (NIR) for minimizing autofluorescence – leads to a characteristic Raman spectrum which is used for particle identification. In other words, SERS nanoparticle probes are labelling agents, similar to dyes used in staining procedures and fluorescence applications.

This chapter intends to summarize current designs of SERS probes (Section 12.2) and their biomedical applications in combination with Raman microscopy (Section 12.3).

12.2
SERS Nanoparticle Probes

12.2.1
Components of a SERS Label

Figure 12.2 displays the building blocks of a SERS label [7, 8]. The minimal configuration comprises Raman reporter molecules adsorbed on the surface of a colloidal SERS substrate. The example shown in Figure 12.2 comprises an organic monolayer on the surface of a Au/Ag nanoshell. A protective encapsulant, such as the glass/silica shell in Figure 12.2, is optional but highly desirable since it stabilizes the actual SERS label and prevents particle aggregation. In this context, it is important to note that the focus in this chapter is on SERS probes <100 nm for microscopic applications. The encapsulated SERS probes can be further functionalized. Since the label itself does not provide selectivity, target-specific binding partners such as antibodies for antigen recognition (Figure 12.2) are required.

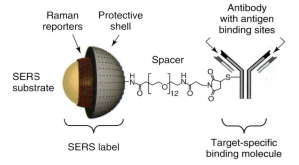

Figure 12.2 Components of a target-specific SERS probe: Colloidal SERS substrate (Au/Ag nanoshell), Raman reporters adsorbed to its surface (organic monolayer), an optional protective encapsulant (glass shell) and a target-specific binding molecule (antibody) attached to the SERS label via a spacer molecule [22].

Figure 12.3 Synthesis of SERS-labelled antibodies: The metal colloid is incubated with Raman labels and optionally encapsulated with a protective shell. The resulting encapsulated SERS label is then conjugated to antibodies via spacer units.

Direct conjugation of the antibody to the metal surface or the Raman reporters is one option. The use of a spacer molecule between the antibody and the SERS label introduces conformational flexibility, which may be supportive or even necessary for proper target recognition via the antigen binding sites.

Figure 12.3 depicts the sequence of chemical modifications by which SERS-labelled antibodies are produced from metal colloids. The addition of Raman reporter molecules to the colloid with subsequent encapsulation leads to a stabilized SERS probe. Modification of the shell with spacers and their conjugation to an antibody finally yields SERS-labelled antibodies.

12.2.2
Choice of Metal Colloid

Parameters such as size, shape and chemical composition determine the optical properties of metal nanoparticles (see also Chapter 1), which is useful to keep in mind for a rational design of sensitive SERS probes. For quantitative SERS,

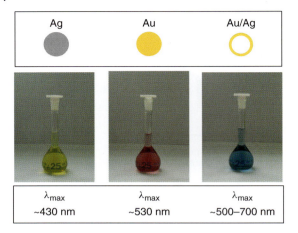

Figure 12.4 Photographs of different noble-metal colloids with a diameter of ~55 nm, together with the approximate wavelength of the plasmon peak in water. In contrast to Ag and Au solid nanospheres, the plasmon resonance of Au/Ag nanoshell occurs is in the red to near-infrared region and can – in addition to the particle diameter – be controlled by the shell thickness.

monodisperse particles are ideal: a narrow size distribution guarantees similar scattering intensities and comparable SERS intensities. Methods for the controlled synthesis of monodisperse metal nanoparticles are therefore highly desirable (see also Chapter 2).

Figure 12.4 contains photographs of ~55 nm colloidal silver nanospheres (λ_{max} ~ 430 nm), gold nanospheres (λ_{max} ~ 530 nm) and nanoshells comprising an alloy of gold and silver (λ_{max} ~ 500–700 nm) in water. Note that the plasmon resonances of the nanoshells can be tuned over a large wavelength range in the red and also NIR for larger shells. The position of the plasmon peak λ_{max} depends on a number of parameters, in particular the size of the nanospheres/shells and the dielectric function of both the metal sphere and the surrounding medium.

The choice of a particular colloidal SERS substrate may depend on external experimental constraints such as the available laser excitation wavelengths as well as the specific properties of the sample. The interest in using plasmonic nanostructures as efficient SERS substrates in combination with red to NIR laser excitation is often triggered by their use in biomedical and biological applications. Minimizing the disturbing autofluorescence of biological specimens leads to an improved (image) contrast, which may be more effective than maximizing absolute SERS signal levels. For optical applications *in vivo*, it is additionally essential to pass tissue such as the skin barrier, taking advantage of the 'biological window' in the NIR region [19, 25].

Figure 12.5 illustrates the tunability of the plasmon band for Au/Ag nanoshells [28, 29] based on Mie calculations [30, 31]. All particles have a diameter of 55 nm and the shell thickness *d* decreases from left to right.

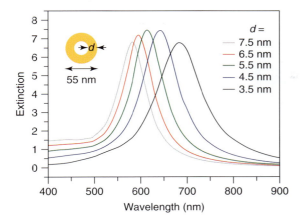

Figure 12.5 Extinction spectra of single Au/Ag nanoshells with a constant diameter of 55 nm but varying shell thickness, calculated by Lorenz–Mie theory with water as the surrounding medium. The plasmon peak shifts to higher wavelengths as the shell becomes thinner.

By using red laser excitation (He–Ne laser, 632.8 nm), it has been shown experimentally that 60 nm gold/silver nanoshells yield about eight times higher SERS intensities compared with gold nanospheres of the same size [22]. In addition to the nanoshells with a solvent core and a gold/silver alloy shell discussed here, also other core/shell structures have been used as SERS substrates: for example, SiO_2/Au, Ag/Au and Au/Ag core/shell particles [32–34]. Mirkin and co-workers, for example, have introduced silver staining of small functionalized gold nanoparticle probes to produce the actual SERS substrate [10].

As pointed out in Chapter 1 (cf. also Figure 1.1), the enhancements for gold and silver in the region >600 nm become similar. Together with the chemical inertness, this makes gold probably the preferred SERS substrate in many biological applications.

12.2.3
Choice of Raman Reporter

In principle, a large variety of molecules may be employed as Raman reporters. Ideally, Raman reporters exhibit (i) high differential Raman scattering cross sections for high signal levels, (ii) a small number of atoms and/or high symmetry leading to a minimal number of Raman bands for dense multiplexing, (iii) low or no photobleaching for signal stability and (iv) chemisorption to the colloidal metal surface for strong binding.

Porter and co-workers, for example, have introduced aryldisulfides as Raman labels because they form self-assembled monolayers (SAMs) on gold surfaces via stable Au–S bonds (Figure 12.6a) [6, 9, 11].

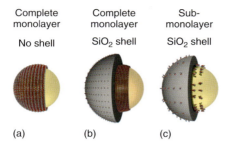

Figure 12.6 Surface coverage and protection of SERS labels with a glass or silica shell. (a) Complete monolayer without protective encapsulation [9, 11]. (b) Complete monolayer coverage with protective silica shell [22]. (c) Sub-monolayer coverage with protective silica shell [13, 14].

Using a SAM of Raman reporter molecules has several advantages [6, 8, 9, 11, 22]. Firstly, a complete monolayer ensures maximum SERS sensitivity since it guarantees the maximum surface coverage with Raman reporter molecules (Figure 12.6a,b) in comparison with sub-monolayer coverage (Figure 12.6c). Secondly, clean and reproducible Raman signatures are obtained because of the uniform orientation of the molecules within the SAM relative to the surface normal of the nanoparticle [22, 24, 27]. Thirdly, spectral interferences with unwanted contributions, due to the co-adsorption of molecules other than Raman labels on the particle surface, are minimized.

Figure 12.7 shows a comparison of the SERS spectra of 4-mercaptobenzoic acid (MBA) and rhodamine 6G (R6G) adsorbed on Au/Ag nanoshells. MBA is a small

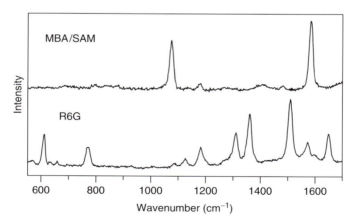

Figure 12.7 SERS spectrum of a self-assembled monolayer (SAM) comprising 4-mercaptobenzoic acid (MBA/SAM) in comparison with rhodamine 6G (R6G, no SAM) on Au/Ag nanoshells. The smaller size of MBA compared to R6G and in particular the uniform orientation of MBA molecules within the SAM in conjunction with the surface selection rules lead to the appearance of only two dominant Raman bands [22].

aromatic molecule, while R6G is a fluorone dye. The smaller size of MBA, but in particular the presence of a SAM in conjunction with the SERS surface selection rules [1], leads to the appearance of only a few dominant Raman bands. The advantage of using dyes, however, is their larger signal due to surface-enhanced resonance Raman scattering (SERRS; see also Chapter 11).

12.2.4
Protection and Stabilization

The encapsulation of SERS probes has several benefits. For instance, the resulting chemical and mechanical stability of the colloidal particles allows particle storage and prevents particle aggregation. Furthermore, the desorption of Raman labels from the metal surface as well as the adsorption of spectrally interfering molecules from the environment to the surface can be completely eliminated.

Various encapsulants are available, including proteins as biopolymers [15], organic polymers [20, 35] and silica [5, 13, 14, 22, 26, 27]. Among them, silica is attractive because of its high mechanical stability and the option for long-term storage.

Mulvaney *et al.* have introduced the concept of silica encapsulation for SERS probes (Figure 12.6c) [13]. Co-adsorption of Raman labels and SiO_2 precursors (typically in a 1 : 20 stoichiometry) leads to a sub-monolayer coverage of the surface with Raman reporter molecules, followed by silica encapsulation with a modified Stöber method [36]. Soon after, Doering and Nie presented a very similar approach to glass-coated nanoparticles [14]. Generally, a glass shell provides chemical and physical stability to the SERS label. For example, a 20 nm thick silica shell leads to a significantly increased lifetime of gold nanospheres in the presence of aqua regia: 3 h in comparison to 15 s for the bare Au colloid [13]. Further advantages are storage stability and protection against mechanical deformation.

Silica-encapsulated SERS probes comprising a complete SAM (Figure 12.6b) combine the chemical stability of a glass shell with the spectroscopic advantages resulting from the maximum and dense surface coverage with Raman reporters in a uniform orientation [22]. In contrast to the sub-monolayer coverage with Raman labels, a complete SAM leads to a higher sensitivity. Using hollow gold/silver nanospheres and red laser excitation (632.8 nm), it has been shown experimentally that a complete SAM yields about 22 times more intense SERS signals compared to sub-monolayer coverage [22].

Figure 12.8 (left) shows the individual preparation steps of silica-encapsulated SERS probes comprising a complete SAM [22]. First, the addition of Raman labels (Ra) to the SERS substrate produces SAM-coated metal nanospheres (**1a**). Layer-by-layer deposition of poly(allylamine hydrochloride) (PAH) and polyvinylpyrrolidone (PVP) onto the SAM leads to the polyelectrolyte-coated SAMs, **1b** and **1c**, respectively. The PVP coating renders the SERS particles vitreophilic. The silica shell in **1d** is then formed upon addition, hydrolysis and finally condensation of tetraethoxy orthosilicate (TEOS).

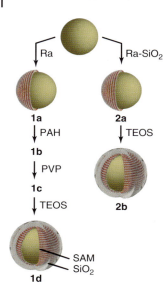

Figure 12.8 Two different routes to SERS labels with a glass shell as an encapsulant and a self-assembled monolayer (SAM) of Raman reporters on the surface of the metal nanoparticle. Left: Three-step synthesis, starting from a SAM **1a** of Raman reporter molecules (Ra) on the surface of Au/Ag nanoshells, subsequent layer-by-layer deposition of the polyelectrolytes PAH (**1b**) and PVP (**1c**) and finally the glass encapsulation using tetraethoxyorthosilicate (TEOS) to **1d** [22]. Right: One-step approach via a SAM **2a** of Raman reporter molecules containing terminal SiO_2 precursors (Ra-SiO_2) and its direct silica encapsulation to **2b** using TEOS [27].

The route to hydrophilic, silica-encapsulated SAMs (**1d**) depicted in Figure 12.8 (left) involves several steps – including different solvents and multiple centrifugations – and is therefore labour and time intensive. In particular, it requires that the SAM is covered by a polyelectrolyte layer prior to silica encapsulation. A faster, simpler and generally applicable route to silica-encapsulated SERS labels is possible via a SAM containing terminal SiO_2 precursors (Figure 12.8, right): both Raman reporter molecule (Ra) and terminal SiO_2-precursor are covalently bound to each other (Ra-SiO_2), that is, both functions are merged into a single molecular unit [27]. In this approach, aminoalkyl-alkoxysilanes such as 3-amino-n-propyltrimethoxysilane (APTMS) may be used as SiO_2 precursors in combination with Raman reporter molecules such as derivatives of mercaptobenzoic acid. The addition of TEOS, its subsequent hydrolysis and condensation then leads to the formation of a silica shell since the SERS labels are already vitreophilic due to the terminal SiO_2 precursor. This strategy to synthesize silica-encapsulated SAMs (Figure 12.8, right) has the advantage that it is much faster and, more importantly, independent of the SAM's surface charge (i.e. the type of particular Raman reporter Ra in Figure 12.8, top) and is therefore ideally universally applicable. However, it requires an additional synthesis step for the Raman reporter–SiO_2 precursor conjugate (Ra-SiO_2 in Figure 12.8, right) [27].

The transmission electron microscopy (TEM) images in Figure 12.9 demonstrate the monodispersity of the resulting glass encapsulated SERS probes [22]. The thickness of the glass shell can be controlled by the amount of TEOS. The 60 nm Au/Ag nanoshells in Figure 12.9 have a ∼10 nm thick (top left) and ∼25 nm thick (bottom left, right) silica shell [22]. Functionalization of the glass surface, for example with amino groups, allows the subsequent conjugation of biomolecules to the silica shell, either directly or via spacers [16, 21, 22, 24].

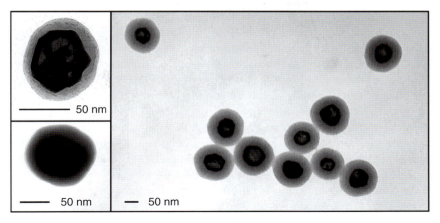

Figure 12.9 Transmission electron microscopy (TEM) images of silica-encapsulated SERS labels with silica shells ~10 nm (top left) and ~25 nm (bottom left, right) thick. The diameter of the gold/silver nanoshells is ~60 nm [22].

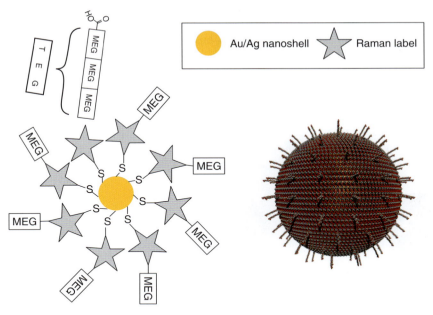

Figure 12.10 Hydrophilic SERS labels for controlled bioconjugation. A complete SAM of arylthiols as Raman reporters is adsorbed onto the surface of the metal colloid. Hydrophilic mono- and triethylene glycols (MEG and TEG, respectively) are conjugated to the Raman reporters and stabilize the SAM. The water soluble SERS label can be conjugated to biomolecules via the terminal carboxy moieties of the longer TEG spacers [24].

Figure 12.10 depicts a second SERS label design which is based on terminal hydrophilic spacers attached to the SAM [24]. This approach combines the stabilization of the SAM by hydrophilic spacers for obtaining water-soluble SERS probes and the option for their controlled bioconjugation. Stabilization of the SAM is achieved by attaching hydrophilic monoethylene glycol (MEG) units with terminal OH groups. A small portion of Raman reporter molecules is conjugated to longer triethylene glycol (TEG) units with terminal COOH moieties for bioconjugation [24]. Four advantages result from this strategy: (i) the colloidal surface is completely covered exclusively by Raman labels for maximum sensitivity; (ii) the entire SERS label is hydrophilic and water soluble due to the MEG/TEG units, which is independent of a particular Raman reporter molecule; (iii) increased steric accessibility of the SAM for bioconjugation via the longer TEG spacers with terminal COOH groups and (iv) the option for controlled bioconjugation by varying the ratio of the two spacer units (TEG-OH : MEG-COOH).

A further advantageous property of the hydrophilic EG units [21, 24] is the minimization of non-specific binding. This is a highly important aspect in many biological and biomedical applications: the binding selectivity is determined by the target-specific binding molecule (e.g. the antibody in Figure 12.2) and should not be diminished by non-specific binding of the labelling agent (e.g. the SERS probe in Figure 12.2) which leads to false-positive results.

The next section discusses selected biomedical applications of SERS probes conjugated to antibodies as target-specific binding molecules.

12.3
Biomedical Applications of SERS Microscopy

SERS microscopy is the combination of SERS labels and Raman microscopy (Figure 12.11). Target-specific probes can be obtained by conjugating SERS particles to recognition molecules such as antibodies. Target localization is achieved by

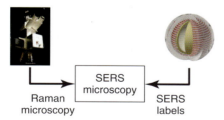

Figure 12.11 SERS microscopy as a novel method of vibrational microspectroscopy for the selective and sensitive localization of targeted molecules. Target-specific SERS probes, obtained after the conjugation of SERS labels to target-specific binding molecules, can be identified and localized in cells and tissues with Raman microscopy.

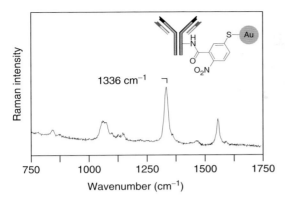

Figure 12.12 Raman spectrum of an SERS-labelled anti-PSA antibody [11, 37] together with its schematic structure (inset). Only Raman bands of the arylthiol as the Raman reporter are observed because of the strong SERS distance dependence.

Raman microspectroscopy, using the characteristic SERS spectrum of the label for identification.

The Raman spectrum of an SERS-labelled antibody together with its schematic structure is shown in Figure 12.12 [11, 37]. The Raman label is a nitro derivative of mercaptobenzoic acid. Strikingly, only three dominant Raman bands are detected, although the number of atoms in the labelled antibody is very large. The band at 1336 cm^{-1} is assigned to the symmetric nitro stretching vibration, while the other two bands can be assigned to phenyl ring modes. In contrast, no spectral contributions from the antibody, for example the amide I band around 1650 cm^{-1} for α-helices, are observed. This selectivity in the enhancement of Raman bands is primarily due to the SERS distance dependence (cf. Chapter 1): only the Raman label moiety is close enough to the nanoparticle surface to experience the drastic near-field enhancement, while the antibody is too distant. A second aspect is the orientation of the Raman label molecules relative to the surface normal: according to the SERS selection rules, z-components of the molecular polarizability tensor experience the largest enhancement (z being the axis parallel to the surface normal) [1].

12.3.1
Immunohistochemistry

The principle of immunohistochemistry using SERS-labelled antibody probes is schematically illustrated in Figure 12.13. The tissue, either frozen or formalin-fixed and paraffin-embedded specimens, contains numerous different proteins and is usually prepared as a micrometre-thick section using a microtome. The target protein is schematically depicted as a filled red circle; other proteins are indicated by a triangle and a square. The aim of immunohistochemistry is to determine the spatial distribution of a target protein and, eventually, even to quantify its abundance. Blocking agents such as certain buffers or bovine serum albumin (BSA) are often employed prior to incubation in order to minimize non-specific binding. The tissue specimen is then incubated with labelled antibodies in order to form the corresponding antigen–antibody complexes. Washing is necessary to

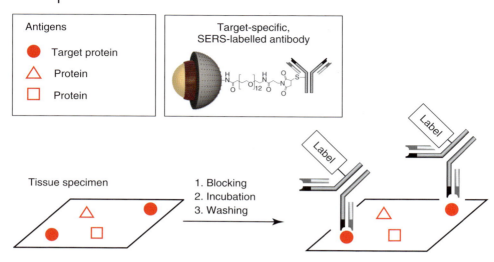

Figure 12.13 Principle of immunohistochemistry for target protein localization. For minimizing non-specific binding, the tissue specimen is first blocked and then incubated with the SERS-labelled antibody as the target-specific probe. The antibody selectively recognizes the antigen via its binding sites and forms the corresponding antigen–antibody complex. Finally, washing removes unbound particles. The antigen can be localized by the characteristic SERS spectrum of the SERS label using a Raman microspectroscopic setup.

remove unbound antibodies. The antigen is located by the characteristic Raman signature of the SERS label.

Figure 12.13 shows the use of labelled primary antibodies. An alternative method (not shown) is the use of unlabelled primary antibodies and subsequent signal amplification by adding labelled secondary antibodies with a high affinity to the primary antibody. This approach has been used in combination with SERS, employing mouse anti-carcinoembryonic antigen as a primary antibody and Malachite Green/anti-mouse/gold particles as the secondary antibody, followed by silver staining [38]. Generally, the binding of several secondary antibodies to a single primary antibody leads to a multiplicative signal enhancement. Antigen quantification, however, is generally difficult because of the amplification step. When both quantitative and multiplexed detection in cells and tissue specimens are of interest, the use of SERS-labelled primary antibodies in SERS microscopy is therefore recommended.

12.3.2
Methodologies in Raman Microspectroscopy

Various methodologies in Raman microspectroscopy are available, and three of them are illustrated in Figure 12.14 [39, 40]. Mapping approaches with point or line focus illumination in combination with an xy-translation stage are most commonly used. Point mapping allows true confocal Raman microscopy and requires lateral

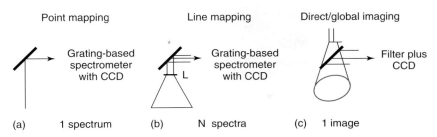

Figure 12.14 Widely used methodologies in Raman microspectroscopy [40]. (a and b) Both point and line mapping are spectra-based approaches using a grating monochromator/CCD detection system. The sample is raster-scanned either in two dimensions (point mapping, one spectrum per spatial position) or one dimension (line mapping, N spectra per spatial position). (c) Direct imaging using wide-field illumination projects the sample plane onto the two-dimensional CCD detector and requires wavelength-selective elements such as dielectric or liquid crystalline tunable filters (global imaging, one image per wavenumber position).

scanning in both x and y directions (Figure 12.14a). In the line mapping approach, scanning is performed perpendicular to the line focus, that is, only along one spatial dimension (Figure 12.14b). This significantly decreases the number of acquisition steps compared to point mapping; however, confocality in the direction of the line focus is lost [40]. Both point and line mapping experiments are usually performed with grating-based monochromators equipped with a CCD. A third, but often less employed methodology is direct or global imaging, using wide-field illumination and dielectric or liquid-crystalline tunable filters as wavelength-selective elements for detecting Raman images at a defined wavenumber position (Figure 12.14c) [40].

In contrast to the direct or global imaging as an image-based methodology, spectra-based mapping approaches require the reconstruction of Raman images. In SERS microscopy, which combines Raman microspectroscopy with target-specific SERS probes, the false-colour image contrast is generated from characteristic SERS label signatures (cf. Figure 12.12), usually using either peak intensities or areas encoded in a false-colour image. False-colour SERS images therefore visualize the distribution and concentration of the corresponding target. Since SERS signal levels are orders of magnitude more intense than conventional Raman scattering, vibrational contributions from the unlabelled tissue are usually not observed.

12.3.3
Immuno-SERS Microscopy for *In vitro* Tissue Diagnostics

The proof of principle for immuno-SERS microscopy using a SERS-labelled primary antibody for target localization in tissue was first demonstrated in 2006 [37]. Localization of prostate-specific antigen (PSA) in tissue specimens of biopsies from patients undergoing prostatectomy was achieved with SERS-labelled primary PSA antibodies, aromatic thiols as Raman reporters and gold/silver nanoshells as colloidal SERS substrate in combination with red laser excitation (He−Ne, 632.8 nm). SERS from the probes was detected in the PSA-(+) epithelium of the

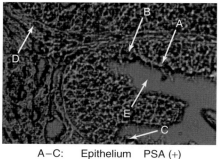
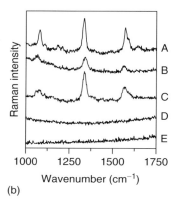
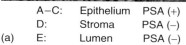

Figure 12.15 Proof of principle for immuno-SERS microscopy using SERS-labelled primary antibodies for tissue diagnostics [37]. Prostate-specific antigen (PSA) was localized in biopsies from patients with prostate cancer. (a) White light image of the prostate tissue section. Three different histological classes can be observed: epithelium, stroma or connective tissue and lumen, in which PSA is either abundant (+) or not (−). (b) Spectra obtained at the locations (A–E) indicated by arrows in the white light image. SERS-labelled anti-PSA antibodies are detected in the PSA-(+) epithelium (locations A–C), exhibiting the characteristic Raman signals of the SERS label. Locations in the PSA-(−) stroma (D) and lumen (E) serve as negative controls where no spectral contributions of the SERS-labelled antibody are detected.

incubated prostate tissue section (Figure 12.15A–C), while locations in the PSA-(−) stroma and lumen served as negative controls (Figure 12.15D–E) [37].

Further applications of immuno-SERS microscopy for cellular and tissue diagnostics with various types of SERS probes appeared soon after this initial study [37]. For example, glass-coated assemblies as SERS probes were used for localizing two targets (HER2 and CD10) on cellular membranes [16]. Au/Ag core–shell nanoparticles were employed for monitoring the overexpression of phospholipase $C\gamma 1$ in human embryonic kidney (HEK 293) cells with SERS imaging [41]. Fluorescent SERS dots [18] composed of silver nanoparticle-embedded silica spheres with fluorescent dye and Raman labels allowed the detection of three cellular proteins which are simultaneously expressed in bronchioalveolar stem cells [42].

The role of small aggregated nanoparticle clusters was elucidated with SERS microscopy in combination with scanning electron microscopy (SEM) in a study of mammalian cell surface receptors. The strongest SERS signals originate from aggregates oriented in the appropriate direction with respect to the polarization of the incident laser [43]. Also, aggregated silver clusters (composite organic inorganic nanoparticles, COINs) were used as SERS probes for PSA localization in prostate tissue sections [44]. The influence of steric hindrance was analysed in a simultaneous two-COIN staining against PSA: both SERS probes were conjugated to an anti-PSA antibody, yielding two distinct groups of SERS-labelled PSA antibodies for two-colour SERS microscopy. The characteristic Raman signatures from both COINs were detected at almost every location in the epithelium, suggesting that steric hindrance from the SERS probes does not represent a major problem. In a subsequent study, the same group presented a detailed comparison of the staining performance with SERS and fluorophore–PSA antibody conjugates on adjacent tissue sections [45]. Staining with COIN- and Alexa-fluorophore labelled antibodies yielded similar results, with a lower staining accuracy but comparable signal intensities for COINs. For decomposition of the overlapping spectral contributions of the individual SERS probes, signal processing algorithms must be applied [46]. In addition to tissue investigations, also single cells were analysed for detecting surface proteins in a two-plex experiment [47].

Selective imaging of PSA in prostate biopsies by SERS microscopy using hydrophilic-spacer SAMs on Au/Ag nanoshells (cf. Figure 12.10) is illustrated in Figure 12.16 [24]. PSA was chosen as a target protein because of its high expression level in prostate tissue and its selective histological abundance in the epithelium of the prostate gland. A bright-field image of an unstained prostate biopsy is shown in Figure 12.16a, together with the region used for Raman point mapping (white box). Three different histological classes can be identified: lumen (L), epithelium (E) and stroma (S). Epithelium is PSA-(+), while (L) and (S) are PSA-(−).

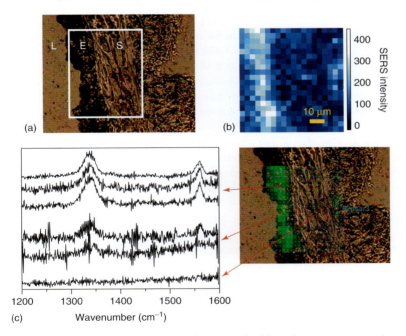

Figure 12.16 SERS imaging for tissue diagnostics. (a) Bright-field microscope image of a prostate tissue section. Different histological classes are indicated: lumen (L), epithelium (E) and stroma (S). The rectangular box shows the region in which spectra were acquired by point mapping. (b and c) The false-colour SERS images show that the characteristic signal of the SERS-labelled antibody is observed selectively in the epithelium. Low non-specific binding is attributed to the hydrophilic ethylene glycols of the SERS probes (cf. Figure 12.10) [24].

The false-colour image in Figure 12.16b is based on the intensity of the Raman marker band at 1336 cm^{-1}. The selective abundance of PSA in the epithelium is evident from this SERS image. Representative SERS spectra recorded at selected locations in the (E), (L) and (S) are shown in Figure 12.16c. In negative control experiments, no binding in the tissue was observed with either bare SERS labels or SERS labels conjugated to BSA. This additionally confirms the specific binding of the SERS-labelled PSA antibody to PSA in the epithelium. The observed minimal non-specific binding is attributed to the hydrophilic ethylene glycol termini of the SAM (cf. Figure 12.10) [24].

12.3.4
Applications *In vivo*

In vivo tumour targeting by SERS probes was first demonstrated by Nie and co-workers in 2008 [19]. Solid gold spheres as SERS substrate covered with crystal violet as the Raman reporter and thiol-modified polyethylene glycols (PEGs) for stabilization were used as the SERS label. The surface coverage with Raman labels was calculated to be about 30%, based on the estimation that about $1.4-1.5 \times 10^4$ Raman reporters and about 3.0×10^4 thiol-PEG molecules are present on the surface of a 60 nm gold particle. Bioconjugation was achieved through the use of a heterofunctional PEG spacer. The SERS probe was conjugated to a single-chain variable fragment antibody for tumour recognition because this fragment binds selectively to the epidermal growth factor receptor overexpressed in many human malignant tumours. These target-specific SERS probes were detected *in vivo* (Figure 12.17a) after injection into the tail vein of a mouse model (Figure 12.17c). Tumour-specific recognition was demonstrated by non-targeted control experiments with bare SERS labels (Figure 12.17b), that is, not conjugated to the recognition element. In comparison with quantum dots emitting in the NIR, the SERS labels yielded about 200 times more intense signal. It is very likely that either aggregates or assemblies of adjacent gold spheres rather than individual gold spheres are responsible for the observed sensitivity in this study.

Soon after this initial study, Gambhir and co-workers demonstrated multiplexed, non-targeted *in vivo* SERS microscopy in mice. Both PEGylated and nonPEGylated gold particles with sub-monolayer coverage of Raman reporter molecules and a silica shell (cf. Figure 12.6c) were employed as SERS probes [48]. Liver pharmacokinetics of these non-targeted probes was investigated after tail vein injection and it was observed that both PEGylated and nonPEGylated labels exhibit no difference in liver accumulation. The authors note that *in vivo* SERS microscopy may complement other imaging (also non-optical) strategies such as radiology. Potential limitations of SERS probes have to be investigated in future studies, addressing issues such as delivery with respect to nanoparticle size, the optimal injected dose and the potential toxicity of the particles.

The photothermal effect associated with metal nanoparticles [49, 50] offers the very attractive option for tumour therapy in addition to diagnostics. This has been demonstrated for nanoshells by Halas and co-workers [49]. Nanorods with a certain

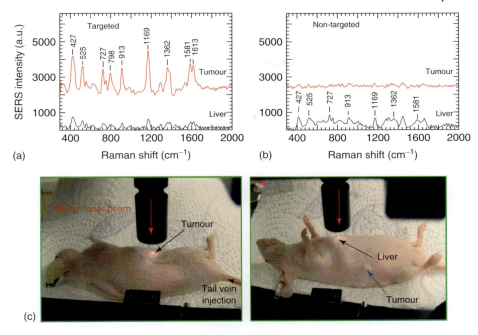

Figure 12.17 *In vivo* cancer targeting with SERS probes. (a,b) SERS spectra obtained from the tumour and the liver using targeted SERS probes (a) and non-targeted particles (b). (c) Photographs showing the locations at which the SERS spectra were obtained using 785 nm laser excitation [19].

size and aspect ratio also exhibit tunable plasmon resonances in the red to NIR [25, 50]. Different visible and NIR dyes have been used to demonstrate non-targeted multiplexing with SERS *in vivo* after subcutaneous injection of nanorods covered with dyes and PEG spacers. In addition to their SERS detection, the ability to be heated photothermally by NIR laser radiation was also tested, demonstrating the combination of SERS spectroscopic detection and remote photothermal heating [25].

12.4 Summary and Outlook

SERS as a molecular vibrational spectroscopic technique can be applied to a vast number of problems in diverse disciplines, including the analytical, biophysical and life sciences. The use of SERS probes as a labelling agent in targeted research is still in its infancy, but has a huge potential for applications in bioanalytics and biomedicine. Central advantages of SERS labels are quantification, sensitivity and dense multiplexing for simultaneous target detection.

First applications of SERS microscopy, the combination of target-specific colloidal SERS probes with Raman microspectroscopy, have only recently been demonstrated. It is realistic to expect that numerous applications for the analysis of cells and tissues will appear within the next few years. Cancer diagnostics will certainly be an important field where this innovative methodology offers significant advantages over existing approaches. However, as pointed out in Chapter 3 in the context of label-free SERS, it is necessary to directly compare and pinpoint the specific advantages and disadvantages of SERS labels compared with other labels and approaches, in particular fluorescence microscopy employing molecular fluorophores and quantum dots.

Both probe development and applications currently are in a rapid growth phase. The acceptance of SERS as a labelling technique will certainly depend on its performance in real-world applications, in particular with respect to reproducibility and sensitivity. Robust and sensitive SERS labels with reproducible signals are a prerequisite and many improvements are necessary to achieve this aim. Otherwise, the general acceptance of this new technology platform may not be accomplished. Using well-defined SERS substrates is as critical as it is in label-free SERS applications. Colloidal SERS particles with uniform size and composition are desired, not complex mixtures of undefined objects. One may envision that large quantities of monodisperse and uniform SERS probes with quantitative single-particle sensitivity will become available in the near future.

Standardization will be required in order to make fair comparisons between different particle probes. Here, we should build on the commonly accepted definition of the enhancement factor and try to extrapolate this, where applicable, to the single-particle level since here the danger of ambiguities due to biased or wrong interpretations is minimized. In addition, it is necessary to establish and share standard protocols for bioconjugation in order to make this technique easily accessible for a wider range of users.

Future applications of SERS probes should explore their sensitivity, multiplexing and size limits *in vitro* and *in vivo*. For all *in vivo* experiments with nanoparticles including SERS labels, in particular animal experiments, biodistribution and toxicity have to be investigated [47, 48]. The interesting option to construct and use hybrid SERS probes in living cells – comprising SERS labels as well as probe molecules for monitoring intracellular parameters such as pH – is discussed in the next chapter by Janina Kneipp.

Acknowledgement

Financial support from the German Research Foundation is acknowledged (DFG; SCHL 594/4-1 and Heisenberg fellowship SCHL 594/5-1). Sincere thanks are due to Magdalena Gellner and Max Schütz for carefully reading the chapter, for their critical comments and for help with preparing several figures. The continuing support and fruitful discussions with Prof. Carsten Schmuck (Duisburg-Essen, Organic Chemistry), Prof. Philipp Ströbel, Prof. Alexander Marx (Mannheim/Heidelberg,

Pathology), PD Dr Patrick Adam (Tübingen, Pathology) and Dr Jens Packeisen (Osnabrück, Pathology) are highly appreciated.

References

1. Aroca, R. (2006) *Surface-Enhanced Vibrational Spectroscopy*, John Wiley & Sons, Inc., New York.
2. Kneipp, K., Moskovits, M. and Kneipp, H. (2006) *Surface Enhanced Raman Scattering: Physics and Applications*, Topics in Applied Physics, Vol. 103, Springer, Berlin.
3. Etchegoin, P. and Le Ru, E. (2009) *Principles of Surface-Enhanced Raman Spectroscopy and Related Plasmonic Effects*, Elsevier, Amsterdam.
4. Stiles, P.L., Dieringer, J.A., Shah, N.C. and Van Duyne, R.P. (2008) Surface-enhanced Raman spectroscopy. *Annu. Rev. Anal. Chem.*, **1**, 601–626.
5. Doering, W.E., Piotti, M.E., Natan, M.J. and Freeman, R.G. (2007) SERS as a foundation for nanoscale, optically detected biological labels. *Adv. Mater.*, **19**, 3100–3108.
6. Porter, M.D., Lipert, R.J., Sioperko, L.M., Wang, G. and Narayanan, R. (2008) SERS as a bioassay platform: fundamentals, design, and applications. *Chem. Soc. Rev.*, **37**, 1001–1011.
7. Schlücker, S. and Kiefer, W. (2009) in *Frontiers in Molecular Spectroscopy* (ed. J. Laane), Elsevier, Amsterdam, pp. 267–288.
8. Schlücker, S. (2009) SERS microscopy: nanoparticle probes and biomedical applications. *ChemPhysChem*, **10**, 1344–1354.
9. Ni, J., Lipert, R.J., Dawson, G.B. and Porter, M.D. (1999) Immunoassay readout method using extrinsic Raman labels adsorbed on immunogold colloids. *Anal. Chem.*, **71**, 4903–4908.
10. Cao, Y.C., Jin, R. and Mirkin, C.A. (2002) Nanoparticles with Raman spectroscopic fingerprints for DNA and RNA detection. *Science*, **297**, 1536–1540.
11. Grubisha, D.S., Lipert, R.J., Park, H.-Y., Driskell, J. and Porter, M.D. (2003) Femtomolar detection of prostate-specific antigen: an immunoassay based on surface-enhanced Raman scattering and immunogold labels. *Anal. Chem.*, **75**, 5936–5943.
12. Cao, Y.C., Jin, R., Nam, J.-M., Thaxton, C.S. and Mirkin, C.A. (2003) Raman dye-labeled nanoparticle probes for proteins. *J. Am. Chem. Soc.*, **125**, 14676–14677.
13. Mulvaney, S.P., Musick, M.D., Keating, C.D. and Natan, M.J. (2003) Glass-coated, analyte-tagged nanoparticles: A new tagging system based on detection with surface-enhanced Raman scattering. *Langmuir*, **19**, 4784–4790.
14. Doering, W.E. and Nie, S. (2003) Spectroscopic tags using dye-embedded nanoparticles and surface-enhanced Raman scattering. *Anal. Chem.*, **75**, 6171–6176.
15. Su, X., Zhang, J., Sun, L., Koo, T.-W., Chan, S., Sundararajan, N., Yamakawa, M. and Berlin, A.A. (2005) Composite organic-inorganic nanoparticles (COINs) with chemically encoded optical signatures. *Nano Lett.*, **5**, 49–54.
16. Kim, J.-H., Kim, J.-S., Choi, H., Lee, S.-M., Jun, B.-H., Yu, K.-N., Kuk, E., Kim, Y.-K., Jeong, D.H., Cho, M.-H. and Lee, Y.-S. (2006) Nanoparticle probes with surface enhanced Raman spectroscopic tags for cellular cancer targeting. *Anal. Chem.*, **78**, 6967–6973.
17. Jin, R., Cao, Y.C., Thaxton, C.S. and Mirkin, C.A. (2006) Glass-bead-based parallel detection of DNA using composite Raman labels. *Small*, **2**, 375–380.
18. Yu, K.N., Lee, S.-M., Han, J.Y., Park, H., Woo, M.-A., Noh, M.S., Hwang, S.-K., Kwon, J.-T., Jin, H., Kim, Y.-K., Hergenrother, P.J., Jeong, D.H., Lee, Y.-S. and Cho, M.-H. (2007) Multiplex targeting, tracking, and imaging of apoptosis by fluorescent surface enhanced Raman spectroscopic dots. *Bioconjug. Chem.*, **2**, 375–380.
19. Qian, X., Peng, X.-H., Ansari, D.O., Yin-Goen, Q., Chen, G.Z., Shin, D.M., Yang, L., Young, A.N., Wang, M.D.

and Nie, S. (2008) In vivo tumor targeting and spectroscopic detection with surface-enhanced Raman nanoparticle tags. *Nat. Biotechnol.*, **26**, 83–90.
20. Yang, M., Chen, T., Lau, W.S., Wang, Y., Tang, Q., Yang, Y. and Chen, H. (2009) Development of polymer-encapsulated metal nanoparticles as surface-enhanced Raman scattering probes. *Small*, **5**, 198–202.
21. McKenzie, F., Ingram, A., Stokes, R. and Graham, D. (2009) SERRS coded nanoparticles for biomolecular labelling with wavelength-tunable discrimination. *Analyst*, **134**, 549–556.
22. Küstner, B., Gellner, M., Schütz, M., Schöppler, F., Marx, A., Ströbel, P., Adam, P., Schmuck, C. and Schlücker, S. (2009) SERS labels for red laser excitation: silica-encapsulated SAMs on tunable gold/silver nanoshells. *Angew. Chem. Int. Ed.*, **48**, 1950–1953.
23. Fernández-López, C., Mateo-Mateo, C., Álvarez-Puebla, R.A., Pérez-Juste, J., Pastoriza-Santos, I. and Liz-Marzán, L.M. (2009) Highly controlled silica coating of PEG-capped metal nanoparticles and preparation of SERS-encoded particles. *Langmuir*, **25**, 13894–13899.
24. Jehn, C., Küstner, B., Adam, P., Marx, A., Ströbel, P., Schmuck, C. and Schlücker, S. (2009) Water soluble SERS labels comprising a SAM with dual spacers for controlled bioconjugation. *Phys. Chem. Chem. Phys.*, **11**, 7499–7504.
25. von Maltzahn, G., Centrone, A., Park, J.-H., Ramanathan, R., Sailor, M.J., Hatton, T.A. and Bhatia, S.N. (2009) SERS-coded gold nanorods as a multifunctional platform for densely multiplexed near-infrared imaging and photothermal heating. *Adv. Mat.*, **21**, 3175–3180.
26. Liu, X., Knauer, M., Ivleva, N.P., Niessner, R. and Haisch, C. (2010) Synthesis of core-shell surface-enhanced Raman tags for bioimaging. *Anal. Chem.*, **82**, 441–446.
27. Schütz, M., Küstner, B., Bauer, M., Schmuck, C. and Schlücker, S. (2010) Synthesis of glass coated SERS nanoparticle probes via SAMs with terminal SiO_2 precursors. *Small*, **6**, 733–737.
28. Sun, Y.G., Mayers, B.T. and Xia, Y.N. (2002) Template-engaged replacement reaction: a one-step approach to the large-scale synthesis of metal nanostructures with hollow interiors. *Nano Lett.*, **2**, 481–485.
29. Gellner, M., Küstner, B. and Schlücker, S. (2009) Optical properties and SERS efficiency of tunable gold/silver nanoshells. *Vib. Spectrosc.*, **50**, 43–47.
30. Mie, G. (1908) Beiträge zur Optik trüber Medien, speziell kolloidaler Metallösungen (Contributions to the optics of turbid media, specifically colloidal metal solutions). *Ann. Phys.*, **25**, 377–445.
31. Bohren, C.F. and Huffman, D.R. (1998) *Absorption and Scattering of Light by Small Particles*, Wiley-VCH Verlag GmbH, Weinheim.
32. Oldenburg, S.J., Westcott, S.L., Averitt, R.D. and Halas, N.J. (1999) Surface enhanced Raman scattering in the near infrared using metal nanoshell substrates. *J. Chem. Phys.*, **111**, 4729–4735.
33. Cui, Y., Ren, B., Yao, J.-L., Gu, R.-A. and Tian, Z.-Q. (2006) Synthesis of Ag core–Au shell bimetallic nanoparticles for immunoassay based on surface-enhanced Raman spectroscopy. *J. Phys. Chem. B*, **110**, 4002–4006.
34. Jana, N.R. (2003) Silver coated gold nanoparticles as new surface enhanced Raman substrate at low analyte concentration. *Analyst*, **128**, 954–956.
35. McCabe, A.F., Eliasson, C., Prasath, R.A., Hernandez-Santana, A., Stevenson, L., Apple, I., Cormack, P.A.G., Graham, D., Smith, W.E., Corish, P., Lipscomb, S.J., Holland, E.R. and Prince, P.D. (2006) SERRS labelled beads for multiplex detection. *Faraday Discuss.*, **132**, 303–308.
36. Stöber, W., Fink, A. and Bohn, E. (1968) Controlled growth of monodisperse silica spheres in the micron size range. *J. Colloid Interface Sci.*, **26**, 62–69.
37. Schlücker, S., Küstner, B., Punge, A., Bonfig, R., Marx, A. and Ströbel, P. (2006) Immuno-Raman microspectroscopy: *in situ* detection of antigens in tissue specimens by surface-enhanced

Raman scattering. *J. Raman Spectrosc.*, **37**, 719–721.

38. Stuart, D.A., Haes, A.J., McFarland, A.D., Nie, S. and Van Dyne, R.P. (2004) Refractive index sensitive, plasmon resonant scattering, and surface enhanced Reman scattering nanoparticles and arrays as biological sensing platforms. *Proc. SPIE - Int. Soc. Opt. Eng.*, **5327**, 60–73.

39. Turrell, G. and Corset, J. (eds) (1996) *Raman Microscopy: Developments and Applications*, Elsevier, San Diego.

40. Schlücker, S., Huffmann, S.W., Schaeberle, M.D. and Levin, I.W. (2003) Raman microspectroscopy: a comparison of point, line, and wide-field imaging methodologies. *Anal. Chem.*, **75**, 4312–4318.

41. Lee, S., Kim, S., Choo, J., Shin, S.Y., Lee, Y.H., Choi, H.Y., Ha, S., Kang, K. and Oh, C.H. (2007) Biological imaging of HEK293 cells expressing PLCγ1 using surface-enhanced Raman microscopy. *Anal. Chem.*, **79**, 916–922.

42. Woo, M.-A., Lee, S.-M., Kim, G., Baek, J., Noh, M.S., Kim, J.E., Park, S.J., Minai-Tehrani, A., Park, S.-C., Seo, Y.T., Kim, Y.-K., Lee, Y.-S., Jeong, D.H. and Cho, M.-H. (2009) Multiplex immunoassay using fluorescent-surface enhanced Raman spectroscopic dots for the detection of bronchioalveolar stem cells in murine lung. *Anal. Chem.*, **81**, 1008–1015.

43. Hu, Q., Tay, L.-L., Noestheden, M. and Pezacki, J.P. (2007) Mammalian cell surface imaging with nitrile-functionalized nanoprobes: biophysical characterization of aggregation and polarization anisotropy in SERS imaging. *J. Am. Chem. Soc.*, **129**, 14–15.

44. Sun, L., Sung, K.-B., Dentinger, C., Lutz, B., Nguyen, L., Zhang, J., Qin, H., Yamakawa, M., Cao, M., Lu, Y., Chmura, A.J., Zhu, J., Su, X., Berlin, A.A., Chan, S. and Knudsen, B. (2007) Composite organic-inorganic nanoparticles as Raman labels for tissue analysis. *Nano Lett.*, **7**, 351–356.

45. Lutz, B., Dentinger, C., Sun, L., Nguyen, L., Zhang, J., Chmura, A.J., Allen, A., Chan, S. and Knudsen, B. (2008) Raman nanoparticle probes for antibody-based protein detection in tissues. *J. Histochem. Cytochem.*, **56**, 371–379.

46. Lutz, B., Dentinger, C.E., Nguyen, L.N., Suni, L., Zhang, J., Allen, A.N., Chan, S. and Knudsen, B.S. (2008) Spectral analysis of multiplex Raman probe signatures. *ACS Nano*, **2**, 2306–2314.

47. Shachaf, C.M., Elchuri, S.V., Koh, A.L., Zhu, J., Nguyen, L.N., Mitchell, D.J., Zhang, J., Swartz, K.B., Sun, L., Chan, S., Sinclair, R. and Nolan, G.P. (2009) A novel method for detection of phosphorylation in single cells by surface enhanced Raman scattering (SERS) using composite organic-inorganic nanoparticles (COINs). *PLoS ONE*, **4**, 1–12.

48. Keren, S., Zavaleta, C., Cheng, Z., de la Zerda, A., Gheysens, O. and Gambhir, S.S. (2008) Noninvasive molecular imaging of small living subjects using Raman spectroscopy. *Proc. Natl. Acad. Sci. U.S.A.*, **105**, 5844–5849.

49. Hirsch, L.R., Stafford, R.J., Bankson, J.A., Sershen, S.R., Rivera, B., Price, R.E., Hazle, J.D., Halas, N.J. and West, J.L. (2003) Nanoshell-mediated near-infrared thermal therapy of tumors under magnetic resonance guidance. *Proc. Natl. Acad. Sci. U.S.A.*, **100**, 13549–13554.

50. Huang, X., El-Sayed, I.H., Qian, W. and El.-Sayed, M.A. (2006) Cancer cell imaging and photothermal therapy in the near-infrared region by using gold nanorods. *J. Am. Chem. Soc.*, **128**, 2115–2120.

13
1-P and 2-P Excited SERS as Intracellular Probe

Janina Kneipp

13.1
From Tags to Probes: Challenges in Intracellular Probing

The increase in Raman cross section from molecules in close proximity to a noble-metal nanostructure [1–5] enables, apart from sensitive detection of analytes, the construction of so-called SERS labels or SERS tags. They consist of metal nanostructures, typically combinations of aggregates of gold or silver nanoparticles and so-called reporter or label molecules or Raman dyes [6–9]. In many cases, these molecules are in fact fluorescent dyes. The advantages of such SERS labels compared to common labels used in fluorescence do not lie necessarily in size or coupling chemistry, but they can be found in the high signal stability and in an enormous multiplexing potential (see also Chapters 11 and 12). As also demonstrated in other chapters of this book, additional gain in sensitivity can result when the excitation occurs in resonance with electronic transitions in the reporter molecules, so that SERRS (see also Chapters 10 and 11) can be employed [10, 11]. The high sensitivity of SERRS labels is paid for with a limited range of excitation wavelengths in label detection. In contrast, excitation of an SERS label out of resonance provides the free choice of excitation wavelength as another significant advantage over fluorescence labels.

Several applications have been reported for non-resonant SERS labels. They exhibit excellent photostability as they do not undergo electronic excitation. Consequently, the excitation wavelength is a free parameter that can be set in the near infrared (NIR), where relatively low-energy photons do not cause photodecomposition of the probed object and where possible interfering cellular autofluorescence is weak. Apart from their application in immunoassays, in DNA detection and for immunostaining of fixed tissue specimens (see also Chapter 12), SERS labels were also introduced into organisms and attached to the surface of cells as targeting agents of specific surface markers [12–15]. Upon introduction into the cellular interior, spectra of Raman labels inside the cells were obtained [16] and also used for imaging [17]. The spectral signature of the reporter is detected, since the reporter molecules are in the proximity of the metal nanostructures.

Surface Enhanced Raman Spectroscopy: Analytical, Biophysical and Life Science Applications. Edited by Sebastian Schlücker
Copyright © 2011 WILEY-VCH Verlag GmbH & Co. KGaA, Weinheim
ISBN: 978-3-527-32567-2

The presence of a specific molecule or compound that is applied to a cell separately before or simultaneously with the metal nanoparticles can also be monitored, for example, a drug accumulating at the cellular membrane [18–20] or a phage binding to a cell [21]. In these cases, the metal substrate acts as a probe of its nanometre-scaled environment rather than a mere label, and extrinsic compounds are detected while they are in contact with the cell. Similarly, the release of molecules from the cell, such as neurotransmitters [22], can be detected.

The first SERS spectra from the cellular interior, generated by unlabelled gold nanoparticles that had been introduced into cells, were reported from the cell line HT29 [23]. These first data indicated a great variety of spectra resulting from all kinds of molecules contained in the cells and provided the first ideas of the complexity of intrinsic cellular information obtained by SERS. Over the last years, more data have been gathered with more and different cellular and probe systems, which revealed the prerequisites that nanoprobes for cellular applications would have to meet [24–26]. They are discussed in the following paragraphs.

13.1.1
Localization and Targeting

The first major point arises from the extreme localization of the signals generated by the SERS probes as opposed to the high level of heterogeneity on the micro and nanoscale encountered by the probes in the cellular interior. The signal that is obtained from an SERS probe comes from a very small volume, and will therefore vary depending on the probe's position and its extremely localized surroundings. As a result, an SERS probe can only provide indirect molecular information about the cell as a whole. This is different from normal Raman experiments in cells, where usually volumes of one to hundreds of femtolitres are probed, containing several organelles and compartments and many molecules [27–30]. In contrast, in an SERS measurement, the probed volume is dictated by the extension of the local fields around the metal nanostructures. In an SERS experiment with a cell, excitation of Raman scattering is still diffraction limited and occurs in the same volumes as in a normal Raman experiment, but the Raman signal that is detected is only the strongly enhanced SERS signal originating from the nanometre-scaled vicinity of the SERS probe. The normal Raman scattering is not detected in the acquisition times and at the laser powers used in these experiments. The extremely high sensitivity that is achieved with the nanoprobes is an enormous advantage, as it provides the only opportunity to obtain Raman spectral information at a time scale that is in accordance with the time scales of important cellular processes such as trafficking and transport in a live cell. In normal Raman experiments, long acquisition times and/or high laser powers are needed, or resonant enhancement has to be employed.

The fact that the signal is extremely localized implies that the position of an SERS probe inside a cell needs to be known. It can be achieved in different ways, one of them being to target the nanoprobes using tools known from cell biology. In general, gold nanoparticles can be designed to take specific routes through a cell. This can be achieved through surface modifications with molecules that target specific sub-structures in the cell such as antibodies or peptides. Routing/targeting can also result from employing different uptake mechanisms of the cells. These depend on the particle properties, the cell lines, as well as external parameters, for example, of the cell culture medium. Finally, the experimentalist also can do his/her part by designing administration regimes that lead to preferred routing. Gold nanoparticles can also be grown inside cultured cells [31]. In a recent study, intrinsic SERS signals were reported from such particles that were intracellularly reduced in epithelial cells [32]. In any case, the position of the probes inside a cell needs to be confirmed in parallel studies by transmission electron microscopy (TEM). Once the pathway of a particular probe type inside a cell is known, one can assume that the SERS signal originating from the probes comes from specific substructures.

13.1.2
Influence of Surroundings on Nanoprobe Aggregation and Stability

The second challenge encountered in the design of intracellular SERS probes is concerned with the enhanced local optical fields that the high SERS enhancement levels were shown to be associated with [33, 34]. The dependence on the local fields implies that the SERS enhancement strongly depends on the morphology (e.g. the size, shape or aggregation) of the nanoparticulate SERS probes. So far, the highest SERS enhancement factors have been obtained by exploiting extremely high field enhancement on aggregates or clusters formed by individual silver or gold nanoparticles [35]. As experiments and theory show, the clusters can vary in size, ranging from dimers [36, 37] to fractal structures [38–40]. Because of the important role of nanoaggregates and clusters for electromagnetic enhancement, the formation of gold nanoparticle aggregates has to be under control in the intracellular SERS experiments. Also here, the importance of imaging the nanoparticles in the biological environment, for example by TEM, cannot be overrated.

As an example, the results of an experiment will be discussed here, where gold nanoprobes were applied in an epithelial cell line as a pulse. Gold nanoparticles were added to cell culture medium and withdrawn again after a defined time of 30 min. Then, in a series of samples taken at different time points, the SERS signal generated in the vicinity of the nanoprobes was observed over time (Figure 13.1). In general, two observables can be indicators of changes in the nanoprobes' environment over time: (i) the spectral signature of the SERS, which contains information on the nm-scaled molecular environment of the gold nanoprobes and (ii) the signal strength of the SERS, which can give information on the SERS enhancement factor and thereby allows for the monitoring of changes in the morphology of the gold nanostructures, such as the formation of aggregates. This relates directly to the problem pointed out at the beginning of this paragraph, namely, the influence of the nanoprobes' environment on their plasmonic properties.

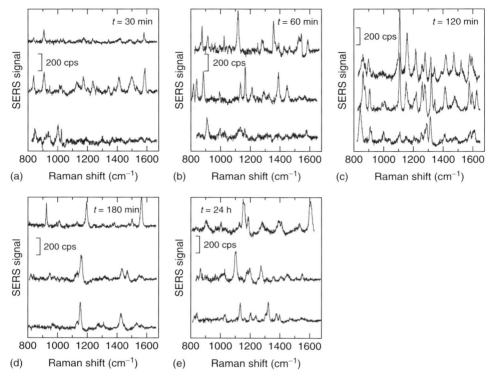

Figure 13.1 Typical surface-enhanced Raman (SERS) spectra from cells of the epithelial cell line IRPT after incubation with gold nanoparticles, excited with $<3 \times 10^5$ W cm^{-2} at 785 nm and collection time 1 s. The incubation time, including a 30 min nanoparticle pulse, is indicated in each panel. The spectra were acquired from living cells in phosphate-buffered saline by raster-scanning over individual cells using a Raman microspectroscopic setup. The positions of the bands are labelled. See Table 13.1 for assignment of the molecular groups/molecules that contribute to the Raman signals. cps, counts per second. (Reprinted with permission from Ref. [25]. Copyright 2006 ACS.)

Figure 13.1 displays typical SERS spectra obtained for each time point in the pulsed experiment described above. Even without an interpretation of the bands, the differences between the spectral signatures indicate that the molecular composition of the endosomal vicinity of the nanoparticles changes over time. To compare potential qualitative alterations (observable (i)), data sets of ∼500–800 spectra were analysed per time point. The number of bands in the spectra that display 90–100% of the maximum signal level of each data set (time point) total 2–8 bands at 30 min and 4–11 bands after 60 min. A significant increase in the number of spectral bands is found at $t = 120$ min. At this time point, the portion of SERS spectra exhibiting 90–100% of the signal level contain 13–19 characteristic bands, resulting from an abundance of different spectral contributions that were not observed for the other time points (for examples see Figure 13.1c). From

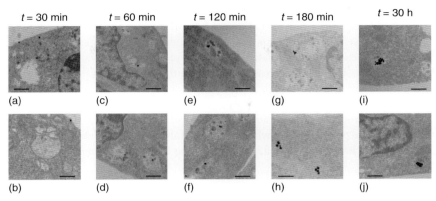

Figure 13.2 Transmission electron micrographs of IRPT cells at different incubation time with the particles. The gold nanoparticles are visible in the cells as black, electron-dense spots. The size of the nanoaggregates varies with incubation time. Nanoclusters of two to three particles are forming after 120 min (e,f), four to six particles after 180 min (g,h), and larger lysosomal nanoaggregates during overnight incubation (i,j) of the cells. The interparticle distance after 180 min is greater than in the other time points, likely due to the enclosure of the particles in multivesicular structures (arrow indicating interparticle space). TEM images were recorded at 80 kV with a Jeol 1011 electron microscope. Scale bars: 500 nm (all except (h)). Scale bar (h): 250 nm. (Reprinted with permission from Ref. [25]. Copyright 2006 ACS.)

the comparison of the 120 min and the 180 min spectra (see also examples in Figure 13.1c,d), we see that there occurs again a decrease in the number of spectral features (three to six bands in the SERS spectra of the 90–100% highest signal level).

Point (ii), the signal strength as a function of gold nanoprobe morphology and/or aggregation, was investigated by a TEM study carried out in parallel samples (Figure 13.2). In our experiments, the spectra obtained after 120 min (Figure 13.1c) exhibit the strongest signals. This improvement in signal strength correlates with the formation of small gold aggregates, at that time mostly dimers and trimers (Figure 13.2e,f). This is in agreement with theoretical estimates showing that extremely high SERS enhancement can exist for two gold nanospheres in close proximity [37]. At later times, when the aggregates grow, the enhancement differs from the optimum enhancement (compare spectra in Figure 13.1c with those in d and e). Also for this observation, a rationale may be found by comparing the TEM images (Figure 13.2): Inter-particle distances, also shown to be important for optimum enhancement to take place [37, 41, 42], in the nanoparticle accumulations after 180 min are greater than in the dimers and trimers (e.g. Figure 13.2e,f), likely due to changes in the cellular sub-environment of the particles.

The presence of gold or silver nanoparticles inside cells is often confirmed by light microscopy. It should be noted that the nanoprobes can be observed in a light microscope only when they have formed large aggregates inside the cells. This means, that they must be present in high concentrations or for a long time of, say, a few hours. At present, the majority of intracellular applications of SERS probes have been carried out under conditions of long incubation times with the probes [17, 43]. In these experiments, it is not clear whether the enhancement is generated by parts of large (sub)micrometre sized clusters or from individual probes or nanoaggregates that are also present. Studies on the behaviour of the probes with respect to cellular ultrastructure and influence on the enhancement that can be expected are urgently needed for a better understanding of the spectra delivered by an SERS probe. The use of adequate biophysical models and characterization of nanoparticle properties in environments relevant for cellular delivery pathways will help us to better understand how intracellular SERS probes are functioning.

13.1.3
Probe Identification

In several cellular processes involved in transport and trafficking, simultaneous investigation of different locations would be favourable. In order to achieve this, a specific probe must identify itself as being of a specific type, for example, with respect to its targeting. Therefore, it would be useful to introduce a probe-specific signature that allows identifying the probe type, but that also enables investigation of the cellular environment. In first studies, we have shown that labelled probes for intracellular studies can be generated [24, 26].

Labelled probes can be identified by the signature of a reporter molecule. At the same time, they deliver information about their environment inside the cell. This is different from a typical SERS tag, which usually does not provide information from other molecules but the reporter. We used biocompatible dye molecules as reporter molecules, such as indocyanine green (ICG) [24], as well as dyes that have been widely studied in cell biology because of their potential as photosensitizers, for example, Rose Bengal (RB) [26]. The absorption maximum of RB lies at 550 nm. In our experiments, we applied NIR excitation at 830 nm, which is not in resonance with the electronic absorption of the dye. This prevents photobleaching or phototoxicity and results in a high stability of the label.

In fluorescence and photochemical applications, the dyes used as reporters here are applied in the micromolar (10^{-6} M) concentration range. SERS probes were constructed from gold nanoparticles and RB as well as the dye crystal violet (CV) that contained the reporter at a concentration of only 10^{-9} M, corresponding to ~10 dye molecules per gold nanoparticle. This low concentration is different from reporter concentrations in usual SERS tags or SERS labels mentioned in the beginning, which are much higher. In most of these cases, SERS labels exhibited full surface coverage of Raman reporter molecules on the metal nanoparticles [10, 44–46].

The characteristic fingerprint-like pattern of many narrow Raman bands of the reporter molecule provides a high spectral specificity of the labelled probes, so they can be identified also in the presence of the signal from the cellular surroundings. Figure 13.3 displays example spectra of CV–gold nanoprobes inside cells. Several bands that can be assigned to CV are identified, but, in addition, other bands

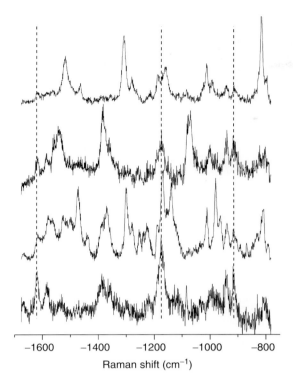

Figure 13.3 SERS spectra measured in 3T3 cells after 2 h incubation with crystal violet (CV) SERS probes. SERS spectra collected in a live cell are composed of Raman lines from the reporter CV along with Raman signatures originating from surrounding cellular molecules. Three Raman lines of CV are marked by the dashed lines. (Reprinted with permission from Ref. [26].)

characteristic of the cellular molecules are also found (Figure 13.3). As illustrated in Figure 13.4, the position of the probes can be imaged by chemical or multivariate mapping. Comparisons can be made regarding the co-localization of signals from the cellular surroundings and the probes' specific signals. The example displayed in Figure 13.4 shows that the molecular composition of different endosomes that contain the labelled probes is not the same. The intensity of the CH_2 deformation mode (1450 cm^{-1}) characteristic of membrane lipids in vesicular structures varies, with positions of high intensity not always coinciding with the presence of the reporter signature. This suggests that not all probes are localized in the nanometre vicinity of a (intracellular) membrane.

As conclusion from these first paragraphs, Figure 13.5 summarizes three major types of SERS probes for intracellular applications. Bare SERS probes, consisting of individual particles or nanoaggregates, deliver information about the cellular biochemistry *in situ* (Figure 13.5a). Labelled probes can be used in targeting approaches with more than one probe type and enable an assignment of intrinsic SERS signatures to specific locations the probes were targeted at (Figure 13.5b).

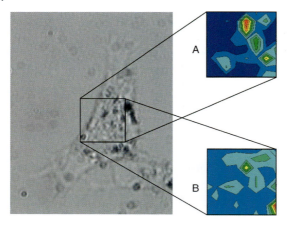

Figure 13.4 SERS images collected for the rectangle (14 × 14 µm² dimension) shown in the bright-field image of a 3T3 cell. Image A is based on the 1174 cm^{-1} Raman line of the crystal violet labelled SERS probe, and image B is based on the intensity of the lipid CH$_2$ deformation mode at ∼ 1470 cm^{-1}. The colours represent relative scattering powers in the scan from red (highest signal) to blue (lowest signal). (Reprinted with permission from Ref. [26].)

A third probe type is a nanosensor, which functions on the basis of a reporter signature. The reporter spectrum changes when the probe environment changes as intrinsic parameter of the cell (Figure 13.5c). An example of such a sensor is the pH nanosensor that will be discussed in the following section. Common to all intracellular nanoprobes is their excitation in the NIR, out of resonance, which keeps experimental conditions flexible, ensures probe stability and avoids phototoxicity. SERS probes enable the chemical characterization of the nanometric vicinity of gold nanoparticles or nanoaggregates and the measurement of vibrational spectra at a sensitivity and lateral resolution unachieved so far in other experiments. This provides the capability of chemical characterization of nanometre-scale units in single live cells. Because of the large effective Raman scattering cross section, SERS probes fulfill the requirements of dynamic, *in vivo* systems – the use of very low laser powers and very short data acquisition times.

13.2
Probing of Intracellular Parameters

In principle, SERS probes can be used to investigate different cellular compartments and, depending on the properties of the nanoprobes, also to probe selected molecular compounds. Differences in the SERS spectra obtained from the molecules in the enhanced local fields of the nanostructures in different cell lines and over time, as well as the direct identification of physiologically relevant molecules, demonstrate that SERS approaches are feasible for the characterization of changing cellular environments and are useful for intracellular applications.

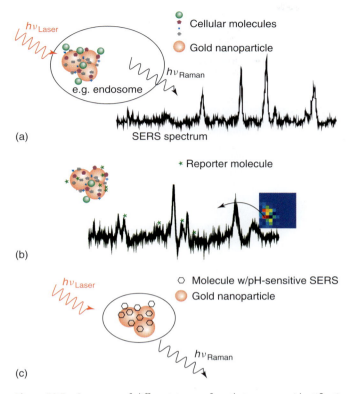

Figure 13.5 Summary of different types of SERS probes for intracellular applications. (a) Probes consisting of gold nanoparticles or nanoaggregates deliver molecular information from their immediate surroundings. (b) Hybrid probes can be constructed from gold nanoparticles and reporter molecules that serve an identification of the probes. Such probes generate signals from both the cell and the reporter. (c) SERS signatures of reporter molecules can vary with the conditions in their nanoenvironment, such as pH. Such probes can function as sensors.

Many studies have been carried out in the endosomal system of different cell lines, both knowingly and unintended. Endosomes form automatically when eukaryotic cells take up nanoprobes in the process of endocytosis. Endocytosis is the most likely way of nanoparticle uptake from the cell culture medium for particle sizes up to 100 nm diameter. An endosome, which is the membrane-bounded environment the particle resides in upon uptake, undergoes a process of maturation from so-called early to late endosome, and finally the lysosome stage. Endosomal maturation is accompanied by drastic changes in the chemical composition. Therefore, the endosomal system constitutes an ideal test environment for SERS nanoprobes and sensors. In addition, elucidation of the chemical composition of endosomes has thus far not been possible without fractionation, purification and other *in vitro* approaches. Therefore, SERS studies are very useful to understand more about endosomal composition and dynamics.

Table 13.1 Wavenumber values of Raman bands observed in endosome SERS spectra of cell lines IRPT and J 774 and their tentative assignments to the classes of molecules and/or vibrational modes.

Raman shift (cm^{-1})	Tentative assignment[a,b]
815	Phosphate: ν(OPO)
827, 850	Proteins, Tyr: δ(CCH) aliphatic, Tyr (ring)
862	Ribose: ν(CC), ring breathing, ν(COC)
899	Ribose-phosphate; saccharides
906	Amino acids
927	Proline: ring ν(CC)
998	Proteins: amide III
1004	Phe, ring breathing
1099	Phosphate: ν(PO$_2^-$), ν(CC), ν(COC), glycosidic link
1106	Proteins, ν(CN)
1118	Proteins: ν(CN)
1133	Proline
1154	ν(CC, CN), ρ(CH$_3$)
1188, 1194, 1204	Nucleotides: base ν(CN), Tyr, Phe
1214, 1240, 1254	T, C, A, ring ν
1254, 1274, 1286	Proteins, lipids: amide III/δ(CH$_2$, CH$_3$)
1313	A/proteins: ring ν/γ_T(CH$_2$, CH$_3$)
1338, 1358	Proteins: γ_T(CH$_2$, CH$_3$), γ_W(CH$_2$, CH$_3$)
1384	Nucleotides, proteins, lipids, δ(CH$_3$) sym.
1414	Amino acids, δ(CH$_3$) asym., ν(COO$^-$)
1427	A, G
1448, 1474	Lipids, proteins: δ(CH$_2$, CH$_3$)
1505, 1518, 1532, 1578	A, C, G
1548, 1563	Proteins: amide II
1582, 1593, 1605	Proteins, Phe, Tyr
1625	Nucleotides, lipids, proteins, ν(C=C) olefinic

[a] Based on Refs. [51–55].
[b] Abbreviations: ν, stretching; δ, deformation; ρ, rocking; γ_T, twisting; γ_T, wagging; sym., symmetrical; asym., asymmetrical; Tyr, tyrosine; Phe, phenylalanine; A, adenine; T, thymine; C, cytosine; G, guanine.
Reprinted with permission from Ref. [25]. Copyright 2006 ACS.

Table 13.1 lists major bands that were observed in Raman spectra measured in endosomes of epithelial cells (compare also Figure 13.1) and indicates that all molecules present are probed, their SERS signatures superimposing in the spectra. In some cases, the SERS spectra measured from endosomes reveal also signatures that can be unequivocally assigned to specific molecules. When a similar experiment to that discussed in Figures 13.1 and 13.2 was carried out in the macrophage cell line J774, a specific spectral signature appeared at all time points, with the exception of the very late lysosomal stage (22 h incubation).

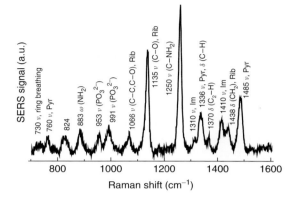

Figure 13.6 One of the typical SERS fingerprints collected from J774 macrophage cells after incubation with gold nanoparticles: excitation wavelength $<3 \times 10^5$ W cm^{-2} at 785 nm, collection time 1 s. This type of spectrum was found in all endosomal stages except the lysosomal. It was identified as the SERS spectrum of AMP and/or ATP. Band assignments based on Ref. [47] are indicated. cps, counts per second; ν, stretching mode; ω, wagging mode; δ, bending mode; Rib, ribose; Pyr, pyrimidine; Im, imidazol.

The spectrum is displayed in Figure 13.6. Almost all bands in this spectrum are characteristic of adenosine phosphate and show contributions from all constituents of this nucleotide (adenine, ribose and phosphate [47]). A comparison with SERS and normal Raman data of pure AMP and ATP from the literature suggested a prevailing contribution from AMP, since triphosphate and diphosphate markers were absent from the spectrum. Interestingly, the acidification and pH regulation of endosomes is achieved by an ATP-dependent proton pump and a Na–K–ATPase [48, 49]. Furthermore, the endosome acidification profile can be modified by cyclic AMP [50]. The outcome of this experiment suggests that, in fact, molecular species involved in the generation of a specific endosomal milieu can be detected and observed by SERS.

As mentioned in the last paragraph, a second parameter that varies drastically in the endosomal system over time is pH [56]. In many cell types, significant acidification compared to earlier endosomal stages takes place in the lysosome. There, to ensure, for example, proper enzymatic function, pH can be well below 5, and even lower (<4) [57]. SERS spectra of 4-mercaptobenzoic acid (pMBA) on silver and gold electrodes and the dependence of the spectra on the pH value have been reported and discussed in terms of adsorption geometry on the metal and on the state of dissociation of the carboxyl group as a consequence of surrounding pH [58]. The pH-dependent SERS spectra of pMBA adsorbed on gold nanoshells bound to a silicon substrate were used to create a pH meter working over the range of 5.8–7.6 pH units [59]. Similar experiments demonstrated that hollow gold nanospheres with pMBA were responsive over a pH range of 3.5–9 [60]. SERS studies on silver nanoparticle clusters functionalized with pMBA showed that the spectrum is sensitive to pH changes in the surrounding solution in the range of 6–8 [16]. The authors concluded from an SERS spectrum measured from these silver particles incorporated into Chinese hamster ovary cells that the pH in the environment of the particles was below 6, which is an observation consistent with the particles being located inside lysosomes [16]. Studies with pMBA on gold nanospheres demonstrated that pH in the endosomal system varied between 6.8 and 5.4 in endosomes of different ages [61]. In these experiments, endosomes of different age and therefore different pH were generated, containing pMBA-functionalized gold nanoparticles. Then, individual cells were raster-scanned in a Raman microspectroscopic setup, and the amount of pH nanosensors was high enough for each probed volume to contain one or more endosomes. Therefore, every sampled spot resulted in an SERS spectrum of pMBA. The pH in the different sampling volumes was determined based on the ratio of two characteristic bands in the spectrum of pMBA, one with varying and one with constant intensity, respectively. The band at 1423 cm^{-1}, which can be assigned to COO^{-} stretching, varies in intensity depending on protonation of the pMBA molecule. The colour map in Figure 13.7 displays the ratio of the SERS signals at 1423 and 1076 cm^{-1} as a function of sampling position in one of the cells, that is, a pH map of the cell. It has to be noted, however, that the pH nanosensor used in this way does not provide the full information about endosomal pH, considering that

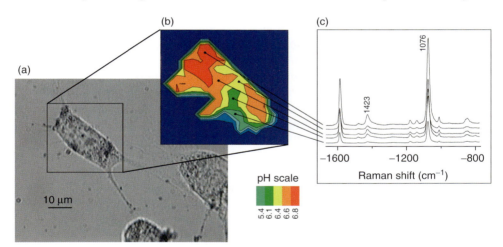

Figure 13.7 Probing and imaging pH values in individual live cells using an SERS nanosensor. (a) Photomicrograph of an NIH/3T3 cell after 4.5 h incubation with the pMBA gold nanosensor. Numerous gold nanoparticles have accumulated in the cell, enabling pH probing in different endosomes over the entire cell based on the SERS signature of pMBA. Lysosomal accumulations can be observed as black spots at the resolution of the light microscope. (b) pH map of the cell displayed as false colour plot of the ratios of the SERS lines at 1423 and 1076 cm^{-1}. The values given in the colour scale bar determine the upper end value of each respective colour. Scattering signals below a defined signal threshold (i.e. where no SERS signals exist) appear in dark blue. (c) Example spectra from regions of different pH. The spectra were collected in 1 s each using 830 nm CW excitation (3 mW). (Reprinted with permission from Ref. [61]. Copyright 2007 ACS.)

pH in lysosomes is significantly lower than 5.4. The pH range below 5.4 cannot be further resolved, since the SERS signals that were used were not sensitive to changes in this acidic range. As is shown in the following section, this situation changes when the concept of the SERS pH sensor is extended to two-photon excitation.

13.3
Surface-Enhanced Hyper Raman Scattering and Its Potential in Studies of Cells

Two-photon excitation is gaining rapidly in interest and significance in microscopy and biospectroscopy. Of particular advantage for the microspectroscopic characterization of cells and tissues are (i) the greatly reduced possible phototoxicity and stress to the sample due to longer wavelengths excitation and (ii) the confinement of the two-photon interaction to the focus of the laser beam. In the following text, two-photon excited SERS, namely, surface-enhanced hyper-Raman scattering (SEHRS) is discussed as a new intracellular probing approach.

Hyper Raman scattering (HRS) is a two-photon excited Raman scattering process and therefore results in Raman signals shifted relative to the doubled energy of the excitation laser [62, 63]. HRS follows symmetry selection rules different from regular one-photon Raman scattering. Therefore, the spectral information obtained in HRS is complementary to the information content of other vibrational (normal Raman and IR) spectra and could therefore be of use for a number of bioanalytical applications. For example, HRS can probe IR-active vibrations that are usually not evident in Raman spectra and in addition reveals so-called silent modes, which are vibrations that are seen neither in Raman nor in IR absorption spectra.

Unfortunately, as a nonlinear incoherent Raman process, HRS is an extremely weak effect with scattering cross sections on the order of 10^{-65} cm^4 s, 35 orders of magnitude smaller than cross sections of 'normal' (one-photon excited) Raman scattering and $\sim$15 orders of magnitude below typical two-photon absorption cross sections. These extremely small cross sections have so far precluded application of HRS as practical spectroscopic tool.

However, HRS benefits to a greater extent from the high local optical fields than normal Raman scattering does in the case of SERS, because of its non-linear dependence on the (enhanced) excitation field [64]. Assuming effective SERS cross sections on the order of 10^{-16} cm^2, which have been reported for several molecules [65–67], effective SEHRS cross sections were found to be on the order of 10^{-46} cm^4 s, which corresponds to 10^4 GM (Goeppert–Mayer) [64]. For comparison, most common fluorescence dyes have two-photon 'action' cross sections (product of the two-photon absorption cross section and fluorescence quantum yield) in the range of 1–300 GM [68]. Webb and co-workers have recently demonstrated cross sections on the order of 10^4 GM for two-photon fluorescence of quantum dots [69]. The effective two-photon cross sections higher than those obtained for two-photon fluorescence as well as the unique capability of probing vibrational modes that are usually not seen in high lateral resolution microscopy or completely absent from

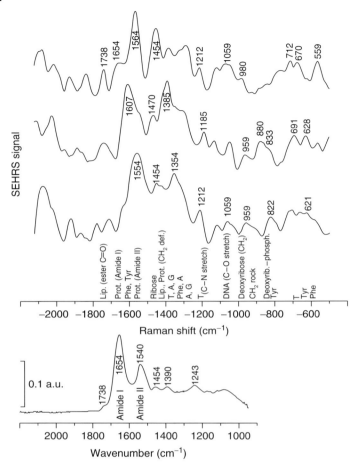

Figure 13.8 SEHRS spectra measured from dry J774 cells after 4 h uptake of gold nanoparticles. Excitation was achieved with 1064 nm mode locked picosecond pulses: average power 30 mW, size of the laser spot ~1 μm, collection time 10 s. The spectra were measured at different, randomly chosen locations in the cell where gold nanoaggregates must have been present in order for SEHRS to occur. For comparison, an example of an IR spectrum of a dried J774 cell on CaF$_2$ substrate is shown below. The IR spectrum was acquired from an ~30 μm diameter spot. (Reprinted with permission from Ref. [64].)

other vibrational spectroscopic data makes SEHRS a promising spectroscopic tool for molecular structural probing in cells.

Since a two-photon excited Raman spectrum follows different selection rules, SEHRS spectra of a cell can contain features that are absent from other Raman and SERS spectra. Examples of such spectra are displayed in Figure 13.8. In addition to other features, a pronounced amide II signal of proteins, usually detected by IR absorption spectroscopy (around 1540 cm^{-1}), is clearly visible here.

An IR spectrum obtained from the cytoplasmic region of the same cell is also displayed (Figure 13.8). IR microspectroscopy is hampered by very low (since diffraction-limited) lateral resolution. In the case of SEHRS, however, IR spectral information can be obtained at the lateral resolution of a Raman, or even better, an SERS experiment. Interestingly, SEHRS observes also spectral bands above 1800 cm^{-1}, which are likely combination modes. The spectral region between 1800 and 2700 cm^{-1} is almost featureless in other vibrational (Raman and IR) spectra of cells and tissues.

The different spectral information in a hyper Raman spectrum also enables an extended applicability of the pH SERS sensor introduced in the last section. There, one-photon excitation limited the use of the intensity change for pH determination to a range of pH 5.4–8. SEHRS spectra of pMBA show the same bands as the SERS spectra, but the relative signal strengths are changed. These differences in the SEHRS signature of pMBA are useful for creating a pH sensor that allows measurements in a more extended range. In an experiment, pH sensors were constructed from p-MBA on silver nanoparticles. In particular, the band associated with the COO$^-$ vibration (on silver shifted with respect to pMBA on gold from 1423 to 1380 cm^{-1}) was used for pH measurements by SERS, as shown, for example, in Figure 13.7. In the SEHRS spectrum it appears at higher signal level compared to the SERS spectra and can therefore be followed down to lower pH values. Additionally, the 1700 cm^{-1} COOH stretching mode appears as pronounced strong band, and shows an opposite dependence on pH (Figure 13.9a). Based on these two signals with opposite pH dependence, an SEHRS sensor can differentiate pH values over a wide range, from pH 2 to 8 [61]. This is of particular interest for measurements in sub-cellular compartments of extreme pH, for example very acidic lysosomes.

In the cellular system, the pH could be determined to be around 4.5 in lysosomes, a value that was in good agreement with measurements by other methods [61]. Example spectra are displayed in Figure 13.9b. It should be noted that SERS-based pH sensing opens up new perspectives for quantitative measurements. Different from other optical methods, no correction regarding cellular background absorption and emission signals is needed. Both SERS and SEHRS provide strong signals also under electronically non-resonant excitation. This avoids photodecomposition of the sensor and allows free selection of the excitation wavelength, optimized for the biological object under study. Other optical pH sensors based on one- or two-photon excited fluorescence signals, including also fluorescence lifetime measurements, in most cases require the application of multiple probes to cover wider pH ranges [70–72]. In contrast to this, the same SERS/SEHRS sensor can operate over a wide range (between pH values of 2 and 8), and thereby enables probing of a variety of cellular compartments.

In summary, although research on intracellular SEHRS probes is still in its infancy, two-photon excited SERS for studies of biological systems may provide new molecular information, combined with the advantages of both excitation in the NIR and the sensitivity and improved lateral resolution of plasmonics-based spectroscopy.

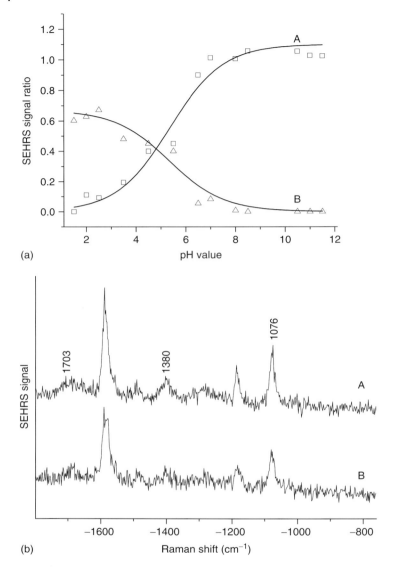

Figure 13.9 Signal ratios at 1380 and 1076 cm^{-1} (a) and 1703 and 1076 cm^{-1} (b) in the SEHRS spectrum of p-MBA on silver nanoparticles plotted as a function of pH demonstrating the operation range of an SEHRS-based pH probe. Typical SEHRS spectra of pMBA on silver nanoaggregates are measured in an NIH/3T3 cell. Spectra were collected after a delay of >10 h delay after termination of the 4.5 h incubation with nanoparticles. Two-photon excitation was achieved with 1064 nm mode-locked ps laser pulses: average power 10 mW, size of the laser spot ∼2 μm, collection time 10 s. The spectral signatures indicate a variation between pH 4.9 and 4.5, consistent with the assumption that the time of the measurement most particles had accumulated inside lysosomes. (Reprinted with permission from Ref. [61]. Copyright 2007 ACS.)

Acknowledgements

I am grateful to my colleagues for laboratory equipment, cell lines, stimulating discussion and collaboration, in particular to Katrin Kneipp, Harald Kneipp, Margaret M. McLaughlin and Dennis Brown of Harvard University Medical School and to Burghardt Wittig of Charité Berlin.

References

1. Otto, A. (1984) in *Light Scattering in Solids IV. Electronic Scattering, Spin Effects, SERS and Morphic Effects* (eds M. Cardona and G. Guntherodt), Springer-Verlag, Berlin, pp. 289–418.
2. Campion, A. and Kambhampati, P. (1998) Surface-enhanced Raman scattering. *Chem. Soc. Rev.*, **27**, 241–250.
3. Moskovits, M. (1985) Surface-enhanced spectroscopy. *Rev. Mod. Phys.*, **57**, 783–826.
4. Persson, B.N.J. (1981) On the theory of surface-enhanced Raman scattering. *Chem. Phys. Lett.*, **82**, 561–565.
5. Kneipp, K., Wang, Y., Kneipp, H., Itzkan, I., Dasari, R.R. and Feld, M.S. (1996) Population pumping of excited vibrational states by spontaneous surface-enhanced Raman scattering. *Phys. Rev. Lett.*, **76**, 2444.
6. Ni, J., Lipert, R.J., Dawson, G.B. and Porter, M.D. (1999) Immunoassay readout method using extrinsic Raman labels adsorbed on immunogold colloids. *Anal. Chem.*, **71**, 4903–4908.
7. Cao, Y.C., Jin, R.C., Nam, J.M., Thaxton, C.S. and Mirkin, C.A. (2003) Raman dye-labeled nanoparticle probes for proteins. *J. Am. Chem. Soc.*, **125**, 14676–14677.
8. Jin, R.C., Cao, Y.C., Thaxton, C.S. and Mirkin, C.A. (2006) Glass-bead-based parallel detection of DNA using composite Raman labels. *Small*, **2**, 375–380.
9. Vo-Dinh, T., Stokes, D.L., Griffin, G.D., Alarie, J., Michaud, E.J., Bunde, T., Kim, U. and Simon, M.I. (1999) Development of a multilabel DNA mapping technique using SERS gene probes. DOE Human Genome Program Contractor-Grantee Workshop VII, 1999, Oakland.
10. Su, X., Zhang, J., Sun, L., Koo, T.W., Chan, S., Sundararajan, N., Yamakawa, M. and Berlin, A.A. (2005) Composite organic-inorganic nanoparticles (COINs) with chemically encoded optical signatures. *Nano Lett.*, **5**, 49–54.
11. Graham, D. and Faulds, K. (2008) Quantitative SERRS for DNA sequence analysis. *Chem. Soc. Rev.*, **37**, 1042–1051.
12. Nithipatikom, K., McCoy, M.J., Hawi, S.R., Nakamoto, K., Adar, F. and Campbell, W.B. (2003) Characterization and application of Raman labels for confocal Raman microspectroscopic detection of cellular proteins in single cells. *Anal. Biochem.*, **322**, 198–207.
13. Kim, J.H., Kim, J.S., Choi, H., Lee, S.M., Jun, B.H., Yu, K.N., Kuk, E., Kim, Y.K., Jeong, D.H., Cho, M.H. and Lee, Y.S. (2006) Nanoparticle probes with surface enhanced Raman spectroscopic tags for cellular cancer targeting. *Anal. Chem.*, **78**, 6967–6973.
14. Qian, X.M., Peng, X.H., Ansari, D.O., Yin-Goen, Q., Chen, G.Z., Shin, D.M., Yang, L., Young, A.N., Wang, M.D. and Nie, S.M. (2008) In vivo tumor targeting and spectroscopic detection with surface-enhanced Raman nanoparticle tags. *Nat. Biotech.*, **26**, 83–90.
15. Lin, Cea. (2008) A new protein assay based on Raman reporter labeled immunogold nanoparticles. *Biosens. Bioelectron.*, **24**, 178–183.
16. Talley, C.E., Jusinski, L., Hollars, C.W., Lane, S.M. and Huser, T. (2004) Intracellular pH sensors based on surface-enhanced Raman scattering. *Anal. Chem.*, **76**, 7064–7068.
17. Wabuyele, M.B., Yan, F., Griffin, G.D. and Vo-Dinh, T. (2005) Hyperspectral surface-enhanced Raman imaging of labeled silver nanoparticles in single cells. *Rev. Sci. Instrum.*, **76**, 063710–063717.

18. Breuzard, G., Piot, O., Angiboust, J.F., Manfait, M., Candeil, L., Del Rio, M. and Millot, J.M. (2005) Changes in adsorption and permeability of mitoxantrone on plasma membrane of BCRP/MXR resistant cells. *Biochem. Biophys. Res. Commun.*, **329**, 64–70.
19. Breuzard, G., Angiboust, J.F., Jeannesson, P., Manfait, M. and Millot, J.M. (2004) Surface-enhanced Raman scattering reveals adsorption of mitoxantrone on plasma membrane of living cells. *Biochem. Biophys. Res. Commun.*, **320**, 615–621.
20. Chourpa, I., Lei, F.H., Dubois, P., Manfait, M. and Sockalingum, G.D. (2008) Intracellular applications of analytical SERS spectroscopy and multispectral imaging. *Chem. Soc. Rev.*, **37**, 993–1000.
21. Souza, G.R., Christianson, D.R., Staquicini, F.I., Ozawa, M.G., Snyder, E.Y., Sidman, R.L., Miller, J.H., Arap, W. and Pasqualini, R. (2006) Networks of gold nanoparticles and bacteriophage as biological sensors and cell-targeting agents. *Proc. Natl. Acad. Sci. U.S.A.*, **103**, 1215–1220.
22. Dijkstra, R.J., Scheenen, W., Dam, N., Roubos, E.W. and ter Meulen, J.J. (2007) Monitoring neurotransmitter release using surface-enhanced Raman spectroscopy. *J. Neurosci. Methods*, **159**, 43–50.
23. Kneipp, K., Haka, A.S., Kneipp, H., Badizadegan, K., Yoshizawa, N., Boone, C., Shafer-Peltier, K.E., Motz, J.T., Dasari, R.R. and Feld, M.S. (2002) Surface-enhanced Raman Spectroscopy in single living cells using gold nanoparticles. *Appl. Spectrosc.*, **56**, 150–154.
24. Kneipp, J., Kneipp, H., Rice, W.L. and Kneipp, K. (2005) Optical probes for biological applications based on surface enhanced Raman scattering from indocyanine green on gold nanoparticles. *Anal. Chem.*, **77**, 2381–2385.
25. Kneipp, J., Kneipp, H., McLaughlin, M., Brown, D. and Kneipp, K. (2006) In vivo molecular probing of cellular compartments with gold nanoparticles and nanoaggregates. *Nano Lett.*, **6**, 2225–2231.
26. Kneipp, J., Kneipp, H., Rajaduraj, A., Redmond, R.W. and Kneipp, K. (2009) Optical probing and imaging of live cells using SERS labels. *J. Raman Spectrosc.*, **40**, 1–5.
27. Uzunbajakava, N., Lenferink, A., Kraan, Y., Volokhina, E., Vrensen, G., Greve, J. and Otto, C. (2003) Nonresonant confocal Raman imaging of DNA and protein distribution in apoptotic cells. *Biophys. J.*, **84**, 3968–3981.
28. Van Manen, H.J., Uzunbajakava, N., Van Bruggen, R., Roos, D. and Otto, C. (2003) Resonance Raman imaging of the NADPH oxidase subunit cytochrome b(558) in single neutrophilic granulocytes. *J. Am. Chem. Soc.*, **125**, 12112–12113.
29. Puppels, G.J., De Mul, F., Otto, C., Greve, J., RobertNicoud, M., Arndt-Jovin, D.J. and Jovin, T. (1990) Studying single living cells and chromosomes by confocal Raman microspectroscopy. *Nature*, **347**, 301–303.
30. Van Manen, H.J. and Otto, C. (2007) Hybrid confocal Raman fluorescence microscopy on single cells using semiconductor quantum dots. *Nano Lett.*, **7**, 1631–1636.
31. Anshup, A., Venkataraman, J.S., Subramaniam, C., Kumar, R.R., Priya, S., Kumar, T.R.S., Omkumar, R.V., John, A. and Pradeep, T. (2005) Growth of gold nanoparticles in human cells. *Langmuir*, **21**, 11562–11567.
32. Shamsaie, A., Jonczyk, M., Sturgis, J., Robinson, J.P. and Irudayaraj, J. (2007) Intracellularly grown gold nanoparticles as potential surface-enhanced Raman scattering probes. *J. Biomed. Opt.*, **12**, 3.
33. Moskovits, M. (2005) Surface-enhanced Raman spectroscopy: a brief retrospective. *J. Raman Spectrosc.*, **36**, 485–496.
34. Kneipp, K., Kneipp, H. and Kneipp, J. (2006) Surface-enhanced Raman scattering in local optical fields of silver and gold nanoaggregates-from single-molecule Raman spectroscopy to ultrasensitive probing in live cells. *Acc. Chem. Res.*, **39**, 443–450.
35. Kneipp, K., Wang, Y., Kneipp, H., Perelman, L.T., Itzkan, I., Dasari, R.R. and Feld, M.S. (1997) Single molecule

detection using surface-enhanced Raman scattering (SERS). *Phys. Rev. Lett.*, **78**, 1667.
36. Inoue, M. and Ohtaka, K. (1983) Surface enhanced Raman scattering by metal spheres. I. Cluster effect. *J. Phys. Soc. Jpn.*, **52**, 3853–3864.
37. Xu, H.X., Aizpurua, J., Kall, M. and Apell, P. (2000) Electromagnetic contributions to single-molecule sensitivity in surface-enhanced Raman scattering. *Phys. Rev. E*, **62**, 4318–4324.
38. Stockman, M.I., Shalaev, V.M., Moskovits, M., Botet, R. and George, T.F. (1992) Enhanced Raman scattering by fractal clusters: scale-invariant theory. *Phys. Rev. B*, **46**, 2821.
39. Shalaev, V.M. (1996) Electromagnetic properties of small-particle composites. *Phys. Rep.*, **272**, 61–137.
40. Yamaguchi, Y., Weldon, M.K. and Morris, M.D. (1999) Fractal characterization of SERS-active electrodes using extended focus reflectance microscopy. *Appl. Spectrosc.*, **53**, 127–132.
41. Li, K.R., Stockman, M.I. and Bergman, D.J. (2003) Self-similar chain of metal nanospheres as an efficient nanolens. *Phys. Rev. Lett.*, **91**, 227402.
42. Goulet, P.J.G., dos Santos, D.S., Alvarez-Puebla, R.A., Oliveira, O.N. and Aroca, R.F. (2005) Surface-enhanced Raman scattering on dendrimer/metallic nanoparticle layer-by-layer film substrates. *Langmuir*, **21**, 5576–5581.
43. Stokes, R.J., McKenzie, F., McFarlane, E., Ricketts, A., Tetley, L., Faulds, K., Alexander, J. and Graham, D. (2009) Rapid cell mapping using nanoparticles and SERRS. *Analyst*, **134**, 170–175.
44. Grubisha, D.S., Lipert, R.J., Park, H.Y., Driskell, J. and Porter, M.D. (2003) Femtomolar detection of prostate-specific antigen: an immunoassay based on surface-enhanced Raman scattering and immunogold labels. *Anal. Chem.*, **75**, 5936–5943.
45. Schlücker, S.B.K. et al. (2006) Immuno-Raman microspectroscopy: in situ detection of antigens in tissue specimens by surface enhanced Raman scattering. *J. Raman Spectrosc.*, **37**, 719–721.
46. Brown, L.O. and Doorn, S.K. (2008) Optimization of the preparation of glass-coated, dye-tagged metal nanoparticles as SERS substrates. *Langmuir*, **24**, 2178–2185.
47. Sanchezcortes, S. and Garciaramos, J.V. (1992) SERS of AMP on different silver colloids. *J. Mol. Struct.*, **274**, 33–45.
48. VanDyke, R.W. (1993) Acidification of rat-liver lysosomes - quantitation and comparison with Endosomes. *Am. J. Physiol.*, **265**, C901–C917.
49. Grabe, M. and Oster, G. (2001) Regulation of organelle acidity. *J. Gen. Physiol.*, **117**, 329–344.
50. Van Dyke, R.W. (2000) Effect of cholera toxin and cyclic adenosine monophosphate on fluid-phase endocytosis, distribution, and trafficking of endosomes in rat liver. *Hepatology*, **32**, 1357–1369.
51. Thomas, G., Prescott, B. and Olins, D. Jr. (1977) Secondary structure of histones and DNA in chromatin. *Science*, **197**, 385–388.
52. Peticolas, W.L., Patapoff, T.W., Thomas, G.A., Postlewait, J. and Powell, J.W. (1996) Laser Raman microscopy of chromosomes in living eukaryotic cells: DNA polymorphism in vivo. *J. Raman Spectrosc.*, **27**, 571–578.
53. Hartman, K.A., Clayton, N. and Thomas, G.J. (1973) Studies of virus structure by Raman spectroscopy.1. R17 virus and R17 RNA. *Biochem. Biophys. Res. Commun.*, **50**, 942–949.
54. Small, E.W. and Peticolas, W.L. (1971) Conformational dependence of Raman scattering intensities from polynucleotides. *Biopolymers*, **10**, 69–88.
55. Parker, F.S. (1983) *Applications of Infrared, Raman, and Resonance Raman Spectroscopy in Biochemistry*, Plenum Press, New York and London.
56. Murphy, R.F., Powers, S. and Cantor, C.R. (1984) Endosome Ph measured in single cells by dual fluorescence flow-cytometry - rapid acidification of insulin to Ph-6. *J. Cell Biol.*, **98**, 1757–1762.
57. Montcourrier, P., Mangeat, P.H., Valembois, C., Salazar, G., Sahuquet, A., Duperray, C. and Rochefort, H. (1994) Characterization of very acidic

phagosomes in breast-cancer cells and their association with invasion. *J. Cell Sci.*, **107**, 2381–2391.

58. Michota, A. and Bukowska, J. (2003) Surface-enhanced Raman scattering (SERS) of 4-mercaptobenzoic acid on silver and gold substrates. *J. Raman Spectrosc.*, **34**, 21–25.

59. Bishnoi, S.W., Rozell, C.J., Levin, C.S., Gheith, M.K., Johnson, B.R., Johnson, D.H. and Halas, N.J. (2006) All-optical nanoscale pH meter. *Nano Lett.*, **6**, 1687–1692.

60. Schwartzberg, A.M., Oshiro, T.Y., Zhang, J.Z., Huser, T. and Talley, C.E. (2006) Improving nanoprobes using surface-enhanced Raman scattering from 30-nm hollow gold particles. *Anal. Chem.*, **78**, 4732–4736.

61. Kneipp, J., Kneipp, H., Wittig, B. and Kneipp, K. (2007) One- and two-photon excited optical ph probing for cells using surface-enhanced Raman and hyper-Raman nanosensors. *Nano Lett.*, **7**, 2819.

62. Ziegler, L.D. (1990) Hyper-Raman spectroscopy. *J. Raman Spectrosc.*, **21**, 769–779.

63. Denisov, V.N., Mavrin, B.N. and Podobedov, V.B. (1987) Hyper-Raman scattering by vibrational excitations in crystals, glasses and liquids. *Phys. Rep. -Rev. Sec. Phys. Lett.*, **151**, 1–92.

64. Kneipp, J., Kneipp, H. and Kneipp, K. (2006) Two-photon vibrational spectroscopy for biosciences based on surface-enhanced hyper-Raman scattering. *Proc. Natl. Acad. Sci. U.S.A.*, **103**, 17149–17153.

65. Michaels, A.M., Nirmal, M. and Brus, L.E. (1999) Surface enhanced Raman spectroscopy of individual rhodamine 6G molecules on large Ag nanocrystals. *J. Am. Chem. Soc.*, **121**, 9932–9939.

66. Kneipp, K., Wang, Y., Kneipp, H., Itzkan, I., Dasari, R.R. and Feld, M.S. (1996) Population pumping of excited vibrational states by spontaneous surface-enhanced Raman scattering. *Phys. Rev. Lett.*, **76**, 2444–2447.

67. Nie, S. and Emory, S.R. (1997) Probing single molecules and single nanoparticles by surface-enhanced Raman scattering. *Science*, **275**, 1102–1106.

68. Zipfel, W.R., Williams, R.M. and Webb, W.W. (2003) Nonlinear magic: multiphoton microscopy in the biosciences. *Nat. Biotechnol.*, **21**, 1368–1376.

69. Larson, D.R., Zipfel, W.R., Williams, R.M., Clark, S.W., Bruchez, M.P., Wise, F.W. and Webb, W.W. (2003) Water-soluble quantum dots for multiphoton fluorescence imaging in vivo. *Science*, **300**, 1434–1436.

70. Bizzarri, R., Arcangeli, C., Arosio, D., Ricci, F., Faraci, P., Cardarelli, F. and Beltram, F. (2006) Development of a novel GFP-based ratiometric excitation and emission pH indicator for intracellular studies. *Biophys. J.*, **90**, 3300–3314.

71. Hanson, K.M., Behne, M.J., Barry, N.P., Mauro, T.M., Gratton, E. and Clegg, R.M. (2002) Two-photon fluorescence lifetime imaging of the skin stratum corneum pH gradient. *Biophys. J.*, **83**, 1682–1690.

72. Lin, H.J., Herman, P., Kang, J.S. and Lakowicz, J.R. (2001) Fluorescence lifetime characterization of novel low-pH probes. *Anal. Biochem.*, **294**, 118–125.

14
Surface- and Tip-Enhanced CARS
Taro Ichimura and Satoshi Kawata

14.1
Introduction

High-order non-linear optical effects contribute to spatial confinement of photons into a volume smaller than the size of excitation light spot. The intensity of the nonlinear effects is proportional to the high-order powers (square, cube, etc.) of the excitation light intensity. The signal emission has a narrower distribution than the intensity distribution of the excitation field. This leads to the reduction of the effective volume of light–matter interaction, and accordingly improves spatial resolution of optical spectroscopy, microscopy and fabrication, which has been so far reported for far-field optics [1, 2].

Near-field optical interaction also gets benefited from the nonlinear effects. In near-field spectroscopy and microscopy [3, 4], in addition to the spatial confinement of photons due to the near-field effects, light–matter interaction can be further confined to a tiny volume at the very end of a near-field probe. When combined with surface-enhanced and tip-enhanced optical spectroscopy, which use localized surface plasmon polaritons (LSPPs) for local enhancement of the light field, the nonlinear effects assist the improvement of the signal sensitivity (intensity and/or contrast) as well as spatial resolution. Because of the nonlinear responses, even a small enhancement of the excitation field could lead to a huge enhancement of the emitted signal, allowing a reduction of the far-field background.

This chapter introduces a combination of nonlinear Raman scattering and surface/tip-enhanced spectroscopy. We describe the fundamentals of the principal and practical instrumentation. Microscopic imaging with high spatial resolution will also be shown and discussed.

14.2
CARS : Coherent Anti-Stokes Raman Scattering

For surface- and tip-enhanced nonlinear Raman spectroscopy, coherent anti-Stokes Raman scattering (CARS) spectroscopy is employed, which is a type of nonlinear Raman spectroscopy [5], now widely used for laser scanning microscopy [6]. The

Surface Enhanced Raman Spectroscopy: Analytical, Biophysical and Life Science Applications. Edited by Sebastian Schlücker
Copyright © 2011 WILEY-VCH Verlag GmbH & Co. KGaA, Weinheim
ISBN: 978-3-527-32567-2

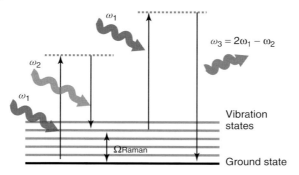

Figure 14.1 Energy diagram of the CARS process.

CARS spectroscopy uses three incident fields including a pump field (ω_1), a Stokes field (ω_2; $\omega_2 < \omega_1$) and a probe field ($\omega_1' = \omega_1$), and induces a nonlinear polarization at the frequency of $\omega_3 = \omega_1 - \omega_2 + \omega_1' = 2\omega_1 - \omega_2$. Figure 14.1 shows an energy diagram for the CARS process. The nonlinear polarization of CARS is given by

$$P_{CARS}^{(3)}(\omega_3 = 2\omega_1 - \omega_2) = \chi^{(3)} E_1(\omega_1) E_2^*(\omega_2) E_1(\omega_1) \tag{14.1}$$

where $\chi^{(3)}$ represents the third-order nonlinear susceptibility, $E_1(\omega_1)$ and $E_2(\omega_2)$ are the electric fields for excitation light. $E_2^*(\omega_2)$ denotes the complex conjugate of $E_2(\omega_2)$. Here, the electric fields are regarded as scalars for simplicity, while the quantities are basically vectors. The nonlinear susceptibility is expressed by the vibration-resonant term ($\chi_R^{(3)}$) and the non-resonant term ($\chi_{NR}^{(3)}$).

$$\chi^{(3)} = \chi_R^{(3)} + \chi_{NR}^{(3)} = \frac{A}{\Omega - (\omega_1 - \omega_2) - i\Gamma} + \chi_{NR}^{(3)} \tag{14.2}$$

The coefficient of the fraction, A, represents a constant related to the strength of the vibration, Ω denotes a frequency of one of molecular vibrations of a given sample, and Γ corresponds to the spectral bandwidth of the same vibration mode. When the frequency difference of ω_1 and ω_2 ($\omega_1 - \omega_2$) coincides with Ω, the anti-Stokes Raman signal is resonantly generated. $\chi_{NR}^{(3)}$ is a contribution from transition process, which does not undergo the vibration state. In particular, the process that undergoes the $2\omega_1$ state may be resonant or pre-resonant to an electronic state, resulting in strong contribution to the susceptibility given by Equation 14.2.

On the basis of Equations 14.1 and 14.2, one can obtain a CARS spectrum by plotting the CARS signal intensity with sweeping ω_2. The CARS spectrum gives essentially identical information with spontaneous Raman spectra, as the selection rule for CARS is the same as spontaneous Raman scattering [7].

One of the important features of the CARS spectroscopy is coherent interference. Owing to constructive interference, strong intensity can be observed even though the signal arises from the anti-Stokes Raman scattering. The CARS polarization is coherently driven by the incident fields so that the induced polarizations, which can be assumed as oscillating dipoles, are spatially coherent. The detected signal of such a coherent radiation is essentially proportional to the square of the number of dipoles or the excited volume. The CARS field at a detector can be expressed as an integration of spatially distributed dipoles multiplied by an appropriate Green's function [8]. In a particular condition when the CARS intensity is not weakened by the phase retardation over an integrated volume, for example, the volume is much smaller than the propagation wavelength of the CARS field, or, the CARS fields constructively interfere at the detector direction (phase matching), the signal intensity can be simply expressed as

$$I_{CARS} \propto \sum_{i,j}^{n} p_i p_j^* \approx n^2 |p|^2 = n^2 |\chi^{(3)}|^2 |E_1|^4 |E_2|^2 \tag{14.3}$$

where n, p_i and p denote the effective number of dipoles, the complex amplitude of the ith dipole and the averaged dipole, respectively. Equation 14.1 is substituted for the intensity of the averaged dipole. This quadratic dependence strongly contributes to the CARS intensity. On the other hand, spontaneous Raman scattering is an incoherent process, so that the Raman fields from two different dipoles do not interfere. Thus, the detected intensity is proportional to the number of dipoles, as

$$I_{Raman} \propto \sum_{i}^{n} p_i p_i^* \approx n|p|^2 = n|\chi^{(1)}|^2 |E|^2 \tag{14.4}$$

where $\chi^{(1)}$ denotes the linear susceptibility of the Raman scattering and $|E|^2$ the intensity of the excitation field. When the number of molecules is large to a certain extent and the intensity of the excitation fields are strong enough, the CARS intensity becomes higher than the spontaneous Raman scattering intensity.

14.3
Local Enhancement of CARS by Metallic Nanostructures

When a molecule is in a nanoscale close vicinity of a metal nanostructure under irradiation of the ω_1 and ω_2 lasers, a CARS polarization can be induced at the molecule by the locally enhanced fields. This phenomenon is referred to as surface-enhanced coherent anti-Stokes Raman scattering (SECARS). When the nanostructure is a nano-sized tip, it is distinctly called tip-enhanced coherent anti-Stokes Raman scattering (TECARS).

Experimental measurement of SECARS was first reported by Liang et al. [9] in 1994. They observed enhanced CARS signals from several organic solvents mixed with silver colloids. In their experiment, in order to satisfy the phase matching condition, two beams were crossed on a sample at a small angle that resulted in a 10 μm coherence length. The density of the silver colloid was adjusted such that several particles lay within the coherence length.

It was, in general, believed that in CARS spectroscopy the propagation angles of incident electric fields have to fulfil the phase matching condition, $k_{CARS} = 2k_1 - k_2$, to induce CARS polarization [5]. However, when the CARS polarization is induced in a volume smaller than the wavelength of the CARS field, the phase matching condition is automatically satisfied. In the small volume, the induced polarization can oscillate in phase, and the wavevector of the CARS field loses the relation with the incident excitation field. This concept has been commonly noticed in laser scanning CARS microscopy [6, 10–12], in which excitation beams are focused into a volume smaller than wavelengths by a high-NA objective lens.

This idea is also applicable to the near-field excitation of CARS. We successfully observed a SECARS spectrum of adenine molecules with isolated gold nanoparticles [13]. The details of the experiment will be shown in the following sections. Through these reports, the possibility of the local enhancement of CARS by metal nanostructures has been verified.

The advantage of the SECARS and TECARS is spatial confinement of effective interaction volume. If one assumes the spatial distribution of the excitation fields as a Bessel or a Gaussian function, the spatial distribution of the CARS field turns out to be much narrower due to the nonlinearities (Figure 14.2a). Here, a system consisting of a nanosphere and a single molecule is considered, where the sizes of both are much smaller than the wavelength of light. In this case, the near field generated around the nanosphere can be regarded as a field generated by an oscillating dipole. The intensity of the field is inversely proportional to the third power of the distance from the dipole ($1/R^3$), where R is the distance from the dipole center [8]. Then, the distribution of CARS polarization is proportional to the third power of $1/R^3$, that is, $1/R^9$. In contrast, spatial distribution of spontaneous Raman scattering polarization is simply proportional to $1/R^3$. Actually, the metallic nanostructure scatters the CARS signal with another enhancement of signal. In consequence, spatial distribution of CARS and spontaneous Raman are proportional to the twelfth power ($1/R^{12}$) and the sixth power ($1/R^6$) of the distance, respectively. This is the major difference between the CARS and spontaneous Raman. This difference is clearly illustrated by a numerical calculation result. Figure 14.2b shows electric field intensity distribution around a conical tip of silver, calculated with the finite-difference time-domain (FDTD) method.

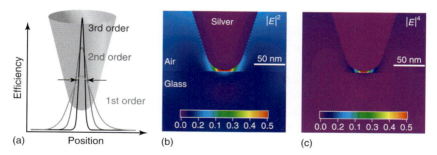

Figure 14.2 Spatial confinement of near-field signal due to the high-order nonlinear optical effects. (a) Schematic illustration. (b) Calculated electric field intensity distribution obtained by the FDTD method, and (c) the square of (b). In this calculation, the tip end is 2 nm away from a surface of a glass substrate. A continuous light (λ = 785 nm) is introduced with an incident angle of 45° from the glass side. The colour scale is normalized by the maximum intensity for each figure.

Figure 14.2c shows the distribution of the square of the electric field intensity shown in Figure 14.2b. Higher confinement is obviously seen in Figure 14.2c. This result guarantees the spatial resolution improvement of CARS spectroscopy and microscopy.

In terms of field enhancement, field enhancement is effective to each of the fields including the CARS field. The nonlinear polarization of SECARS and TECARS is expressed by

$$P^{(3)}_{CARS}(\omega_3) = L(\omega_3) \left\{ \chi^{(3)} \left[L(\omega_1) E_1(\omega_1) \right] \left[L(\omega_2) E_2(\omega_2) \right]^* \left[L(\omega_1) E_1(\omega_1) \right] \right\}$$
$$= L(\omega_3) L(\omega_1) L(\omega_2)^* L(\omega_1) \left[\chi^{(3)} E_1(\omega_1) E_2^*(\omega_2) E_1(\omega_1) \right] \quad (14.5)$$

where $L(\omega_i)(i = 1, 2, 3)$ is a complex number referred to as *local field factor*, which represents the factor of enhancement at a given wavelength at a particular position. The polarization of spontaneous Raman scattering, however, can be expressed by

$$P^{(3)}_{RS}(\omega_2) = L(\omega_2) \left\{ \chi^{(1)} \left[L(\omega_1) E_1(\omega_1) \right] \right\}$$
$$= L(\omega_2) L(\omega_1) \times \left[\chi^{(1)} E_1(\omega_1) \right] \quad (14.6)$$

Eventually, the enhancement factor for CARS signal has higher-order dependence. It results in not only the increase of the signal intensity but also relative reduction of far-field background, which is the CARS/Raman signal generated at an area far from the enhanced spot. The far-field background makes near-field optical images difficult to interpret, and was previously discussed in tip-enhanced Raman imaging [14, 15]. In TECARS microscopy, the near-field contribution becomes dominant to the far-field contribution, allowing one to interpret the obtained images in a simpler way.

Consequently, one can expect three advantages in the use of nonlinear optical effect, (i) improvement of spatial resolution, (ii) enhancement of signal intensity and (iii) reduction of far-field background.

14.4
Surface-Enhanced CARS

14.4.1
Experimental System for SECARS Measurements

For our SECARS experiment, a CARS microscope equipped with a microlens array scanner was used for fast image acquisition [16, 17] (Figure 14.3). The excitation laser beams are split into ∼100 beamlets by the microlens array, and the same number of foci are formed on the specimen. By rotating the microlens array disk, each focal spot on the specimen is scanned simultaneously. This enables observation of CARS images of the specimen within a short time compared to single-focus scanning CARS microscopy. For excitation of CARS, two mode-locked Ti:sapphire lasers (pulse duration: 5 ps, repetition rate: 80 MHz) are used. The ω_1 and ω_2 beams are collinearly combined in time and space, and introduced to

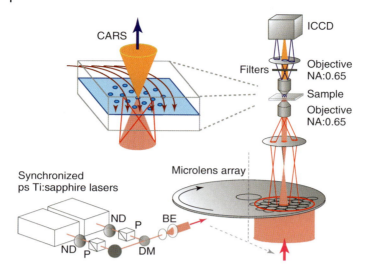

Figure 14.3 Schematic of a far-field CARS microscope using microlens array scanner.

the microlens array scanner. The beamlets are collimated again and are incident upon an oil-immersion lens (NA: 0.65). Multiple numbers of foci illuminating a specimen are produced. Each spot is separated from adjacent spots by approximately 4 μm, preventing interference between neighbouring focal fields from causing degradation of spatial resolution.

The setup adopts the transmission configuration with parfocally opposed two objective lenses. The CARS signal can be detected by an image-intensified CCD camera. The excitation lasers have to be rejected by a set of band-pass filters. The total OD of the filters for CARS emission is ∼0.4, and the ODs for the ω_1 and ω_2 beams are larger than 9–10. The residual light from the ω_1 and ω_2 beams passing through the filters was smaller than the noise level of the detector.

14.4.2
SECARS of Adenine Nanocrystals

Spherical gold nanoparticles (AuNPs) with diameter of 60 nm were used to enhance CARS. The particles were dispersed and fixed on an aminopropyltriethoxysilane-coated glass substrate through electrostatic interaction between the negative charges on the particles and the positive charges of amino groups at the surface. Other methods using polylysine and thiolcompounds are also available for immobilizing the particles on surfaces. Figure 14.4a shows a topographic image of the AuNPs fixed on the substrate obtained by a tapping mode AFM. The height of each particle is ∼60 nm corresponding to its diameter. Adenine molecules dissolved in water (1.35×10^{-2} wt %) were cast and dried on the AuNPs-immobilized substrate. The thickness of the adenine molecule layer is

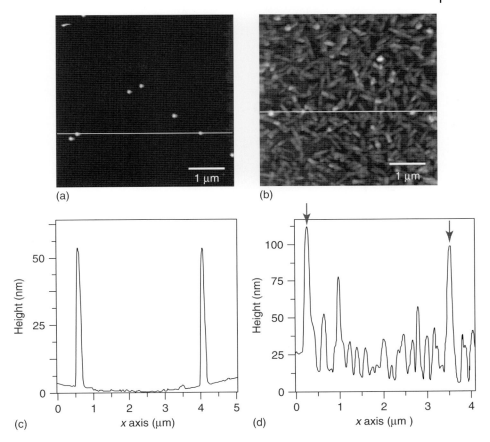

Figure 14.4 AFM images of the sample used for the SE-CARS imaging, measured (a) after immobilization of AuNPs and (b) after dispersion of adenine molecules. (c) and (d) are cross-sections of the lines indicated in (a) and (b), respectively.

about 20 nm. Figure 14.4(b) shows an AFM image of the sample after the cast of adenine molecules, and indicates most of the gold particles are well isolated while the adenine molecules are distributed densely.

Figure 14.5 shows a Raman spectrum obtained with a conventional Raman spectrometer (excitation wavelength: 532 nm). The strong Raman peaks at 719.7 cm^{-1} and 1329.5 cm^{-1} indicated by arrows are assigned to the ring breathing mode of a whole molecule and to the ring stretching mode of the diazole, respectively. The latter peak is adopted in this section and the next section for TECARS imaging of adenine.

The pump laser frequencies, ω_1 and ω_2, were tuned to 12 731.7 cm^{-1} (λ: 785.4 nm) and 11 402.5 cm^{-1} (λ: 877.0 nm) resulting in the Raman shift of 1329.2 cm^{-1}, which

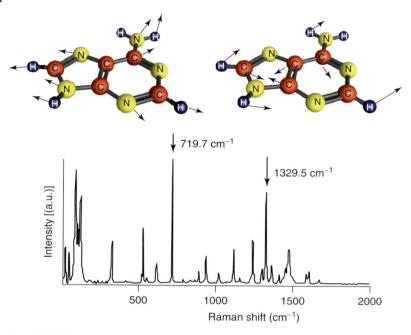

Figure 14.5 Spontaneous Raman spectrum of adenine molecules. Shown above two strong peaks (719.7 cm^{-1} and 1329.5 cm^{-1}) are the vibrational motions. The peak at 1329.5 cm^{-1} was employed for the SECARS experiment.

corresponds to the right peak indicated in Figure 14.5. Accordingly, the anti-Stokes Raman signal at 14 060.9 cm^{-1} (λ: 711.2 nm) was generated and observed.

Figure 14.6 shows the SECARS images of the Raman band of 1329.2 cm^{-1} obtained with an exposure time of 30s. The scan area is 31 µm × 31 µm. The average pump laser intensities at ω_1 and ω_2 before the microlens array were 323 and 259 mW, respectively. Several bright spots (indicated by arrows) can be seen in the image, which are strong CARS signals from isolated AuNPs. Only one particle lies in each spot, according to the AFM image in Figure 14.4a. When the pulses were temporally desynchronized by a phase shifter, the CARS signal from the particles mostly vanished as shown in Figure 14.6b. The slightly remaining signal in Figure 14.6b can be attributed to multi-photon-excited autofluorescence of adenine molecules and white light continuum emission of the particles, which do not require the pulse synchronization. The CARS intensities in Figure 14.6a are quite different from other particles. Furthermore, the number density of bright spots found in Figure 14.6a is much less than that of gold particles seen in Figure 14.4. These phenomena are attributed to the concept of 'hot spot', as seen in surface-enhanced Raman scattering (SERS) experiments [18]. It is caused by the difference in the geometry of the particle structure and the severe conditions for strong field enhancement. Because of the third-order nonlinearity of the CARS

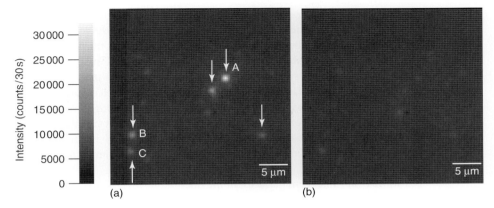

Figure 14.6 CARS images of adenine molecules, which are obtained when the two laser pulses are temporally (a) synchronized and (b) not synchronized. The bright spots are the AuNPs enhancing the surrounding field and emitting strong CARS signals. The scan area is 31 μm × 31 μm consisting of 100 pixels × 100 pixels. The exposure time of the image-intensified CCD camera is 30 s.

process, the hot spot effect becomes further prominent compared to the linear Raman scattering.

By sweeping the Stokes laser frequency ω_2 from 11 431.7 cm^{-1} (874.8 nm) to 11 381.0 cm^{-1} (878.7 nm), while keeping the pump laser frequency ω_1 at 12 731.7 cm^{-1} (785.4 nm), SECARS spectra were measured. Figure 14.7 shows an observed SECARS spectra of adenine molecules obtained at three different AuNPs indicated by the capital letters (A∼C) in Figure 14.6a. The thick solid line is

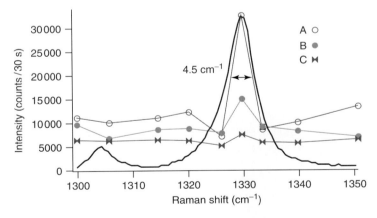

Figure 14.7 Enhanced CARS spectra of adenine molecules at the three different AuNPs labeled with A–C in Figure 14.6. The lines with marks are the spectra from each particle and the thick solid line shows a close-up of the peak in the spontaneous Raman spectrum shown in Figure 14.5.

a close-up of the Raman spectrum peak at 1329.5 cm^{-1} shown in Figure 14.5. The SECARS spectra from particle A and B are in good agreement with the spontaneous Raman spectra. On the other hand, C does not strongly exhibit the resonance enhancement at the Raman shift of 1329.5 cm^{-1}. This can be attributed to the strong non-resonant background signal of either the adenine molecules or gold. The non-resonant signal of gold is caused by the nonlinearity of the free electron oscillation, which locally induces optical four-wave mixing (FWM) [19]. Owing to the presence of coherent interference between the resonant and nonresonant signals (Equation 14.1), there should be a peak shift in the CARS spectra [7], though it is not observed in Figure 14.7. This may be because of the rough step of wavelength sweeping of the excitation laser.

To estimate the enhancement factor in the microscope images, Figure 14.6b was subtracted from Figure 14.6a to remove the contributions of the optical background signals and detector noise. The corrected image contains only the contributions of CARS and nonresonant FWM. The CARS intensity within a bright spot is integrated and regarded as a total intensity from a single particle since the spatial resolution of our microscope is larger than the size of the single particles. The enhancement factor by a single gold particle is defined as the ratio of the total intensity from a particle to the mean CARS intensity without particles.

Figure 14.8a shows enhancement factors for several single gold particles in a SECARS image. The enhancement factor for single particles reached up to 2000. Figure 14.8b shows a CARS image, which contains a very strong spot exhibiting enhancement factor of ~6000. This brightness originates from, not a single particle but, a dimer structure consisting of two attaching spheres, which was found in the AFM image of the particle shown in Figure 14.8c. This is analogous to a well-known fact that dimer structures give huge enhancement of Raman scattering [20, 21].

14.4.3
SECARS of Single-Walled Carbon Nanotubes

The local enhancement effect of CARS spectrum for single-walled carbon nanotubes (SWCNTs) was also examined. The SWCNTs with purities of 90% were

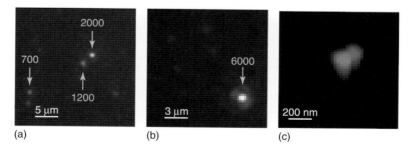

Figure 14.8 Estimated enhancement factors. (a and b) SECARS images with the enhancement factors. (c) An AFM image of the AuNP observed in (b).

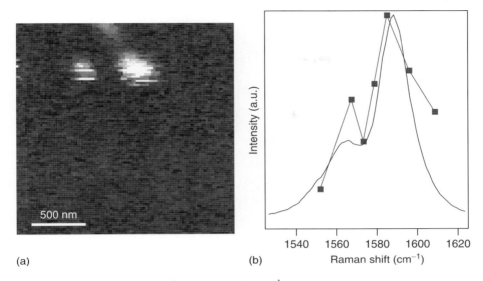

Figure 14.9 (a) SECARS image of SWCNTs at 1584 cm^{-1}. (b) SECARS spectrum (line with boxes) shown with a spontaneous Raman spectrum (solid line).

produced by high-pressure carbon monoxide (HiPCO) technique. Individual bundles of SWNTs could be exfoliated from the aggregates (1 mg) after exposure to 65% HNO$_3$ with ultra-sonication for 3 h. After being diluted with ethanol, the solution (200 ml) was sonicated for 3 h, and spincoated onto a cover glass [22]. Figure 14.9a shows a SECARS image of SWCNTs with AuNPs at 1584 cm^{-1}, corresponding to the strongest peak of typical SWCNTs called *G-band* [23]. The bright spots are the SWCNTs attached to the gold particles and emitting the SECARS signals. In this measurement, the total excitation laser power was set to be ~50 mW. The SWCNTs in the area surrounding the particles cannot be seen in Figure 14.9a, because the CARS of single SWCNTs without enhancement are not strong enough to be detected. An SECARS spectrum of SWCNTs is shown in Figure 14.9 with a spontaneous anti-Stokes Raman spectrum (solid line). The two peaks are obviously seen in the SECARS spectrum and correspond well with the spontaneous Raman spectrum.

14.5
Tip-Enhanced CARS

14.5.1
Experimental System for TECARS Microscopy

For TECARS imaging, we used an inverted microscope and a contact mode AFM mounted on the microscope stage, as shown in Figure 14.10 [24]. The same

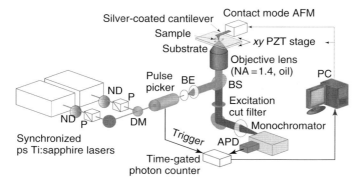

Figure 14.10 An experimental system of tip-enhanced CARS microscopy. See the text for detail.

laser system as used in the SECARS experiments in the previous section was employed for excitation of CARS. The repetition rate of the excitation lasers is controlled by an electro-optically modulated pulse picker. The ω_1 and ω_2 beams are collinearly combined in time and space, and introduced into the microscope with an oil-immersion objective lens ($NA = 1.4$) focused onto the sample surface. The AFM controls a metallic tip around the focal area. Metallic tips can be prepared in several possible ways such as vapour deposition to a silicon cantilever [25], electrodeless plating on a silicon cantilever [26], electrochemical etching of a metal wire [27] and attaching a metal nanoparticle on an end of a tapered glass fibre [28]. For the following experiments, we prepared silver tips by the vapour deposition process. CARS was enhanced by bringing the tip close to the sample. A raw TECARS signal usually contains a far-field contribution coming from the whole focal area, together with a near-field contribution coming from a small part of the sample adjacent to the tip.

The backscattered CARS emission is collected with the objective lens and detected with an avalanche-photodiode based photon-counting module through an excitation-cut filter and the monochromator. The pulse signal from the APD are counted by a time-gated photon counter synchronously triggered with the pulse picker, which effectively reduces the dark counts down to almost 0 counts/s. Scanning the sample stage, while keeping the tip at the focused spot, can acquire two-dimensional TECARS images of a specific vibrational mode with a high spatial resolution.

14.5.2
TECARS Imaging of DNA Molecules

DNA imaging by the TECARS microscopy was demonstrated. Two types of samples were prepared, (i) DNA clusters and (ii) DNA network. For preparation of DNA clusters, an aqueous solution of single-stranded DNA (poly(dA-dT)) (250 µg ml^{-1}) was cast and dried on a coverslip at room temperature with a fixation time of

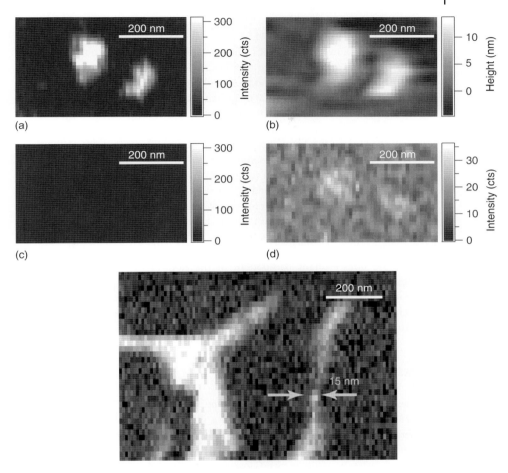

Figure 14.11 TECARS images of DNA molecules. (a) TECARS image at 1337 cm^{-1} of adenine, and (b) the simultaneously obtained topographic image. (c) TECARS image at the off-resonant frequency (1278 cm^{-1}). (d) The same image as (c) shown with a different grey scale. (e) TECARS image of the DNA network structure. For all images, the number of photons counted in 100 ms was recorded for one pixel.

~24 h. The dimensions of the clusters are typically ~20 nm in height and ~100 nm in width. For a DNA network, an aqueous solution of double-stranded DNA (poly(dA-dT)-poly(dA-dT)) (250 µg ml^{-1}) was mixed with MgCl$_2$ (0.5 mM) solution, and the DNA solution was then cast on a coverslip and blow-dried after a fixation time of ~2 hours [29]. The DNA chains were self-assembled to form a network structure. The Mg^{2+} ions electrostatically connect phosphorus atoms of DNA and oxygen atoms of the glass. The DNA network consists of bundles of DNA double-helix filaments aligned parallel on the glass substrate. The height of the bundle structures is ~2.5 nm, which is comparable to the diameter of single-DNA double-helix. CARS imaging was performed at 1337 cm^{-1}, which coincides with a Raman mode of adenine by tuning the excitation frequencies ω_1 and ω_2 to be 12 710 cm^{-1} (λ_1: 786.77 nm) and 11 373 cm^{-1} (λ_2: 879.25 nm), respectively.

Figure 14.11 shows the TECARS images of the DNA clusters obtained by our system. Figures 14.11a and b are the TECARS image at the on-resonant wave number difference (1337 cm^{-1}) and the simultaneously acquired topographic AFM image. The DNA clusters of ~100 nm diameter are visualized in Figure 14.11a. The two DNA clusters with distance of ~160 nm are obviously distinguished by the tip-enhanced CARS imaging. This indicates that the CARS imaging successfully achieved super-resolving capability beyond the diffraction limit of light. For comparison, an imaging of the same area was performed at 1278 cm^{-1} where the sample DNA does not have a unique Raman peak. Figure 14.11c shows that the CARS signals mostly vanished at the off-resonant position. Figures 14.11a and c verify that DNA molecules emit vibrationally resonant CARS at the specific frequency. However, there remains some slight signal increase at the clusters at the off-resonant frequency, as seen in Figure 14.11d, which is the same as Figure 14.11c but is shown with a different grey scale. This can be caused by both the non-resonant component of the nonlinear susceptibility of DNA (Equation 14.2) [5] and a topographic artefact [30]. Figure 14.11e shows a CARS image of the DNA network structure at 1337 cm^{-1}. The network structure of the DNA chains was clearly visualized. The full width of half maximum of the finest structure in the image (indicated by arrows) is found to be ~15 nm.

14.5.3
TECARS Imaging of CNTs

Bundles of SWCNTs of semiconductor-type were used for CARS imaging [19]. The SWCNT bundles were dispersed on a glass substrate by spin-coating. The frequencies of the two beams were set to be ω_1 = 12 744 cm^{-1} (λ_1: 784.70 nm) and ω_2 = 11 163 cm^{-1} (λ_2: 895.84 nm) such that the frequency difference is 1581 cm^{-1} corresponding to the G-band of SWCNTs [23]. Figure 14.12a and b show a simultaneously obtained topographic image and a tip-enhanced CARS image of the SWCNT bundles, respectively. Figure 14.12c shows a far-field CARS image obtained without the tip. Figure 14.12d shows the cross-sections of the tip-enhanced and far-field CARS signals of the line indicated by the arrows. By sweeping the frequency of the ω_2 beam from 11 521 cm^{-1} (868.00 nm) to 11 112 cm^{-1} (899.90 nm) a TECARS spectrum and the corresponding far-field CARS spectrum were obtained, as shown in Fig. 14.12e, where a spontaneous anti-Stokes Raman spectrum is also drawn for comparison. The CARS spectra are in good agreement with the spontaneous Raman spectrum, and the characteristic band shape of the semiconductor-type of the SWCNTs are clearly seen in the tip-enhanced CARS spectrum. In the tip-enhanced CARS image in Figure 14.12b, only the several bundles aligned parallel to the polarization of the incident fields are selectively visualized, while lots of randomly oriented bundles are found in the topographic image in Figure 14.12a. This selectivity can be explained by a combination of two possible mechanisms. First, since the G-band is a longitudinal

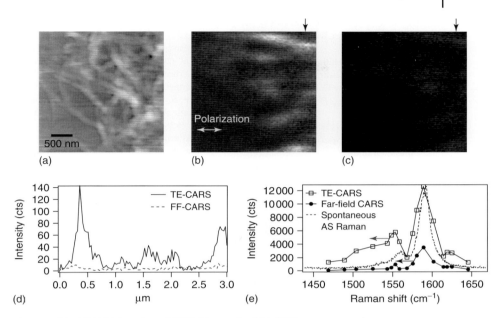

Figure 14.12 TECARS imaging of SWCNTs. (a) A topographic image of the SWTs dispersed on a glass substrate. (b) A TECARS image obtained simultaneously with the topographic image in (a). (c) A far-field CARS image obtained without the tip. (d) The cross-sections of the line indicated by the arrows in (b) and (c). (e) A TECARS spectrum, a far-field CARS spectrum and spontaneous anti-Stokes (AS) Raman spectrum of an SWCNT bundle. The average powers of the ω_1 beam and ω_2 beam are 12 µW and 6 µW at the 800 kHz repetition rate except for the spectrum measurement in (e), which used the powers of 26 µW and 13 µW.

mode susceptible to the electric fields parallel to the tubes [23], only those several SWCNTs parallel to the polarization of the incident field can be strongly excited. Second, the variety of the SWCNTs, which have many different chiralities are contained in the sample, and only some have electronic resonance frequencies matching with either of the excitation fields. Therefore, there can be a large difference in the CARS signal intensity, tube by tube. It is verified from all the results shown in Figure 14.12 that our TECARS microscope effectively amplified the CARS signal of the SWCNTs and can obtain vibrational images of SWCNTs.

References

1. Denk, W., Strickler, J.H. and Webb, W.W. (1990) Two-photon laser scanning fluorescence microscopy. *Science*, **248**, 73–76.
2. Kawata, S., Sun, H.-B., Tanaka, T. and Takada, T. (2001) Finer features for functional microdevices. *Nature*, **412**, 697–698.
3. Kawata, S. (ed.) (2001) *Near-field Optics and Surface Plasmon Polaritons*, Springer, Berlin.
4. Kawata, S and Shalaev, V.M. (eds) (2007) *Tip Enhancement (Advances in Nano-optics and Nano-photonics)*, Elsevier Science.

5. Shen, Y.R. (1984) *The Principles of Nonlinear Optics*, John Wiley & Sons, Inc, New York.
6. Zumbusch, A., Holton, G.R. and Xie, X.S. (1999) Three-dimensional vibrational imaging by Coherent anti-Stokes Raman scattering. *Phys. Rev. Lett.*, **82**, 4142–4145.
7. Levenson, M.D. (1988) *Introduction to Nonlinear Laser Spectroscopy*, Academic press, Orlando.
8. Jackson, J.D. (1998) *Classical Electrodynamics*, 3rd edn, John Wiley & Sons, Inc, New York.
9. Liang, E.J., Weippert, A., Funk, J.M. et al. (1994) Experimental observation of surface-enhanced coherent anti-Stokes Raman scattering. *Chem. Phys. Lett.*, **227**, 115–120.
10. Hashimoto, M., Araki, T. and Kawata, S. (2000) Molecular vibration imaging in the fingerprint region by use of coherent anti-Stokes Raman scattering microscopy with a collinear configuration. *Opt. Lett.*, **25**, 1768–1770.
11. Hashimoto, M. and Araki, T. (2001) Three-dimensional transfer functions of coherent anti-Stokes Raman scattering microscopy. *J. Opt. Soc. Am. A*, **18**, 771–776.
12. Volkmer, A., Cheng, J.-X. and Xie, X.S. (2001) Vibrational imaging with high sensitivity via Epi-detected Coherent Anti-Stokes Raman Scattering (E-CARS) microscopy. *Phys. Rev. Lett.*, **87**, 023901.
13. Ichimura, T., Hayazawa, N., Hashimoto, M. et al. (2003) Local enhancement of coherent anti-Stokes Raman scattering by isolated gold nanoparticles. *J. Raman Spectrosc.*, **34**, 651–654.
14. Hayazawa, N., Inouye, Y., Sekkat, Z. and Kawata, S. (2001) Near-field Raman scattering enhanced by a metallized tip. *Chem. Phys. Lett.*, **335**, 369–374.
15. Mehtani, D., Lee, N., Hartschuh, R.D. et al. (2005) Nano-Raman spectroscopy with side-illumination optics. *J. Raman Spectrosc.*, **36**, 1068–1075.
16. Minamikawa, T., Hashimoto, M., Fujita, K. et al. (2009) Multi-focus excitation coherent anti-Stokes Raman scattering (CARS) microscopy and its applications for real-time imaging. *Opt. Express*, **17**, 9526–9536.
17. Kobayashi, M., Fujita, K., Kaneko, T. et al. (2002) Second-harmonic-generation microscope with a microlens array scanner. *Opt. Lett.*, **27**, 1324–1326.
18. Nie, S. and Emory, S.R. (1997) Probing single molecules and single nanoparticles by surface-enhanced Raman scattering. *Science*, **275**, 1102–1106.
19. Hayazawa, N., Ichimura, T., Hashimoto, M. et al. (2004) Amplification of coherent anti-Stokes Raman scattering by a metallic nano-structure for a high resolution vibration microscopy. *J. Appl. Phys.*, **95**, 2676–2681.
20. Le Ru, E.C., Meyer, M. and Etchegoin, P.G. (2006) Proof of single-molecule sensitivity in surface enhanced Raman scattering (SERS) by means of a two-analyte technique. *J. Phys. Chem. B*, **110**, 1944–1948.
21. Michaels, A.M., Jiang, J. and Brus, L. (2000) Ag nanocrystal junctions as the site for surface-enhanced Raman scattering of single rhodamine 6G molecules. *J. Phys. Chem. B*, **104**, 11965–11971.
22. Bower, C., Kleinhammes, A., Wu, Y. and Zhou, O. (1998) Intercalation and partial exfoliation of single-walled carbon nanotubes by nitric acid. *Chem. Phys. Lett.*, **288**, 481–486.
23. Duesberg, G.S., Loa, I., Burghard, M. et al. (2000) Polarized Raman spectroscopy on isolated single-wall carbon nanotubes. *Phys. Rev. Lett.*, **85**, 5436–5439.
24. Ichimura, T., Hayazawa, N., Hashimoto, M. et al. (2004) Tip-enhanced coherent anti-Stokes Raman scattering for vibrational nanoimaging. *Phys. Rev. Lett.*, **92**, 220801.
25. Hayazawa, N., Inouye, Y., Sekkat, Z. and Kawata, S. (2000) Metallized tip amplification of near-field Raman scattering. *Opt. Commun.*, **183**, 333–336.
26. Saito, Y., Murakami, T., Inouye, Y. and Kawata, S. (2005) Fabrication of silver probes for localized plasmon excitation in near-field Raman spectroscopy. *Chem. Lett.*, **34**, 920–921.
27. Hartschuh, A., Sánchez, E.J., Xie, X.S. and Novotny, L. (2003) High-resolution near-field Raman microscopy of

single-walled carbon nanotubes. *Phys. Rev. Lett.*, **90**, 095503.
28. Anger, P., Bharadwaj, P. and Novotny, L. (2006) Enhancement and quenching of single-molecule fluorescence. *Phys. Rev. Lett.*, **96**, 113002.
29. Tanaka, S., Cai, L.T., Tabata, H. and Kawai, T. (2001) Formation of two-dimensional network structure of DNA molecules on Si substrate. *Jpn. J. Appl. Phys.*, **40**, L407–L409.
30. Hecht, B., Bielefeldt, H., Inouye, Y. *et al.* (1997) Facts and artifacts in near-field optical microscopy. *J. Appl. Phys.*, **81**, 2492–2498.

Index

a

absorption of a material 4
acetaminophen 131
adenosine monophosphate (AMP) 78
adsorption factor, in Raman enhancement 244
Ag-coated Au particles 50–51
Ag colloid monolayers 157
aggregated colloids 105
aldrin (ALD) 114
alexa-fluorophore labelled antibodies 277
alkanethiols 44
alkylating agents 142
α,ω-aliphatic diamines (ADs)
– AD–Ag NP systems 115
– AD–metal NP systems 113–114
– detection of 111–115
amine modifications 245–246
para-aminobenzoic acid (p-ABA) 167, 179
3-amino-n-propyltrimethoxysilane (APTMS) 270
6-amino-2-naphthoic acid 179
amino-terminated alkane thiols 80
amphiphiles 221
analgesics 130–138
Anatomical Therapeutic Chemical Classification 130
anisotropic silver nanoplates 55
anthracyclines 142
anti-bonding resonance 15
anticarcinogenics 142–151
antimalarial drugs 139
antimetabolites 142
antimutagenics 142–151
antioxidant vitamins 142
antipyretics 130–138
Apaf-1 protein 237
aromatic thiols 44

aspirin 131–133, 138
Au/Ag nanoshell 264, 266
Au–Ag hybrid devices 235
Au–Li Ag–Ag film substrates 157
Au–Li Ag substrates 157

b

benzene 165
benzene ring 44
benzo[c]phenanthrene (BcP) 106
benzotriazoles 245
bimetallic nanoparticle SERS substrate 50–51
biomimetic devices 221
biotin 45
black hole quenchers (BHQs) 246
Blue-ray technology 181
bovine serum albumin (BSA) 47
bulk materials, optical properties of 2

c

calixarenes, detection of 106–111
capillary-driven test stripes 176–178
capillary electrophoresis SERS (CE–SERS) systems 157–161
carbon nanotube composites, detection of 118–119
cardiolipin 237
centrifugal platform 180–181
cetylpyridinium salts 47
cetylquinolinium salts 47
cetyltrimethylammoniumbromide 55
chemical effect (CM), SERS 39
chemical enhancements 31
chemical reducing agents 41
chemisorption 135
chemotherapeutic drugs 142

Surface Enhanced Raman Spectroscopy: Analytical, Biophysical and Life Science Applications. Edited by Sebastian Schlücker
Copyright © 2011 WILEY-VCH Verlag GmbH & Co. KGaA, Weinheim
ISBN: 978-3-527-32567-2

chloroform   133
chloroquine   139–142
chrisene (CHR)   106
citrate   45
citrate-reduced silver colloids   163
cocaine   132
coherent anti-Stokes Raman scattering (CARS) spectroscopy
– coherent interference feature   306–307
– enhancement factor   314
– incident fields   306
– intensity   306–307
– local enhancement   307–309
– near-field excitation   308
– nonlinear susceptibility   306–307
– polarization   306–308
– spatial distribution of fields   308
– surface-enhanced   309–315
– third-order nonlinearity of   312–313
– tip-enhanced   315–319
coinage metals   19–20, 243
– optical properties of   2
colloidal SERS substrates   179
colloid silver nanoparticles   42
confocal Raman microspectrometer   157
core–shell metal nanoparticles   41
coronene (COR)   106
coupled plasmon resonances   15–17
cyclohexane   204
cytochrome c (Cyt c)
– with DOPG complexes   224–225
– dynamics of interfacial processes   235–236
– electron transfer processes at coated electrodes   225–228
– horse heart (HH)   233
– immobilized form of   224–225
– interfacial electric fields and the biological functions of   237–239
– $K_{B2/B1}$ values   224–225
– overall interfacial processes of   238
– oxidized complex of   238
– positively charged region of   222
– protein–liposome complex, formation of   222
– rapid-scan SEIRA spectra of   234
– redox function of   237–238
– RR spectrum of   222–223
– stationary SERR spectra of   229
– TR SERR experiments, with Soret-band excitation   228, 236
cytochrome c oxidase (CcO)   222

d

DABCYL   246
dacarbazine   132
DCEC calixarene   106
defocussing method   203–204
density functional theory (DFT)   130
diacetylmorphine   162
dialkyl sulfides   44
dibenzoanthracene (DBA)   106
dielectric function (ω)   2
– of Ag and Au   2–4
– of a (lossless) metal   3
dielectrophoresis (DEP)   181
Diels Alder cycloaddition   246
dihydrocodeine   162
2,5-dimethoxy-4-bromoamphetamine (DOB)   75
4,6-dinitrocresol   163
2,4-dinitrophenol   163
dioleoyl-phosphatidylglycerol (DOPG)   222
dioxepine   162
dipicolinic acid (DPA)   78
dithiocarbamate calixarene (DTCX)   108
– DTCX-functionalized Ag NPs   109, 111
– marker parameters   109
DNA detection techniques, quantitative
– amine modifications   245–246
– assays   256–259
– dye labels   247–248, 255
– fluorescence readout approaches   249–250
– fluorescent labelling probes   245
– fluorescent probes   241–242
– of hospital-acquired infections (HAIs)   258
– of human immunodeficiency virus type-1 (HIV-1) DNA sequence   257
– limits of detection (LoDs)   246, 254–255
– multiplexing capabilities, review of   253–256
– multivariate analysis (MVA) approach   255
– non-fluorescent labels   246, 249
– propargylamine–deoxyuridines, incorporation of   251
– Raman reporter molecules   245–248
– real-time polymerase chain reaction (rt-PCR)   241
– sensitive detection   252–255
– SERRS DNA probes   248–251
– SERS detection of glucose   244
– surface-enhanced resonance Raman scattering (SERRS)   242–245
– Taqman approach   250

– use of nanoparticles in   243–244
– using closed-tube formats   242
droplet-based microfluidics, for SERS   183–187
droplet drying   78–79
drug absorption process, in human body   136

e

$|E|^4$ approximation   27–28
e-beam lithography   166, 168
electrochemical cyclic voltammetry   40
electrochemical surface-enhanced Raman scattering (EC-SERS)
– applications   204–213
– benzene adsorption and reaction, study of   204–206
– biological application of   216
– cell culturing, study of   215
– cell design   198–199
– characterization   192
– cytochrome c on a DNA-modified gold surface, study of   208–211
– detection of dopamine   211
– detection sensitivity   199
– discrimination of mutations in DNA sequences, study of   212–213
– electrochemical (EC) field   193–194
– electrode materials and excitation energy dependence   194–195
– electrode potential   193
– electrolyte solution and solvent dependence   195
– EM and EC enhancements   195–197
– experimental techniques   197–204
– in fabrication of well-ordered nanostructured electrode surfaces   214–215
– features   192–197
– identification   192
– integration with microfluidic devices   216
– measurement on bio-related systems   203–204
– NADH, study of the adsorption behavior of   207–208
– overview   191–192
– oxidation and reduction cycles (ORCs)   199–201
– photon-driven CT states   196
– perspectives   213–216
– SERS-active electrode surfaces, study of   199–201
– single-stranded and double-stranded DNA on gold surfaces, study of   208
– spectral characters   194
– for studying biological molecules   206–211
– substrate cleaning   201–202
– surface plasmon resonance (SPR) effect   194
– and surface Raman enhancements   192
electrokinetic effects, basic   181
electrokinetic platform   181–183
electromagnetic effect (EM), SERS   39
– fluorescence. see fluorescence signals, enhancement of
– SERS   22–23, 27–28, 31
electron microscopies   58, 105
electron transfer (ET) reactions   219
– in CcO   237–239
– of cytochrome c, at coated electrodes   225–228
– electric field effects   232–234
– and protein orientation dynamics   231–232
– and protein structural changes   234–235
– relaxation time τ_{relax} 229–231
electro-osmotic flow (EF)   181
electrophoresis (EP)   181
'emission leg' fluorescence   28
ensemble SERS/SERRS   88
erythrosin B   158–159
ethylbenzene   165

f

finite-difference time-domain (FDTD) method   308
flavanoids   142
flow management   175
fluidic anchors   180
fluorescence process   26
fluorescence quenching   28–29
fluorescence signals, enhancement of
– absorption   23–24
– vs Raman processes   24–26
– vs SERS   29–31
fluorescent labelling probes   245
fluorophores   245
5-fluorouracil   132, 142–150
foams, microchannel systems   185
Fourier transform (FT) techniques   46

g

G-band   315, 318
gold (Au)
– absorption of   4
– branched nanoparticles   52
– dielectric functions of   2–4

gold (Au) (contd.)
– growth reaction in HAuCl$_4$ 43
– high enhancement region   22
– imaginary part of $\varepsilon(\lambda)$ 3–4
– LFIEF on the surface   6, 19
– mirrors   5
– nanoparticle colloids   243
– nanoparticles   41, 43–44, 104, 183
– nanospheres   266
– nanostructures   89
– performance   18–19
– polymer-encapsulated dimers of nanoparticles   49
– poly(N-vinyl-2-pyrrolidone) (PVP) capped dendritic nanoparticles   52
– reflectance   4
– self-assembled monolayers (SAMs)   221
– SERS-active electrode   200
– spherical nanoparticles (AuNPs)   310
gold nanotags   178

h

haematin, resonance spectra of   139
haemozoin   139–140
high-pressure carbon monoxide (HiPCO) technique   315
hospital-acquired infections (HAIs)   241, 258
host–guest interaction mechanisms   105
hot spots   20–21, 56, 87, 97, 112, 312–313
– spatial localization of   22
humic acids (HAs), detection of   119–122
hydrazine   41
hydrazine dihydrochloride   42
4-hydroxyacetanilide   134
hydroxylamine hydrochloride   41
hyper Raman scattering (HRS)   297

i

ibuprofen   131
immunohistochemistry, using SERS-labelled antibody probes   273–274
index of refraction n(ω)   2
indium tin oxide (ITO) electrodes   79
indocyanine green (ICG)   290
interband electronic transitions   4
intracellular probes
– challenges   285–286
– cross sections corresponding to 10^4 GM (Goeppert–Mayer)   297
– identification of   290–292
– influence of surroundings   287–290
– IR-active vibrations   297
– localization and targeting   286–287

– macrophage cell line J774   294–295
– parameters   292–297
– pH values   295–296, 299
– spectra of pMBA   299
– surface-enhanced hyper-Raman scattering (SEHRS) approach   297–300
– using gold or silver nanoparticles   289–290
– values of Raman bands   294
iodide   202
isotachophoretic (ITP) focussing method   167
isotopomers   77

k

keto–enol tautomerism   144

l

lab-on-a-chip technology   173–175
Langmuir–Blodgett films   32
Langmuir–Blodgett method, for SERS/SERRS substrates   151
– approach   90–91
– to biologically relevant systems   91–92
– experimental details   93–94
large-scale integration (LSI) platform   178–180
laser-ablated metal colloids   44–45
laser-induced fluorescence (LIF) detection   167
laser-induced silver substrate (LISS)   160
laser scanning microscopy   305
Lee–Meisel silver colloids   55–56
local field intensity enhancement factor (LFIEF), at a surface   5–6
– at ω_L 27–28
– at ω_S 27
– and power law   22
– between cylinders   15–17, 20–21
– caused by a coupled plasmon resonance   19–20
– gold (Au)   6, 19
– long-tail distributions   23
– maximum   20
– silver (Ag)   6, 19
– spatial distribution   20–21
localized surface plasmon polaritons (LSPPs)   305
localized surface plasmon resonance (LSPR)   39, 87
locked nucleic acid (LNA)   256
long-tail distributions   21–23
lossless Drude model   3–4
lucigenin (LG)   115

m

Malachite Green/anti-mouse/gold particles 274
maleimide dye 246
marker bands 222
Maxwell's equations 20
mecA gene 242
mefloquine 139
membrane model 225
membrane-based electron relays 219
mercaptanes 221
4-mercaptobenzoic acid (MBA) 268, 270
mercaptopropanesulfonic acid (MPS) 81
Met80 ligand 223
metal colloids, use in SERS 40
metallic cylinder
– electrostatic approximation 7–9
– local field enhancements 11
– localized surface plasmon resonances of 9–10
– shape effects 12–15
– size effects 12
metallic sphere
– local field enhancements 11
– localized surface plasmon resonances of 10–11
– shape effects 12–15
– size effects 12
metals, optical properties of 2–4
methadone 162
methicillin-resistant *Staphylococcus aureus* (MRSA) 242
3,4-methylenedioxy-*N*-methylamphetamine (MDMA) 75
microfluidics, for SERS
– capillary-driven test stripes 176–178
– centrifugal platform 180–181
– droplet-based 183–187
– electrokinetic platform 181–183
– large-scale integration (LSI) platform 178–180
– optofluidic CD platform 181
– polydimethyl siloxane (PDMS) microchannels 178–180
micro total analysis systems, concept 174
micro-Raman scattering experiments 94
micro-Raman system 203
Mie theory 20
model membranes 225
molecule-specific Raman fingerprints 242
monoethylene glycol (MEG) units 272
mutagens 142

n

N3-deprotonated tautomer 148
$NaBH_4$ solution 41
NaCl-activated citrate-reduced silver colloids 161
nanoparticle SERS substrates, metal-based 103
– Ag/Au core–shell nanoparticles 50–51
– aggregation of 47–50
– aromatic molecules 44
– bimetallic 50–51
– characterization 57–58
– charge-transfer effect and 'active site' effect 48
– chemical reaction 41–44
– colloidal gold particle aggregation on SERS, effect of 48
– colloidal spherical 41–47
– dicationic ADs on 114
– in different shapes 52–57
– dimensions 40
– dimers 49
– factors influencing 41–42
– functionalized 105
– hydrosol activation 47
– laser ablation and photoreduction method for preparing 44–45
– mercaptoacetic acid-capped spherical silver nanoparticles 42
– near-infrared (NIR) excitation 46
– preparation of 40–41, 200–201
– self-assembly of diamines, effect 114–115
– size effects 45
– stability of 40–41, 47
– synthesizing of spherical metal nanoparticles 41
– on unfunctionalized solid substrates 58–60
– use of ADs as linkers 113
nanostructure-enhanced Raman scattering 40
nanotags 263
narcotic drugs 131
near-field optical interaction 305
near-infrared (NIR) excitation, of nanoparticles 46
nicotine 77
2-nitrophenol 163
4-nitrophenol 163
non-fluorescent labels 246, 249
nonlinear Raman spectroscopy. *see* coherent anti-Stokes Raman scattering (CARS) spectroscopy

non-steroidal anti-inflammatory drugs (NSAIDs)   131

o

oligonucleotide probe   246
optical techniques   2
oxidation–reduction cycle   40

p

pain relievers. *see* analgesics
paracetamol   131–134
paracetamol sinus solutions   135
PDMS microdevices   179–180
persistent organic pollutants (POPs)   103
ortho-phenanthroline   179
pharmaceutical compounds, SERS analysis   129–130
– acetaminophen   131
– analgesics   130–138
– anticarcinogenics   142–151
– antimalarial drugs   139
– antimutagenics   142–151
– antipyretics   130–138
– aspirin   131–134, 138
– band positions   147
– bonds   148
– *β*-carotene   150–151
– chloroquine–haem complex   139–140
– cocaine   132
– dacarbazine   132
– 5-fluorouracil   132, 142–150
– ibuprofen   131
– mefloquine–haematin interaction mechanism   140
– N3–H deformation   147
– narcotic drugs   131
– non-steroidal anti-inflammatory drugs (NSAIDs)   131
– paracetamol   131–134
– of paracetamol sinus solutions   134–135
– phenobarbital   132
– at pH values   136–137, 148
– pK_a value of a base   144
– preferred medium   131
– using ordinary aqueous silver colloids, issues   131–132
– using poly(methyl methacrylate) (PMMA) plastic optical fibre   132
phenobarbital   132
phosphoramidite   246
photobleaching   32, 242
photonic materials, metals as
– Ag *vs* Au   18–19
– coinage metals   19–20
– coupled plasmon resonances   15–17
– electrostatic approximation   7–9
– gap effects   16
– local field enhancements   11
– localized surface plasmon resonances of   9–11
– long-tail distributions of enhancement   21–23
– polarization along tips   17
– shape effects   12–15
– size effects   12
– tip-enhanced Raman scattering (TERS)   17–18
– transition metals   19–20
photosensitizers   290
phthalocyanines   246
physisorption   135
piezo-controllers   17
planar surfaces   4–7
plant alkaloids   142
plasma frequency, of metals   3
Plasmodium falciparum   139
plasmonics   17
plasmon resonances
– shape effects   12–15
– size effects   12
2-plex system   257
5-plex system   254, 255
polychlorinated pesticides (PCPs)   104
polycyclic aromatic hydrocarbons (PAHs)   103
– assays   106
– detection using HA-functionalized systems   120
– layer-by-layer deposition of   269
– limit of detection (LOD)   110
– pyrene (PYR)   106
– selective recognition of   107
– sensitivity of detection by SERS   111
– SERS intensity of pollutants   108
polydimethyl siloxane (PDMS)   159
– microchannels   178–180
poly-(L-lysine)   244
polymerase chain reaction (PCR)   241
poly(vinyl alcohol)   47
poly(vinylpyrrolidone)   47
positive real numbers   2
propanethiol (PT)   81
protein–protein ET processes   219
pyridine   45, 77, 156, 193, 243
4-(2-pyridylazo)resorcinol   160

q

quantitative SERS analysis
– batch-to-batch reproducibility problems 74
– choice of enhancing media 72–73
– of colloidal suspensions 71–72
– disordered materials 72
– within a gel matrix 73–74
– internal standards 74–78
– microfluidic system 73
– plasmonic materials 72–73
– reproducibility 74–78
– root mean square (RMS) error 77
– selectivity 78–82
– shell life 73–74
– of solid materials 72–73
– stability 73–74
quinine 139–140

r

radiation efficiency 28–29
Raman process 24–26
Raman reporter molecules 245–248
reflectance 4
reflection process, of metals 7
resorufin 179
respiratory syncytial virus (RSV) 178
rhodamine 6G (R6G) 160, 179, 268
– molecules 48
– SERS spectra of 76
Rose Bengal (RB) 290
rubicene (RUB) 106

s

SAM–protein interface, electric field strength 221
scattering process 24
SEIRA spectra, of calixarenes 106–107
selective detection of trace pollutants, by SERS-based sensors
– α,ω-aliphatic diamines (ADs) 111–115
– on assembled metallic single-walled nanotube 118
– C–C skeletal stretching modes 112
– calixarenes 106–111
– carbon nanotube composites 118–119
– contact hosts 115–119
– diquat (DQ) and paraquat (PQ) 117
– of endosulfan 117
– of functionalized metal NPs 105
– humic acids (HAs) 119–122
– inclusion hosts 106–115
– limit of detection (LOD) 109–110
– lucigenin (LG) 115–117
– occlusion hosts 119–122
– using aggregated colloids 105
– viologen dication (VGD) species 115–117
self-assembledmonolayers (SAMs) 104, 267–268
– of ADs 112
– of arylthiols 271
– silica-encapsulated 270
separation techniques, using SERS
– capillary electrophoresis SERS (CE–SERS) systems 157–161
– combination of western blotting and SERS detection 166–167
– e-beam lithography 166, 168
– of flowing liquids 161–162
– gas chromatography (GC) 164–165
– HPLC–SERS study 161–163
– HPTLC–SERS study 165
– isotachophoretic (ITP) focussing method 167
– laser-induced fluorescence (LIF) detection 167
– liquid chromatography (LC) 161–164
– of SERRS-active substrates 160
– thin layer chromatography (TLC) 163, 165–166
– using modified SERS substrates 165
SERS-active nanostructures 182
SERS enhancement factor 22–23
– of dimers 50
$|E|^4$ approximation 27–28
– vs fluorescence enhancement 29–31
– gold nanoparticles 44
– magnitude of 32–33
– multiple-molecule 33
– in non-optimized conditions 33
– in resonant conditions 30
– single-molecule 28, 32–33
– in trace-level or single-molecule regime 88
SERS labels 263
SERS microscopy, biomedical application of
– immunohistochemistry 273–274
– immuno-SERS microscopy for tissue diagnostics 275–278
– mapping approaches 274–275
– methodologies in Raman microspectroscopy 274–275
– silica-encapsulated SERS probes 269–271
– target-specific SERS probe 265
– in vivo applications 278–279
SERS nanoparticle probes
– components 264–265
– gold spheres as 278
– metal colloids 265–267

SERS nanoparticle probes (contd.)
– protection and stabilization   269–272
– Raman reporters   267–269
SERS uncertainty principle   33
silica thin layer chromatography (TLC) plate   160
silver (Ag)
– aggregation of nanoparticle   49
– dielectric functions of   2–4
– electron transfer processes   228
– functionalization of NPs   109, 111, 113–114
– imaginary part of ε (λ)   3–4
– island films   149–151
– LFIEF on the surface   6, 19
– nanoparticle colloids   42, 243
– nanoparticle via photoreduction of $AgNO_3$ 45
– nanoparticles   41–42, 46
– nanospheres   266
– nanostructures   89
– performance   18–19
– polyol synthesis of silver nanoparticles   55
– real index of refraction   4
– reflectance   4
– self-assembled monolayers (SAMs)   221
– self-assembly of ADs on NPs   113
– SERS-active electrode   200
– spherical nanoparticles   52
silver-quantum dots (Ag-QDs) nanoparticles   163
single-molecule surface-enhanced Raman scattering (SM-SERS)   87–88
– basic elements needed   89
– degree of enhancement needed   89
– Langmuir–Blodgett method for   90–94
– octadecyl rhodamine B (R18) samples   97–99
– phospholipid, tagged   94–97
– Raman cross section   89–90
– requirements   89–90
single wall carbon nanotube (SWCNT) composites   118
size effect, in SERS activity   45
– metallic cylinder   12
sodium acrylate   43
sodium borohydride   41
sodium citrate   41
sodium dodecyl sulfate   47
spatial averaging   32
spatial localization, of resonance   20
spectrally modified fluorescence   31
spermine   244
sputtered Ag substrates   157

stability of enhancing media   73–74
standard boundary conditions, for electromagnetic fields   4
straight plug-flow concept   184
surface-enhanced coherent anti-Stokes Raman scattering (SECARS)   307
– of adenine nanocrystals   310–314
– advantages   308
– experimental system   309–310
– nonlinear polarization of   309
– of single-walled carbon nanotubes   314–315
– spectra of adenine molecules   313
– spectrum of SWCNTs   315
– Stokes laser frequency   313
– surface-enhanced hyper-Raman scattering (SEHRS)   297–300
surface-enhanced fluorescence (SEF)   2
surface-enhanced infrared absorption (SEIRA)   105
surface-enhanced resonance Raman scattering (SERRS)   30, 242
– multiplexing   253
– surfaces facilitating   242–245
surface-immobilized SERS substrates   179
surface plasmon resonance (SPR) spectroscopy   2
surfactant-stabilized sample droplets   185
SWCNT/PYR systems   118
SyBr Green   242

t

TAMRA dye   253
TCEC calixarene   106
terephthalic acid   179
tetraethoxy orthosilicate (TEOS)   269
tetramethyl orthosilicate (TMOS)   132
tetra-4-n-methylpyridyl porphyrin (TMPyP)   81–82
thiols   245
– alkyl   104
– amino-terminated alkane   80
– linkers   104
thiophenol, SERS spectra of   72
tip-enhanced coherent anti-Stokes Raman scattering (TECARS)   307
– advantages   308
– of carbon nanotubes   318–319
– of a DNA network   317–318
– experimental system   315–316
– G-band of SWCNTs   318
– imaging of DNA molecules   316–318
– metallic tips, preparation of   316

– near-field contribution 309
– nonlinear polarization of 309
tip-enhanced Raman scattering (TERS) 17–18
Tollens' reagent 72
toluene 165
topoisomerase inhibitors 142
trace-level SERS 88
trans-1,2-bis(4-pyridyl)ethylene and N,N-dimethyl-4-nitrosoaniline 157
transition metals 19–20
transparency region, of optical properties 24
triphenylene (TP) 106
TRITC–DHPE system 94–97
TR SERR experiments, with Soret-band excitation 228
tyrosine kinase inhibitors 142

u
uracil compounds 142
UV laser irradiation 45

UV–visible absorption spectrometer 161

v
viologen dication (VGD) species, detection of 115–117
vitamin A 134, 142

w
Watson–Crick geometry 144
western blotting and SERS detection 166–167

x
X-ray diffraction (XRD) technique 58
X-ray energy dispersive spectrometry (EDS) 58
X-ray photoelectron spectroscopy (XPS) 58
o,m,p-xylenes 165

z
zirconia 132